Arnold M. Kosevich

The Crystal Lattice

Arnold M. Kosevich

The Crystal Lattice

Phonons, Solitons, Dislocations, Superlattices

Second, Revised and Updated Edition

WILEY-VCH

WILEY-VCH Verlag GmbH & Co. KGaA

Author

Arnold M. Kosevich
B. Verkin Institute for Low Temperature Physics and Engineering
National Academy of Sciences of Ukraine
310164 Kharkov, Ukraine
e-mail: kosevich@ilt.kharkov.ua

Library of Congress Card No.: applied for.

British Library Cataloging-in-Publication Data:
A catalogue record for this book is available from the British Library.

Bibliographic information published by Die Deutsche Bibliothek
Die Deutsche Bibliothek lists this publication in the Deutsche Nationalbibliografie; detailed bibliographic data is available in the Internet at <http://dnb.ddb.de>.

Printed in the Federal Republic of Germany
Printed on acid-free paper

Satz Uwe Krieg, Berlin
Printing Strauss GmbH, Mörlenbach
Bookbinding Litges & Dopf Buchbinderei GmbH, Heppenheim

ISBN-13: 978-3-527-40508-4
ISBN-10: 3-527-40508-9

Contents

The Crystal Lattice: Phonons, Solitons, Dislocations, Superlattices, Second Edition. Arnold M. Kosevich

ISBN: 3-527-40508-9

Prefaces

Preface to the First Edition

The design of new materials is one of the most important tasks in promoting progress. To do this efficiently, the fundamental properties of the simplest forms of solids, i. e., single crystals must be understood.

Not so long ago, materials science implied the development, experimental investigation, and theoretical description, of primarily construction materials with given elastic, plastic and resistive properties. In the last few decades, however, new materials, primarily crystalline, have begun to be viewed differently: as finished, self-contained devices. This is particularly true in electronics and optics.

To understand the properties of a crystal device it is not only necessary to know its structure but also the dynamics of physical processes occurring within it. For example, to describe the simplest displacement of the crystal atoms already requires a knowledge of the interatomic forces, which of course, entails a knowledge of the atomic positions.

The dynamics of a crystal lattice is a part of the solid-state mechanics that studies intrinsic crystal motions taking into account structure. It involves classical and quantum mechanics of collective atomic motions in an ideal crystal, the dynamics of crystal lattice defects, a theory of the interaction of a real crystal with penetrating radiation, the description of physical mechanisms of elasticity and strength of crystal bodies.

In this book new trends in dislocation theory and an introduction to the nonlinear dynamics of 1D systems, that is, soliton theory, are presented. In particular, the dislocation theory of melting of 2D crystals is briefly discussed. We also provide a new treatment of the application of crystal lattice theory to physical objects and phenomena whose investigation began only recently, that is, quantum crystals, electron crystals on a liquid-helium surface, lattices of cylindrical magnetic bubbles in thin-film ferromagnetics, and second sound in crystals.

In this book we treat in a simple way, not going into details of specific cases, the fundamentals of the physics of a crystalline lattice. To simplify a quantitative descrip-

The Crystal Lattice: Phonons, Solitons, Dislocations, Superlattices, Second Edition. Arnold M. Kosevich

ISBN: 3-527-40508-9

tion of physical phenomena, a simple (scalar) model is often used. This model does not reduce the generality of qualitative calculations and allows us to perform almost all quantitative calculations.

The book is written on the basis of lectures delivered by the author at the Kharkov University (Ukraine). The prerequisites for understanding this material are a general undergraduate-level knowledge of theoretical physics.

Finally, as author, I would like to thank the many people who helped me during the work on the manuscript.

I am pleased to express gratitude to Professor Paul Ziesche for his idea to submit the manuscript to WILEY-VCH for publication, and for his aid in the realization of this project.

I am deeply indebted to Dr. Sergey Feodosiev for his invaluable help in preparing a camera-ready manuscript and improving the presentation of some parts of the book. I am grateful to Maria Mamalui and Maria Gvozdikova for their assistance in preparing the computer version of the manuscript. I would like to thank my wife Dina for her encouragement.

I thank Dr. Anthony Owen for his careful reading of the manuscript and useful remarks.

Kharkov July 1999 *Arnold M. Kosevich*

Preface to the Second Edition

Many parts of this book are not very different from what was in the first edition (1999). This is a result of the fact that the basic equations and conclusions of the theory of the crystal lattice have long since been established. The main changes ("reconstruction") of the book are the following

1. All the questions concerning one-dimensional (1D) crystals are combined in one chapter (Chapter 1). I consider the theory of a 1D crystal lattice as a training and proving ground for studying dynamics of three-dimensional structures. The 1D models allow us to formulate and solve simply many complicated problems of crystal mechanics and obtain exact solutions to equations not only of the linear dynamics but also for dynamics of anharmonic (nonlinear) crystals.

2. The second edition includes a new chapter devoted to the theory of elastic superlattices (Chapter 5). A new class of materials, namely, phonon and photon crystals has recently been of the great interest, and I would like to propose a simple explanation of many properties of superlattices that were studied before and known in the theory of normal crystal lattices.

3. New sections are added to the new edition concerning defects in the crystal lattice.

Finally, I would like to thank the people who helped me in the preparation of the manuscript.

I am indebted to Dr. Michail Ivanov and Dr. Sergey Feodosiev for their advise in improving the presentation of some parts of the book. I express many thanks to Alexander Kotlyar for his invaluable help in preparing the figures and electronic version of the manuscript. The author is grateful to Oksana Charkina for assistance in preparing the manuscript. I would like to thank my wife Dina for her encouragement.

Kharkov March 2005 *Arnold M. Kosevich*

Part 1 Introduction

The Crystal Lattice: Phonons, Solitons, Dislocations, Superlattices, Second Edition. Arnold M. Kosevich

ISBN: 3-527-40508-9

0
Geometry of Crystal Lattice

0.1
Translational Symmetry

The crystalline state of substances is different from other states (gaseous, liquid, amorphous) in that the atoms are in an ordered and symmetrical arrangement called the crystal lattice. The lattice is characterized by space periodicity or *translational symmetry*. In an unbounded crystal we can define three noncoplanar vectors $\boldsymbol{a}_1$, $\boldsymbol{a}_2$, $\boldsymbol{a}_3$, such that displacement of the crystal by the length of any of these vectors brings it back on itself. The *unit vectors* $\boldsymbol{a}_\alpha$, $\alpha = 1, 2, 3$ are the shortest vectors by which a crystal can be displaced and be brought back into itself.

The crystal lattice is thus a simple three-dimensional network of straight lines whose points of intersection are called the crystal lattice[1]. If the origin of the coordinate system coincides with a site the position vector of any other site is written as

$$\boldsymbol{R} = \boldsymbol{R}_n = \boldsymbol{R}(\boldsymbol{n}) = \sum_{\alpha=1}^{3} n_\alpha \boldsymbol{a}_\alpha, \qquad \boldsymbol{n} = (n_1, n_2, n_3), \tag{0.1.1}$$

where n_α are integers. The vector $\boldsymbol{R}$ is said to be a *translational vector* or a *translational period* of the lattice. According to the definition of translational symmetry, the lattice is brought back onto itself when it is translated along the vector $\boldsymbol{R}$.

We can assign a translation operator to the translation vector $\boldsymbol{R}(\boldsymbol{n})$. A set of all possible translations with the given vectors $\boldsymbol{a}_\alpha$ forms a *discrete group of translations*. Since sequential translations can be carried out arbitrarily, a group of transformations is commutative (Abelian). A group of symmetry transformations can be used to explain a number of qualitative physical properties of crystals independently of their specific structure.

Now consider the geometry of the crystal lattice. The parallelepiped constructed from the vectors corresponding to the translational periods is called a *unit cell*. It is

1) The lattice sites are not necessarily associated with the positions of the atoms.

The Crystal Lattice: Phonons, Solitons, Dislocations, Superlattices, Second Edition. Arnold M. Kosevich

ISBN: 3-527-40508-9

clear that the unit vectors, and thus the unit cell, may be chosen in different ways. A possible choice of unit cell in a planar lattice is shown in Fig. 0.1. As a rule, the unit cell is chosen so that its vertex coincides with one the atoms of the crystal. The lattice sites are then occupied by atoms, and vectors $\boldsymbol{a}_\alpha$ connect the nearest equivalent atoms.

By arranging the vectors $\boldsymbol{a}_1, \boldsymbol{a}_2, \boldsymbol{a}_3$ in the correct sequence, it is easy to see that the unit cell volume $V_0 = \boldsymbol{a}_1[\boldsymbol{a}_2, \boldsymbol{a}_3]$. Although the main translation periods are chosen arbitrarily, the unit cell volume still remains the same for any choice of the unit vectors.

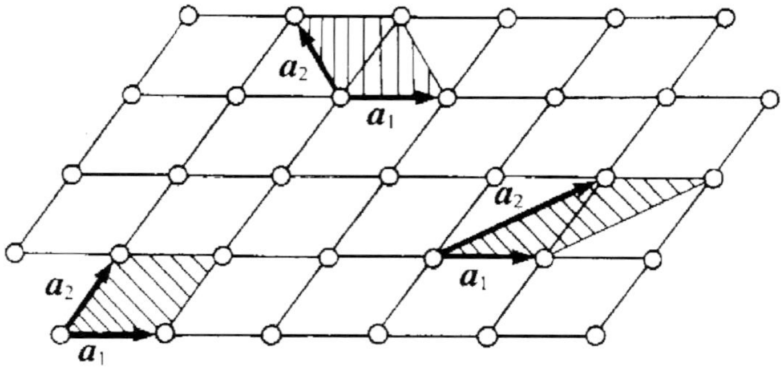

Fig. 0.1 Choice of unit cells (dashed) in a two-dimensional lattice.

The unit cell contains at least one atom of each of the types that compose the crystal[2]. Since the atoms of different type are distinguished not only by their chemical properties but also by their arrangement in the cell, even in a crystal of a pure element there can be more than one type of atom. If the unit cell consists of only one type of atom it is called *monatomic*, otherwise it is *polyatomic*. A monatomic lattice is also often called *simple* and a polyatomic lattice *composite*. Table salt (NaCl) containing atoms of two types is an example of a polyatomic crystal lattice (Fig. 0.2), and 2D lattice composed of atoms of two types is presented also in Fig. 0.3a. A polyatomic crystal lattice may also consist of atoms of the same chemical type. Figure 0.3b shows a highly symmetrical diatomic planar lattice whose atoms are located at the vertices of a hexagon.

The differences between simple and composite lattices lead to different physical properties. For example, the vibrations of a diatomic lattice have some features that distinguish them from the vibrations of a monatomic lattice.

We would like to emphasize that the unit cell of a crystal involves, by definition, all the elements of the translation symmetry of the crystal. By drawing the unit cell one can construct the whole crystal. However, the unit cell may not necessarily be symmetrical with respect to rotations and reflections as the crystal can be. This is clearly seen in Fig. 0.3 where the lattices have a six-fold symmetry axis, while the unit cells do not.

2) We note that the contribution to a cell of an atom positioned in a cell vertex equals 1/8, on an edge 1/4 and on a face 1/2.

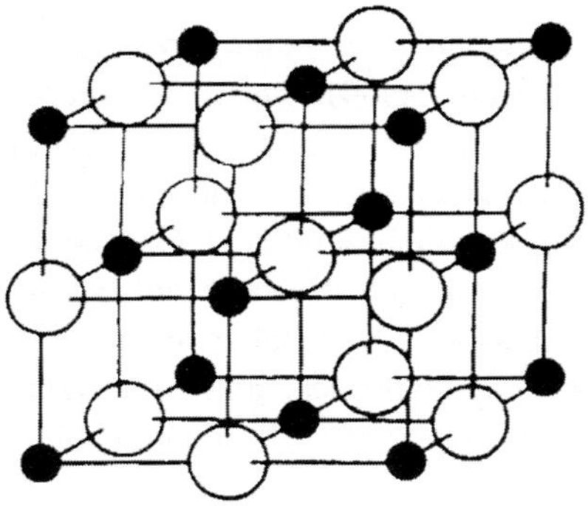

Fig. 0.2 NaCl crystal structure (○ - Na, ● - Cl).

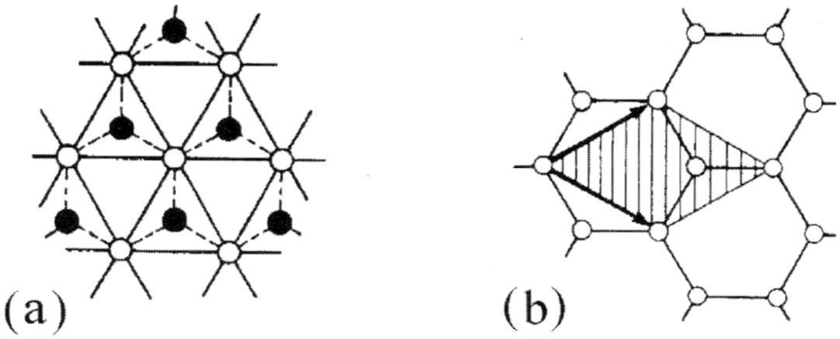

Fig. 0.3 Hexagonal 2D diatomic lattice composed of atoms (*a*) of different types and (*b*) of the same type. The unit cell is hatched.

0.2 Bravais Lattice

The Bravais lattice is the set of all equivalent atoms in a crystal that are brought back onto themselves when they are displaced by the length of a unit vector in a direction parallel to a unit vector. Bravais and monatomic lattices are usually coincident. A polyatomic lattice, however, consists of several geometrically identical interposed Bravais lattices.

The Bravais lattice of a polyatomic crystal is often more symmetrical than the crystal lattice itself. It contains all the elements of the crystal symmetry and may also have additional symmetry elements. For example, a planar crystal may have three-fold symmetry (Fig. 0.3a) whereas its Bravais lattice may have six-fold symmetry. The Bravais lattice has inversion centers at all of the sites, whereas the crystal lattices (necessarily polyatomic) do not necessarily have such a symmetry element.

The Bravais lattices are classified according to the symmetry of rotations and reflections. Seven *symmetry groups* or *space groups* are defined. Each of the lattices of a given group has an inversion center, a unique set of axes and symmetry planes. Each space group is associated with a polyhedron whose vertices correspond to the nearest sites of the corresponding Bravais lattice and that has all the symmetry elements of the space group. The polyhedron is either a parallelepiped or a prism.

In the most symmetrical Bravais lattice, the cube is used as the symmetry "carrier", and the lattice is called *cubic*. A *hexagonal lattice* is characterized completely by a regular hexahedral prism, the Bravais rhombohedron lattice by a rhombohedron, (i. e., the figure resulting when a cube is stretched along one of its diagonals), etc. A rectangular prism with at least one square face has *tetragonal* symmetry.

Within a given space group an additional subdivision into several types of Bravais lattices can be made. The type of Bravais lattice depends on where the lattice sites are located: either only at the vertices of the polyhedrons or also on the faces or at the center. We distinguish between the following types of Bravais lattice: primitive (P), base-centered (BC), face-centered (FC) and body-centered (BC) lattices.

The lattice of NaCl in Fig. 0.2 gives an example of a cubic lattice. A plane diatomic lattice with the 3-fold symmetry axes is shown in Fig. 0.3a, however, its Bravais lattice has 6-fold symmetry axes; a hexagonal lattice with the 6-fold symmetry axes is presented in Fig. 0.3b.

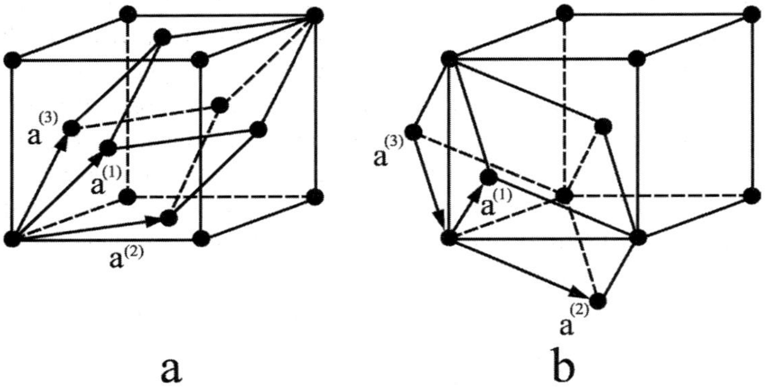

Fig. 0.4 Unit cells with translation vectors inside the cubic unit cells (a) of the FCC lattice and (b) of the BCC lattice.

It should be noted that the unit cell is not a principal geometrical figure being the "carrier" of all rotation elements of symmetry in the case of *centered* lattices. In order to demonstrate this fact a situation of the atoms in the single cube of BC-cubic and FC-cubic lattices is shown in Fig. 0.4a and 0.4b where the unit cells of these lattices are presented as well.

Naming the cubic, hexagonal and tetragonal lattices we have thereby counted the lattices possessing axes of 2-, 3-, 4- and 6-fold symmetry. Naturally, the question arises what types of the symmetry axes are compatible with the translational symmetry of a spatial lattice. It appears that the symmetry axes of the 2-, 3-, 4- and 6-fold only can exist in the unbounded spatial lattice (see *Problems* at the and of the chapter).

0.3
The Reciprocal Lattice

In order to describe physical processes in crystals more easy, in particular wave phenomena, the crystal lattice constructed with unit vectors $\boldsymbol{a}_\alpha$ in real space is associated with some periodic structure called the *reciprocal lattice*. The reciprocal lattice is constructed from the vectors $\boldsymbol{b}_\beta$, $\beta = 1, 2, 3$, related to $\boldsymbol{a}_\alpha$ through

$$\boldsymbol{a}_\alpha \boldsymbol{b}_\beta = 2\pi\delta_{\alpha\beta}, \quad \alpha, \beta = 1, 2, 3,$$

where $\delta_{\alpha\beta}$ is the Kronecker delta. The vectors $\boldsymbol{b}_\beta$ can be simply expressed through the initial translational vectors $\boldsymbol{a}_\alpha$:

$$\boldsymbol{b}_1 = \frac{2\pi}{V_0}[\boldsymbol{a}_2, \boldsymbol{a}_3], \quad \boldsymbol{b}_2 = \frac{2\pi}{V_0}[\boldsymbol{a}_3, \boldsymbol{a}_1], \quad \boldsymbol{b}_3 = \frac{2\pi}{V_0}[\boldsymbol{a}_1, \boldsymbol{a}_2].$$

The parallelepiped constructed from $\boldsymbol{b}_\beta$ is called the unit cell of a reciprocal lattice. It is easy to verify that the unit cell volume in the reciprocal lattice is equal to the inverse value of the unit cell volume of the regular lattice:

$$\Omega_0 = \boldsymbol{b}_1[\boldsymbol{b}_2, \boldsymbol{b}_3] = \frac{(2\pi)^3}{V_0}.$$

Note that the reciprocal lattice vectors have dimensions of inverse length. The space where the reciprocal lattice exists is called *reciprocal space*. The question arises: what are the points that make a reciprocal space? Or in other words: what vector connects two arbitrary points of reciprocal space?

Consider a wave process associated with the propagation of some field (e. g., electromagnetic) to be observed in the crystal. Any spatial distribution of the field is, generally, represented by the superposition of plane waves such as

$$\psi_q = e^{i\boldsymbol{q}\boldsymbol{r}},$$

where $\boldsymbol{q}$ is the wave vector whose values are determined by the boundary conditions.

However, in principle the vector $\boldsymbol{q}$ takes arbitrary values. The dimension of the wave vector coincides with the dimension of inverse length, and the continuum of all possible wave vectors forms the reciprocal space. Thus, the reciprocal space is the three-dimensional space of wave vectors.

By analogy to the translation vectors of the regular lattice (0.1.1), we can also define translation vectors in reciprocal space:

$$\boldsymbol{G} \equiv \boldsymbol{G}(\boldsymbol{m}) = \sum_{\alpha=1}^{3} m_\alpha \boldsymbol{b}_\alpha, \quad \boldsymbol{m} = (m_1, m_2, m_3), \tag{0.3.1}$$

where m_α are integers. The vector $\boldsymbol{G}$ is called a *reciprocal lattice vector*.

It can be seen that simple lattices in reciprocal space correspond to simple lattices in real space for a given Bravais space group. The reciprocal lattice of FC Bravais lattices (rhombic, tetragonal and cubic) is a body-centered lattice and vice versa. A lattice with a point at the center of the base has a corresponding reciprocal lattice also with a point at the center of the base.

In addition to the unit cell of a reciprocal lattice, one frequently constructs a "symmetry" cell. This cell is called the *Brillouin zone*. We choose a reciprocal lattice site as origin and draw from it all the vectors $\boldsymbol{G}$ that connect it to all reciprocal lattice sites. We then draw planes that are perpendicular to these vectors and that bisect them. If $\boldsymbol{q}$ is a vector in a reciprocal space, these planes are given by

$$\boldsymbol{q}\boldsymbol{G} = \frac{1}{2}G^2. \tag{0.3.2}$$

The planes (0.3.2) divide all of reciprocal space into a set of regions of different shapes (Fig. 0.5a).

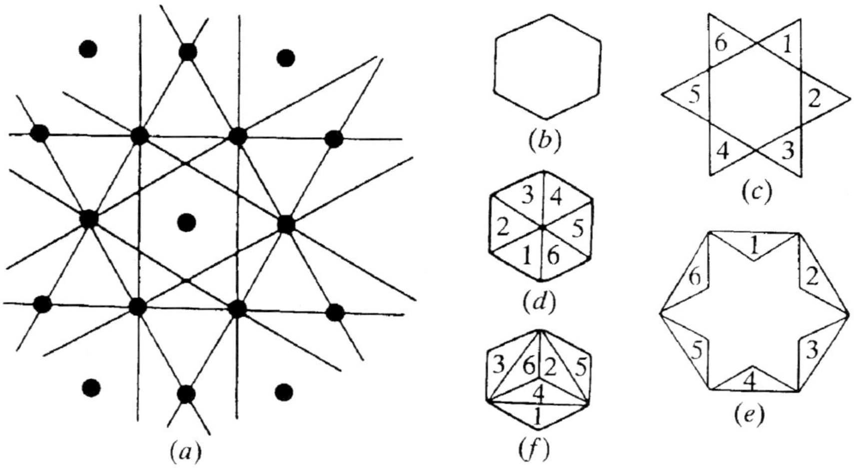

Fig. 0.5 Brillouin zones of hexagonal crystal: (*a*) construction of zones; the point in the middle is the origin, the lines drawn are the planes perpendicular to and bisecting the vectors connecting the origin with all other lattice sites (not shown); (*b*) the first zone; (*c*) the six parts of the second zone; (*d*) reduction of the second zone to the first; (*e*) the six parts of the third zone; (*f*) reduction of the third zone to the first.

The region containing the origin is called the first Brillouin zone. The regions of the reciprocal space that directly adjoin it make up the second zone and the regions bordering that are the third Brillouin zone, etc. The planes given by (0.3.2) are the *boundaries of the Brillouin zones*.

The regions of higher Brillouin zones can be combined into a single figure, identical to the first zone (Fig. 0.5d, f). Thus, any zone can be reduced to the first one. The concept of a reduced zone is convenient because it requires knowledge of the geometry of the first Brillouin zone only.

Mathematical relations between quantities in real and reciprocal space are entirely symmetrical with respect to these spaces and, formally, the lattices constructed with two sets of three vectors $\boldsymbol{a}_\alpha$ and $\boldsymbol{b}_\beta$ are reciprocal to one another. That is, if one is defined as the lattice in real space, the other is its reciprocal. It should be noted, however, that the physical meaning of these spaces is different. For a crystal, one initially defines the crystal lattice as the lattice in real space.

The concept of a reciprocal lattice is used because all physical properties of an ideal crystal are described by functions whose periodicity is the same as that of this lattice. If $\phi(\boldsymbol{r})$ is such a function (the charge density, the electric potential, etc.), then obviously,

$$\phi(\boldsymbol{r}+\boldsymbol{R}) = \phi(\boldsymbol{r}), \tag{0.3.3}$$

where $\boldsymbol{R}$ is a lattice translation vector (0.1.1). We expand the function $\phi(\boldsymbol{r})$ as a three-dimensional Fourier series

$$\phi(\boldsymbol{r}) = \sum_{\boldsymbol{q}} \phi_{\boldsymbol{q}} e^{i\boldsymbol{q}\boldsymbol{r}}, \tag{0.3.4}$$

where it is summed over all possible values of the vector $\boldsymbol{q}$ determined by the periodicity requirement (0.3.3)

$$\sum_{\boldsymbol{q}} \phi_{\boldsymbol{q}} e^{i\boldsymbol{q}\boldsymbol{r}} e^{i\boldsymbol{q}\boldsymbol{R}} = \sum_{\boldsymbol{q}} \phi_{\boldsymbol{q}} e^{i\boldsymbol{q}\boldsymbol{r}}. \tag{0.3.5}$$

Equation (0.3.5) can be satisfied if

$$e^{i\boldsymbol{q}\boldsymbol{R}} = 1, \quad \boldsymbol{q}\boldsymbol{R} = 2\pi p, \tag{0.3.6}$$

where p is an integer. To satisfy (0.3.6) it is necessary that

$$\boldsymbol{q}\boldsymbol{a}_\alpha = 2\pi p_\alpha, \quad \alpha = 1,2,3, \tag{0.3.7}$$

where p_α are the integers.

The solution to (0.3.7) for the vector $\boldsymbol{q}$ has the form

$$\boldsymbol{q} = m_1\boldsymbol{b}_1 + m_2\boldsymbol{b}_2 + m_3\boldsymbol{b}_3. \tag{0.3.8}$$

It follows from (0.3.8) that the vector $\boldsymbol{q}$ is the same as that of the reciprocal lattice: $\boldsymbol{q} = \boldsymbol{G}$ where $\boldsymbol{G}$ is determined by (0.3.1).

Thus, any function describing a physical property of an ideal crystal can be expanded as a Fourier series (0.3.4) where the vector $\boldsymbol{q}$ runs over all points of the reciprocal lattice

$$\phi(\boldsymbol{r}) = \sum_{\boldsymbol{G}} \phi_{\boldsymbol{G}} e^{i\boldsymbol{G}\boldsymbol{r}}. \tag{0.3.9}$$

Since there is a simple correspondence between the real and reciprocal lattices there should also be a simple correspondence between geometrical transformations in real and reciprocal space. We illustrate this correspondence with an example widely used in structural analysis. Consider the vector $\boldsymbol{r}$ such that

$$\boldsymbol{G}\boldsymbol{r} = 2\pi p, \qquad (0.3.10)$$

where p is the integer and $\boldsymbol{G}$ is a reciprocal lattice vector. Equation (0.3.10) describes a certain plane in the crystal. It is readily seen that this is a crystal plane, i. e., the plane running through an infinite set of Bravais lattice sites. Since the constant p may take any value, (0.3.10) describes a family of parallel planes. Thus, each vector of a reciprocal lattice $\boldsymbol{G} = \boldsymbol{G}(\boldsymbol{m})$ corresponds to a family of parallel crystal planes (0.3.10) rather than to a single plane. The distance between adjacent planes of the family is $d_B = 2\pi/G$, where G is the length of a corresponding vector of a reciprocal lattice. Three quantities m_1, m_2, m_3 in these relations can always be represented as a triplet of prime numbers p_1, p_2, p_3 (i. e., assume that p_1, p_2, p_3 have no common divisor except unity). These three numbers (p_1, p_2, p_3) are called the *Miller indices.*

0.4
Use of Penetrating Radiation to Determine Crystal Structure

We consider the transmission of a field (X-rays, beams of fast electrons or slow neutrons) through a crystal. We assume the distribution of the field in space to be described by a scalar function ψ that in vacuo obeys the equation

$$\epsilon\psi + c^2\Delta\psi = 0,$$

where for electromagnetic waves ϵ is the frequency squared ($\epsilon = \omega^2$) and c the light velocity, or in the case of electrons and neutrons they are the energy and the inverse mass ($c = \hbar^2/2m$). The crystal atoms interact with the wave, generating a perturbation. This perturbation is taken into account in the above equation by an additional potential

$$\epsilon\psi + c^2\Delta\psi + U(\boldsymbol{r})\psi = 0. \qquad (0.4.1)$$

The potential $U(\boldsymbol{r})$ has the same periodicity as the crystal (for example, it may be proportional to the electric charge density in a crystal).

We now consider how the periodic potential can affect the free wave

$$\psi_k = e^{i\boldsymbol{k}\boldsymbol{r}}, \quad c^2k^2 = \epsilon. \qquad (0.4.2)$$

We assume that U is weak, i. e., we can use perturbation theory (this is a reasonable assumption in many real systems). Let the wave (0.4.2) be incident on a crystal and scattered under the effect of the potential U. In the Born approximation, the amplitude of the elastically scattered wave with wave vector $\boldsymbol{k}'$ is proportional to the integral

$$U(\boldsymbol{k}', \boldsymbol{k}) = \int U(\boldsymbol{r})e^{-i(\boldsymbol{k}'-\boldsymbol{k})\boldsymbol{r}}\, dV, \qquad (0.4.3)$$

which is the matrix element of the potential U. The scattering probability, i. e., the probability for the wave (0.4.2) to be transformed to a wave

$$\psi_{k'} = Ae^{ik'r}, \quad A = \text{constant}, \quad c^2k'^2 = \epsilon, \tag{0.4.4}$$

is proportional to the squared matrix element (0.4.3).

To calculate the integral (0.4.3) we use an expansion such as (0.3.9) for the periodic function $U(r)$:

$$U(k', k) = \sum_G U_G \int e^{i(G-k'+k)r}\, dV. \tag{0.4.5}$$

In an unbounded crystal (0.4.5) is reduced to

$$U(k', k) = (2\pi)^3 \sum_G U_G \delta(k' - k - G). \tag{0.4.6}$$

It is clear that the incident wave (0.4.2) with the wave vector $\boldsymbol{k}$ can be transformed only into the waves whose wave vector is

$$k' = k + G, \tag{0.4.7}$$

where $\boldsymbol{G}$ is any reciprocal lattice vector.

In elastic scattering the wave frequency (or the scattered particle energy) does not change, so that

$$k'^2 = k^2. \tag{0.4.8}$$

The relations (0.4.7), (0.4.8) are called the *Laue equations* and are used in the analysis of X-ray diffraction and the electron and neutron elastic scattering spectra in crystallography. By fixing the direction of the incident beam and measuring the directions of the scattered waves, one can determine the vectors $\boldsymbol{G}$, i. e., the reciprocal lattice. From these it may be possible to reproduce the crystal structure.

To simplify (0.4.7), (0.4.8) further, we first establish their relation to the reciprocal lattice. We take the scalar product of (0.4.7) and take into account (0.4.8):

$$k'G = -kG = \frac{1}{2}G^2. \tag{0.4.9}$$

Comparing (0.4.9) and (0.3.2), it can be seen that only those beams whose wave-vector ends lie on the Brillouin zone boundaries (the origin of the waves vectors is at the center of the Brillouin zones) are reflected from the crystal.

We denote the angle between the vectors $\boldsymbol{k}$ and $\boldsymbol{k}'$ by 2θ. Then from (0.4.8) we obtain the relation

$$G = 2k \sin\theta. \tag{0.4.10}$$

As was shown above, the length of the vector $\boldsymbol{G}$ is inversely proportional to the distance d between the nearest planes of atoms to which this vector is perpendicular

$$G = \frac{2\pi n}{d}, \tag{0.4.11}$$

where n is the integer. Substituting (0.4.11) into (0.4.10) and introducing the wavelength of the incident radiation $\lambda = 2\pi/k$ we obtain

$$n\lambda = 2d \sin\theta. \tag{0.4.12}$$

This relation is known as the *Bragg reflection law*. The diffraction described by (0.4.12) is sometimes referred to as "reflection" from cristal planes.

It should be noted that this simplest *geometrical* (or *kinematic*) *theory* of diffraction in crystals is applicable only to scattering in thin crystal samples. It does not include the interaction of the incident and diffracted beams with deeper atomic layers in thick samples.

0.4.1
Problems

1. Prove that if $\boldsymbol{r}$ is the radius-vector of an arbitrary site in the crystal the following equation is valid

$$\sum_{\boldsymbol{G}} e^{i\boldsymbol{G}\boldsymbol{r}} = V_0 \sum_{\boldsymbol{R}} \delta(\boldsymbol{r} - \boldsymbol{R}), \tag{0.4.13}$$

where the summation on the r.h.s. is carried out over all lattice sites and on the l.h.s. over all reciprocal lattice sites.

2. Derive from (0.4.13) the equation

$$\sum_{\boldsymbol{R}} e^{-i\boldsymbol{k}\boldsymbol{R}} = \Omega_0 \sum_{\boldsymbol{G}} \delta(\boldsymbol{k} - \boldsymbol{G}), \tag{0.4.14}$$

where $\boldsymbol{k}$ is the position vector of an arbitrary point in the reciprocal space.

3. Elucidate which symmetry axes can be inherent elements of the symmetry of a lattice.

Hint. Consider two neighboring sites A and B in the plane perpendicular to the symmetry axis (see Fig. 0.6). Perform a rotation by the angle ϕ=$2\pi/n$ about the axis C_n through the point A; after that B occupies position B'. Analogous rotation about B transfers A to A'. Since the sites B' and A' belong to the same lattice the length $B'A'$ should be divisible by the length AB.

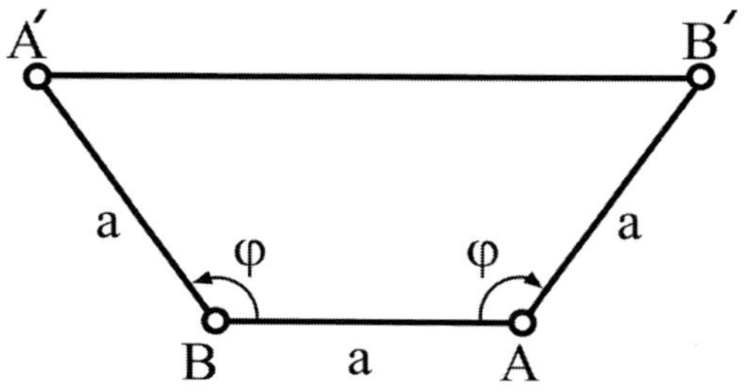

Fig. 0.6 Graphical solution of Problem 3.

Solution. $n = 2, 3, 4, 6$.

Part 2 Classical Dynamics of a Crystal Lattice

The Crystal Lattice: Phonons, Solitons, Dislocations, Superlattices, Second Edition. Arnold M. Kosevich

ISBN: 3-527-40508-9

1
Mechanics of a One-Dimensional Crystal

1.1
Equations of Motion and Dispersion Law

The physics of condensed media concerns itself generally with periodic structures of lower dimensions, in particular, one-dimensional (1D). One-dimensional problems are attractive mainly because of the simplicity of the mathematics and the possibility of obtaining exact solutions in many cases, not only for small (harmonic) vibrations, but also for more complex situations in which the nonlinear (anharmonic) crystal properties may be involved. Qualitative aspects of nonlinear dynamics can be understood most easily in one-dimensional systems; therefore, these systems are still being intensely studied. Finally, some phenomena in three-dimensional crystals can be modelled by one-dimensional problems. Thus mechanics of the one-dimensional crystal can be considered as a spring-board for studying dynamics of three-dimensional structures.

In our study of 1D crystal mechanical vibrations, we focus on the existence of two physical objects each of which can be called a one-dimensional crystal. First, there are periodic linear structures in a three-dimensional space. A long macromolecule of any homopolymer is the best example. We shall call such crystals *linear chain*. The second object is a periodic structure that enables one to study the motion in one-dimensional space. For this purpose, we should imagine the physical phenomena and processes in a "Straight Line Land: Lineland".

We start with equations of mechanics in 1D space, i.e., we consider the one-dimensional crystal itself. Consider a periodical array of particles (atoms, moleculas or other units whose internal structure can be neglected) situated along the x-axis with period a. The position of an arbitrary site is equal to

$$x_n \equiv x(n) = na, \tag{1.1.1}$$

where n is an integer. Suppose Eq. (1.1.1) gives the coordinates of particles ("atoms") in the equilibrium state of the crystal. A real coordinate of an atom differs from

The Crystal Lattice: Phonons, Solitons, Dislocations, Superlattices, Second Edition. Arnold M. Kosevich

ISBN: 3-527-40508-9

the corresponding lattice site (1.1.1) when the atoms are displaced relative to their equilibrium positions. Such a situation appears undoubtedly during vibrations of the crystal. We denote the displacement of the atom with number n in a monatomic (with a single atom per unit cell) from its equilibrium position by $u(n)$.

In order to write dynamic equations of motion of crystal atoms we need to describe an interatomic interaction. First restrict ourselves to the model of nearest-neighbor interaction, i. e., consider the 1D crystal to be similar to the array of small balls connected with elastic springs. Secondly take into account the fact that the relative atomic displacements being considered in crystal dynamics are small compared with a. Therefore, it is natural to begin studying the lattice dynamics with the case of small harmonic vibrations. Assuming the crystal to be in equilibrium at $u(n) = 0$ we can write the potential energy U in the harmonic approximation as

$$U = U_0 + \frac{1}{2}\sum_n \alpha(u_n - u_{n-1})^2, \tag{1.1.2}$$

where U_0 = constant and the summation is over all crystal sites. In the simplest model the interatomic interaction is characterized by only one elastic parameter α.

With the expression for the potential energy (1.1.2) one can easily write down the equation of motion of every atom:

$$m\frac{d^2u}{dt^2} = -\frac{\partial U}{\partial u(n)}, \tag{1.1.3}$$

where m is the atomic mass. Equation (1.1.3) leads to the following dynamic equation

$$\frac{d^2u(n)}{dt^2} = \left(\frac{\omega_{\mathrm{m}}}{2}\right)^2 [u(n+1) - 2u(n) + u(n-1)], \tag{1.1.4}$$

where $\omega_{\mathrm{m}}^2 = 4\alpha/m$.

Take the solution to (1.1.4) in the form

$$u(n) = ue^{ikx(n)}. \tag{1.1.5}$$

The parameter k is analogous to a wave number of vibration and is regarded as a *quasi-wave number*.

The stationary crystal vibrations for which the displacement of all atoms are time dependent only by the factor $e^{-i\omega t}$ are of special interest. For such vibrations substituting (1.1.5) into (1.1.4), we obtain

$$\omega^2 u - \frac{1}{2}\omega_{\mathrm{m}}^2(1 - \cos ak)u = 0,$$

hence

$$\omega^2 = \frac{1}{2}\omega_{\mathrm{m}}^2(1 - \cos ak) = \omega_{\mathrm{m}}^2 \sin^2\frac{ak}{2}. \tag{1.1.6}$$

In mechanics the dependence of frequency on the wave number is called the *dispersion law* or *dispersion relation*. Thus, (1.1.6) gives the dispersion relation

$$\omega^2 = \omega^2(k)$$

for the lattice vibrations.

From (1.1.6) we note that the dispersion law determines the frequency as a periodic function of the quasi-wave number with a period of a reciprocal lattice

$$\omega(k) = \omega(k + G), \quad G = \frac{2\pi}{a}.$$

This periodicity is the basic distinction between the dispersion law of crystal vibrations and that of continuous medium vibrations, since the monotonic wave-vector dependence of the frequency is typical for the latter. The difference between the quasi-wave number k and the ordinary wave number is also observed in the fact that only number k values lying inside one unit cell of a reciprocal lattice ($-\pi/a < k < \pi/a$) correspond to physically nonequivalent states of a crystal.

When the lattice period a tends to zero, the Brillouin zone dimension becomes infinitely large and we return to the concept of momentum and its eigenfunctions in the form of plane waves.

To clarify the available restrictions on the region of physically nonequivalent $\mathbf{k}$ values we note that $k = 2\pi/\lambda$ always, where λ is the corresponding wavelength. We consider, for simplicity, a one-dimensional crystal (a linear chain) with a period a for which the reciprocal lattice "vector" $G = 2\pi/a$. Choose the interval $-\pi/a \leq k \leq \pi/a$ as the reciprocal lattice unit cell. The limiting value of the quasi-wave number $k = \pi/a$ will then respond to the wavelength $\lambda = 2a$. It follows from the physical meaning of wave motion that this wavelength is the minimum in the crystal, since we can observe the substance motion only at points where material particles are located. A wave of this length is shown as a solid curve in Fig. 1.1 (the dark points are the equilibrium positions of particles, the light ones are their positions at a certain moment of the motion). A wave with wave number larger than the limiting one reciprocal lattice period namely, $k = \pi/a + 2\pi/a = 3\pi/a$, is shown as a dashed line. Both waves correctly reproduce the crystal motion but the introduction of the wavelength $\lambda = 2a/3$, carrying no additional information on the particle motion, is not justified physically.

We now propose a short analysis of the dispersion relation of the one-dimensional crystal. According to (1.1.6), possible vibration frequencies fill band $(0, \omega_{\mathrm{m}})$ where ω_{m} is the upper boundary of the band of possible vibration frequencies. To continue the analysis one needs to know a spectrum of quasi-wave number values inside the Brillouin zone. In order to define such a spectrum consider the one-dimensional crystal containing N atoms ($N \ll 1$) and having the length L ($L = Na$). The spectrum mentioned depends on the boundary conditions at the crystal ends. We formulate the simplest boundary conditions supposing that the atomic chain is closed up into a ring

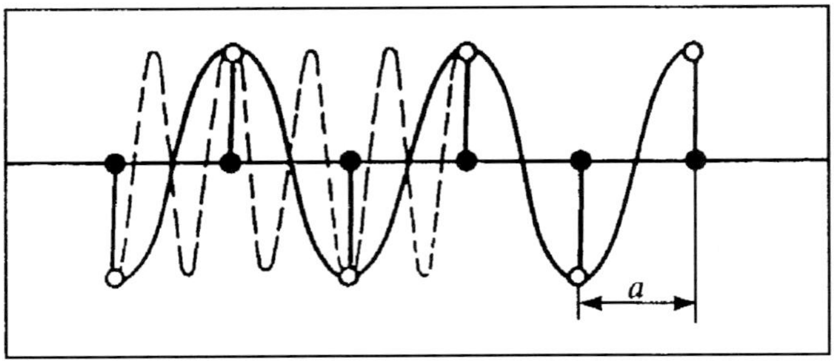

Fig. 1.1 Profile of the wave with $\lambda = 2a$ (*solid line*) and the wave with $3\lambda = 2a$ (*dashed line*).

and N+n atom coincides with the n-th atom:

$$u(x_n) = u(x_n + Na). \tag{1.1.7}$$

The cyclicity condition (1.1.7) is called the *Born–Karman condition.* Combining (1.1.7) and (1.1.5), we obtain

$$ka = \frac{2\pi}{N}p, \quad \text{or} \quad k = \frac{2\pi}{L}p, \tag{1.1.8}$$

where $p = 0, 1, 2, \ldots, N$. Usually it is convenient to choose symmetrical conditions

$$p = 0, \pm 1, \pm 2, \ldots, \pm\frac{N+1}{2}.$$

Since in the macroscopic case ($N \gg 1$) the discrete values of the k numbers are divided by the very small interval $\Delta k \sim \frac{1}{L}$ and the spectrum of k values can be regarded as quasi-continuous. Therefore one can analyze the dispersion relation considering the frequency as a continuous function of quasi-wave number.

Let us begin from the vibrations with small k. The 1D-crystal long-wave vibrations ($ak \ll 1$) have the ordinary acoustic frequency spectrum

$$\omega^2 = s^2k^2 \quad \text{or} \quad w = sk, \tag{1.1.9}$$

where s is a sound velocity ($2s = a\omega_m$).

To describe the dispersion law at the upper edge of the band of possible frequencies (when $\omega_m - \omega \ll \omega_m$) it is convenient to introduce $q = k - \pi/(2a)$ and consider $|q| \ll 1$. Then the atom displacements assume the form

$$u(n) = u_0 e^{ikan} = (-1)^n u_0 e^{iqan}. \tag{1.1.10}$$

It is interesting to note that the neighbor atoms vibrate with the opposite phases in the limiting case $\omega \to \omega_m$. And the following dependence is obtained for the dispersion law

$$\omega = \omega_m - \frac{1}{2}\gamma q^2, \qquad \gamma = \omega_m \left(\frac{a}{2}\right)^2. \tag{1.1.11}$$

A dispersion relation of the type (1.1.11) is known as a *quadratic dispersion law.*

According to (1.1.6) the dependence of the frequency on k is characterized by the monotonic plot in the interval $(-\pi/a < k < \pi/a)$. However, this fact is a result of using a model of the nearest-neighbor interaction. Taking into account the interaction of the next neighbors leads to a possibility to obtain a nonmonotonic dependence of ω on k with a diagram similar to Fig. 1.2 (see Problem 1 at the end of the section). Therefore maxima inside the Brillouin zone and minima on its boundaries can appear on a graphical representation of the dispersion relation.

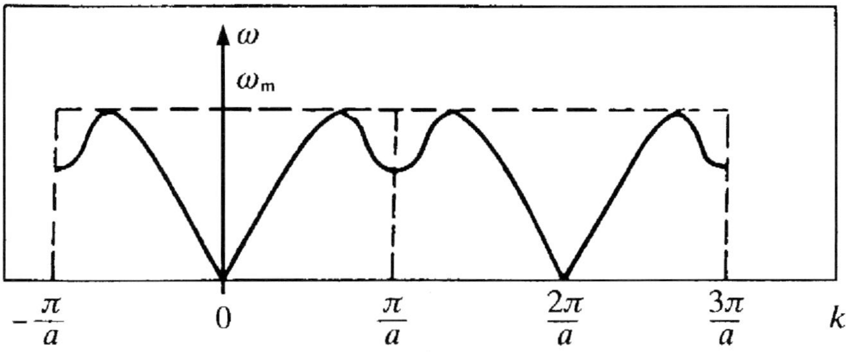

Fig. 1.2 One-dimensional dispersion diagram.

Having discussed the behavior of the frequence spectrum inside a band of possible free vibrations in an unbounded 1D crystal and at the edges of the band one should say that vibrations with frequencies outside of this band are possible in a bounded chain (see Problem 2 at the end of the section).

Note in conclusion that the acoustic dispersion relation (1.1.9) is a natural consequence of the long-wave approximation for an equation of motion of the crystal. Actually, in such an approximation the function $u(n)$ of a discrete argument n can be considered as a continuous function of the argument $x \equiv x_n = na$ and the following expansion can be used

$$u_{n\pm 1} - u_n = \pm a u' + \frac{1}{2} a^2 u'' + \cdots, \tag{1.1.12}$$

where

$$u' = \frac{\partial u(n)}{\partial x}, \quad u'' = \frac{\partial^2 u(n)}{\partial x^2}, \tag{1.1.13}$$

and so on. Then (1.1.4) can be transformed into the following differential equation in partial derivatives

$$\frac{\partial^2 u(x)}{\partial t^2} - s^2 \frac{\partial^2 u(x)}{\partial x^2} = 0, \tag{1.1.14}$$

where x is a continuous coordinate that determines the position in a 1D crystal.

Obviously, the acoustic dispersion law (1.1.6) is typical for the solutions to the wave equation (1.1.14).

Another form of a dynamic differential equation in partial derivatives is typical for vibrations with frequencies close to the upper edge of the frequency spectrum ($\omega_m - \omega \ll \omega_m$). Taking into account (1.1.10), in this region the following approximation can be used

$$u(n) = (-1)^n v(na), \tag{1.1.15}$$

where $v(x)$ is a continuous function of x. Now, using the expansion (1.1.18) for the function $v(na) = v(x)$, one can transform (1.1.4) into the following

$$\frac{\partial^2 v(x)}{\partial t^2} + \omega_{\mathrm{m}}^2 v(x) + s^2 \frac{\partial^2 v(x)}{\partial x^2} = 0. \tag{1.1.16}$$

Obviously, the dispersion relation (1.1.11) follows from (1.1.16).

Proceeding further from quasi-continuity of the spectrum of k values we change the summation over the discrete values of a quasi-wave number for the integration. Taking (1.1.8) into account it is easy to obtain the rule governing this transition to the integration

$$\sum_k f(k) = \frac{L}{2\pi} \int f(k)\, dk, \tag{1.1.17}$$

where the integration is carried out over the interval of a single unit cell in k-space (or the Brillouin zone).

Having analyzed the spectrum of eigenvalues of the crystal vibrations let us consider a set of eigenfunctions of this problem. The crystal eigenvibrations (1.1.5) are numbered by k. Introduce normal vibrations in the form

$$\phi_k(n) = \frac{1}{\sqrt{N}} e^{ikna}, \tag{1.1.18}$$

which provides the normalization condition

$$\sum_n \phi_k^*(n) \phi_k'(n) = \delta_{kk'}. \tag{1.1.19}$$

The normal eigenvibrations (1.1.19) are often called the *normal modes* of the vibrations.

The set of normal modes allows us to construct easily the so-called Green function for crystal vibrations. According to the definition the *Green function* for vibrations of the unbounded chain in the site representation has the form

$$G_{\omega^2}(n, n') = \sum_k \frac{\phi_k^*(n)\phi_k(n')}{\omega^2 - \omega^2(k)} = \frac{1}{N} \sum_k \frac{e^{ika(n-n')}}{\omega^2 - \omega^2(k)}. \tag{1.1.20}$$

The function $G_\varepsilon(n)$ where $\varepsilon = \omega^2$ is called the Green function of stationary crystal vibrations. Expression (1.1.20) can be rewritten as

$$G_\varepsilon(n) = \frac{1}{N} \sum_k G(\varepsilon, k) e^{ikna}; \tag{1.1.21a}$$

$$G(\varepsilon, k) = \frac{1}{\varepsilon - \omega^2(k)}, \tag{1.1.21b}$$

determining the Green function in the (ε, k) representation.

It is not difficult to calculate the Green function for the vibrations with the dispersion law (1.1.6):

$$G_\varepsilon(n) = \frac{i}{\sqrt{\varepsilon(\omega_m^2 - \varepsilon)}} \begin{cases} e^{-iakn}, & n < 0; \\ e^{iakn}, & n > 0, \end{cases} \tag{1.1.22}$$

where $ak = 2\arcsin(\sqrt{\varepsilon}/\omega_m) = 2\arcsin(\omega/\omega_m)$.

1.1.1
Problems

1. Find the dispersion law for a 1D crystal vibrations taking into account interactions between nearest neighbors and next-but-one neighbors. Find a possibility of a nonmonotonous dependence $\omega = \omega(\mathbf{k})$ within the interval $0 < ak < \pi$.

Solution. The required dispersion relation

$$\omega^2 = \frac{4\alpha_1}{m}\sin^2\frac{ak}{2} + \frac{4\alpha_2}{m}\sin^2 ak, \tag{1.1.23}$$

where α_1 and α_2 are parameters of the elastic interaction between the nearest and next-nearest neighbor atoms, respectively. A required nonmonotonic dependence appears under the condition $4\,|\alpha_2| > \alpha_1$ and the plot of the (1.1.18) becomes similar to Fig. 1.2.

2. Find the wave number values for the frequencies exceeding the maximum frequency of harmonic 1D crystal vibrations with interaction of nearest neighbors only. Interpret the result.

Hint. Proceeding from the fact that in (1.1.4) ω is real, find the complex k values corresponding to $\omega > \omega_{\mathrm{m}}$.

Solution. Complex values

$$k = \pm\pi/a + i\kappa, \tag{1.1.24}$$

where κ is determined by

$$\omega = \omega_{\mathrm{m}} \cosh\frac{a\kappa}{2}. \tag{1.1.25}$$

The solutions exponentially decreasing (or increasing) with the distance can describe the vibrations of a bounded 1D crystal that are localized near its free edge and not penetrating inside.

3. Find the Green function for stationary vibrations, accounting for the interactions of not only the nearest neighbors, when $\omega = \omega(k)$ is a nonmonotonic function in the interval $0 < k < \pi/a$.

1.2
Motion of a Localized Excitation in a Monatomic Chain

It is customary to think that an excitation in the crystal moves with the sound velocity. What meaning can be associated with such an assertion? A concept of "sound" is connected with the acoustic vibrations, i. e., with the long-wave approximation of the crystal dynamics. Motion of small vibrations in the long-wave approximation is described by the wave equation (1.1.14). Any function of the argument $x - st$ is the solution to (1.1.14), and an arbitrary perturbation $u = u(x - st)$ with small enough gradients $a(\partial u/\partial x) \ll 1$ moves with velocity s along the 1D crystal without changing its shape. This is a consequence of the fact that the dispersion relation (1.1.9) is dispersionless, i. e., it determines a wave phase velocity independent of k or ω.

The dispersion law (1.1.6) does not possess such a property and the phase velocities of corresponding waves depend on k, and a localized excitation should expand moving along the crystal. The unique stationary solution to (1.1.4) not being deformed when moving along a 1D crystal is the harmonic wave whose frequency and wave number are related by the dispersion relation (1.1.6). However, it is said generally that a perturbation in a crystal moves with sound velocity, with the character of the perturbation remaining unspecified. We shall clarify the meaning of such an assertion.

We analyze (1.1.4) that coincides with the recursion relation for a Bessel function of the first kind $J_\nu(z)$:

$$\frac{d^2 J_\nu(z)}{dz^2} = \frac{1}{4}\left[J_{\nu+2}(z) - 2J_\nu(z) + J_{\nu-2}(z)\right].$$

It is clear that the solution to (1.1.4) can be expressed directly through Bessel functions. Assuming a 1D crystal of infinite length ($-\infty < n < \infty$) and taking into account the boundedness of displacements, we can write

$$u_n(t) = \text{const} \cdot J_{2(n-p)}(\omega_\text{m} t), \qquad (1.2.1)$$

where p is an arbitrary integer and n an independent integer.

If the initial displacements u_n^0 and the initial velocities v_n^0 of all atoms are given at $t = 0$ the corresponding solution to (1.1.4) for $t > 0$ reads as

$$u_n(t) = \sum_{p=-\infty}^{\infty} u_p^0 J_{2(n-p)}(\omega_\text{m} t) + \sum_{p=-\infty}^{\infty} v_p^0 \int_0^t J_{2(n-p)}(\omega_\text{m}\tau)\, d\tau. \qquad (1.2.2)$$

We assume that at the initial time only one atom is displaced from the equilibrium position ($u_n^0 = u_n\delta_{n0}$, $v_n = 0$). It then follows from (1.2.2) that

$$u_n(t) = u_0 J_{2n}(\omega_\text{m} t), \qquad t > 0. \qquad (1.2.3)$$

The perturbation (1.2.3) has no pronounced propagation front even far from the onset ($n \gg 1$). But it follows from the properties of Bessel functions that the first (major) perturbation maximum comes at the point $n(n \gg 1)$ at a time t_1 determined by the condition $\omega_m t_1 \approx 2n$ (Fig. 1.3). The velocity of motion of a perturbation maximum is $na/t_1 \approx (1/2)\omega_m a = s$, i. e., it is practically the same as the sound velocity in a 1D crystal. Thus, the effective velocity of the transfer of the perturbation is not different from that of the sound velocity, i. e., it is the same as the limiting group velocity that follows from the dispersion relation (1.1.6) or (1.1.9).

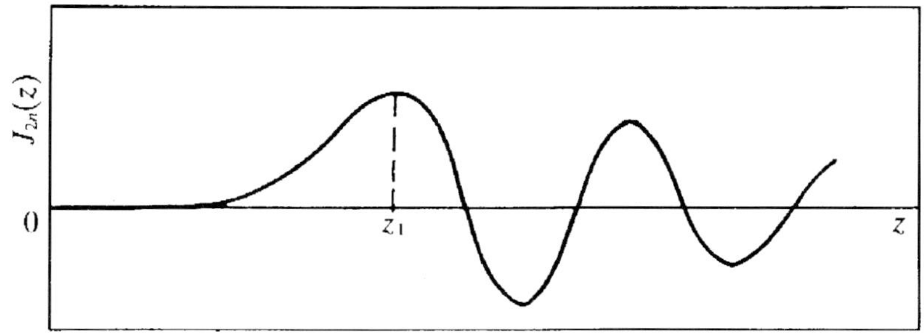

Fig. 1.3 The time dependence of the perturbation coming to the point located at a distance an from the displaced atom $(z = \omega_m t, z_1 = \omega_m t_1 \approx 2n)$.

Having clarified the role of the dispersion in an excitation signal moving along the 1D discrete chain, we now describe the velocity dispersion in the long-wave continuous approximation. It is known that competition of the higher dispersion with nonlinearity is very important in the dynamics of complex media. Hence, deriving a dynamic equation in partial derivatives for the 1D crystal anew, we take into account unharmonic terms in the interaction energy of the nearest neighbors. We assume the potential energy of a crystal to have the form of a sum

$$U = \sum_n \varphi(\xi_n) = \sum_n \varphi(u_n - u_{n-1}); \qquad \xi_n = u_n - u_{n-1}, \tag{1.2.4}$$

so that in the harmonic approximation

$$\varphi(\xi) = \frac{1}{2}\varphi''(0)\xi^2, \qquad \varphi''(0) = \alpha > 0. \tag{1.2.5}$$

Generalizing the equations of motion of a 1D crystal, we assume the function $\varphi(\xi)$ to be different from (1.2.5) and reduced to the parabolic dependence (1.2.5) only at $\xi \to 0$.

The equation of crystal motion with interactions between nearest neighbors only has the form

$$m\frac{d^2u_n}{dt^2} = \varphi'(u_{n+1} - u_n) - \varphi'(u_n - u_{n-1}). \tag{1.2.6}$$

For small relative displacements, one can take

$$\varphi'(\xi) = \varphi''(0)\xi + \frac{1}{2}\varphi'''(0)\xi^2, \tag{1.2.7}$$

including only the so-called *cubic anharmonicity*.

We substitute (1.2.7) into (1.2.6) and compare the result with (1.1.1); it is seen that $\alpha = m\varphi''(0)$. Just this relation connects the elements of the force matrix introduced phenomenologically with the parameter of the interatomic interaction potential.

However, (1.2.6) allows one to avoid the restrictions that arise in considering the harmonic vibrations by means of (1.1.4). To enable a description of inhomogeneous crystal states varying weakly in space we consider (1.2.6) in the long-wave approximation. Assume the characteristic distance Δx of the change in the field of displacements to be large ($\Delta x \gg a$). This makes it possible to pass to a continuum treatment, i. e., to replace the functions of a discrete argument n by the function of a continuous coordinate x and use the expansions

$$\begin{aligned} \xi_{n+1} = & \quad u_{n+1} - u_n = & au' + \frac{1}{2}a^2u'' + \frac{1}{3!}a^3u''' + \frac{1}{4!}a^4u''''; \\ \xi_n = & \quad u_n - u_{n-1} = & au' - \frac{1}{2}a^2u'' + \frac{1}{3!}a^3u''' - \frac{1}{4!}a^4u''''. \end{aligned} \tag{1.2.8}$$

In (1.2.8), the terms with the fourth-order derivatives remain, since the corresponding terms in the equations of motion may compete with terms generated by nonlinearity.

The nonlinear terms in (1.2.6) will be calculated to the first nonvanishing approximation with the lowest orders of the derivatives. Therefore, using (1.2.8), we take

$$\xi_{n+1}^2 = a^2\left(u'\right)^2 + a^3u'u'', \qquad \xi_n^2 = a^2\left(u'\right)^2 - a^3u'u''. \tag{1.2.9}$$

Substitute (1.2.8), (1.2.9) into (1.2.7), (1.2.6) to obtain

$$m\frac{d^2u}{dt^2} = a^2\varphi''(0)\left[\frac{\partial^2 u}{\partial x^2} + \frac{1}{12}a^2\frac{\partial^4 u}{\partial x^4}\right] + a^3\varphi'''(0)\frac{\partial u}{\partial x}\frac{\partial^2 u}{\partial x^2}. \tag{1.2.10}$$

Introduce the notations

$$s^2 = \frac{a^2}{m}\varphi''(0), \quad B^2 = \frac{a^2}{12}s^2, \quad \Lambda = -\frac{a^3}{m}\varphi'''(0), \tag{1.2.11}$$

and write (1.2.10) as a nonlinear wave equation (the *Boussinesq equation*):

$$u_{tt} = s^2u_{xx} + B^2u_{xxxx} - \Lambda^2u_xu_{xx}, \tag{1.2.12}$$

where u_{tt} is the second derivative in time; u_x, u_{xx}, u_{xxxx} are the derivatives of the corresponding order with respect to the coordinate x.

Since, when the atoms approach each other their mutual repulsion increases and when they separate from each other their attraction decreases, it may be assumed that $\varphi''(0) < 0$. This was used in introducing the notation of (1.2.11). Besides that, although a simple relation between the parameters B and s results from the assumption of interaction of nearest neighbors only, it is quite natural to regard the coefficient of $\partial^4u/\partial x^4$ to be positive. Indeed, the dispersion law of harmonic vibration of a 1D crystal is generally such that the group velocity $d\omega/dk$ decreases with increasing k for small k.

For the equation of motion (1.2.12) the long-wave ($ak \ll 1$) dispersion relation for small (harmonic) vibrations has the form

$$\omega^2 = s^2k^2 - B^2k^4, \tag{1.2.13}$$

and the group velocity for $ak \ll 1$ is $v = s - (3/2)(B^2k^2/s)$. Thus, the coefficient discussed should really be positive to provide a decrease in v with k increasing.

An harmonic approximation describes well small crystal perturbations. But in some cases there arises the necessity to describe the motion of crystal atoms, which is accompanied with their large displacements. It is natural to pose the question whether the motion of crystal atoms is possible in which a strong perturbation will move along the crystal without changing the form of this perturbation?

If the displacement gradients connected with this perturbation are small there exists a positive answer to this question within the harmonic approximation. As we have noted, in this case the atom displacement is described by the linear wave equation (1.1.14) whose solution is any double-differentiated function depending on the argument $x - st$ (if the wave runs in the positive direction of the axis x) or $x + st$ (if the wave runs in the opposite direction). We remind readers that s is the sound velocity.

But if the harmonic approximation is insufficient, the answer to the question posed is no longer obvious. We study this question using the *Boussinesq equation* (1.2.12). Since a transfer of any deformation impulse in a 1D crystal is connected with motion of a local compression, the analysis of dynamics of the derivative $p = u_x$ makes sense supposing that the plot of $p(x)$ has a form similar to Fig. 1.4.

Thus, we find a stationary solution to (1.2.12), moving along the axis x, i. e., a solution of the form $u = u(x - Vt)$, where V is an arbitrary parameter (the perturbation velocity). As the desired function is, actually, a function of one argument $\xi = x - Vt$ (1.2.12) is transformed into the ordinary differential equation

$$B^2u_{xxxx} - \Lambda^2u_xu_{xx} - \gamma u_{xx} = 0, \tag{1.2.14}$$

where $\gamma = V^2 - s^2$. We introduce new notations $\alpha = \Lambda/B$, $\beta = \gamma/B^2$ and rewrite (1.2.14) with respect to p:

$$p_{xxx} - \alpha^2pp_x - \beta p_x = 0. \tag{1.2.15}$$

We detract from a specific value of the coefficient B given by the definition (1.2.11) and seek for a formal solution to (1.2.15) for all possible values of β. Integrating (1.2.15) over x twice, taking into account the boundary conditions at infinity, we obtain

$$\left(\frac{dp}{dx}\right)^2 - \frac{1}{3}\alpha^2p^3 - \beta p^2 = 0. \tag{1.2.16}$$

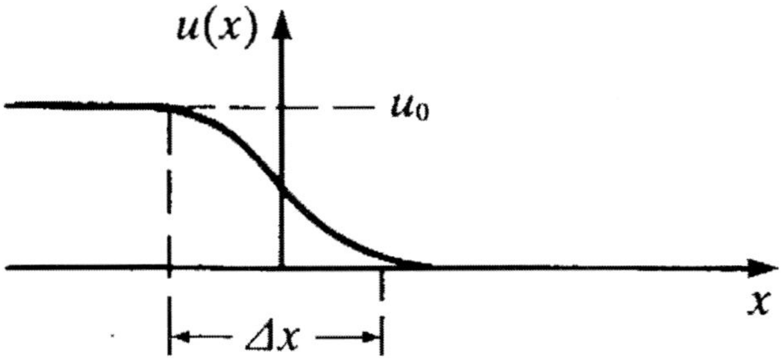

Fig. 1.4 Solitary wave of 1D crystal deformation.

Equation (1.2.16) is easily integrated, and we obtain the solution that is interesting for us

$$p = -p_0 \operatorname{sech}^2 \left[\frac{1}{2} \sqrt{\beta}(x - Vt) \right], \tag{1.2.17}$$

where $p_0 = 3\beta/\alpha^2$. It follows from the solution (1.2.17) that it exists only for $p_0 > 0$, i. e., for $V > s$.

As the quantity p has the meaning of 1D crystal strain, the solution (1.2.17) obtained by us really describes a local compression moving with velocity V along a 1D crystal. A similar solution for the deformation is called a solitary solution, or a *soliton*. This is a singular (isolated) solution to the equation concerned, which may move along the crystal without changing its form.

Rewrite (1.2.17)

$$p = -p_0 \operatorname{sech}^2 \left(\frac{x - Vt}{l} \right), \tag{1.2.18}$$

where l is the soliton width determining the transition region Δx in Fig. 1.4. The value of l can be found from asymptotes of the solution $u \sim \exp(-x/l)$ of a corresponding linear equation (obtained from (1.2.14) or (1.2.15) with $\Lambda = 0$), and it is

$$l = \frac{1}{\sqrt{\beta}} = \frac{B}{\sqrt{V^2 - s^2}}. \tag{1.2.19}$$

The velocity of the motion of the perturbation concerned is connected with the amplitude p_0:

$$V^2 = s^2 + \frac{1}{3}\Lambda^2 p_0. \tag{1.2.20}$$

Thus, the nonlinear differential equation (1.2.14) has the desired solution (1.2.17) only for the velocity determined by (1.2.20).

In conclusion, it should be noted that substituting a specific value $B^2 = a^2s^2/12$ that follows from (1.2.11) into the formula for the overall width l yields: $l = as/\sqrt{12(V^2 - s^2)}$. Thus, the long-wave approximation ($l \gg a$) is actually valid only in a narrow interval of velocities V near s.

1.3
Transverse Vibrations of a Linear Chain

We consider a special linear analog of a simple crystal lattice assuming the atoms to be positioned periodically along a certain line in 3D space. Let a be a lattice constant and n the atom number counted from any point of the chain. We direct the x-axis along the undeformed straight line chain and denote by v the vector of transverse atom displacements (perpendicular to the x-axis), retaining the notation u for the longitudinal component of the displacement vector. The great interest in studying the vibrations of the 1D system proposed is explained by the fact that this problem is an excellent model of the dynamics of homopolymer molecules.

Keeping in mind possible applications to vibrations of the homopolymer molecules we restrict ourselves to the long-wave approximations. In the harmonic approximation the longitudinal and transverse vibrations are independent, and we analyze each form of motion separately. If the atoms are displaced along the x-axis, the elastic energy is determined by their relative displacements. The relative displacement of neighboring atoms is $\xi_n = u_n - u_{n-1}$, and, in the nearest-neighbor approximation, the crystal potential energy equals a sum such as (1.2.4), so that in the harmonic approximation, it is possible to employ only the expansion (1.2.5). The forces generating the potential energy (1.2.4), (1.2.5) provide, between the neighboring atoms, a certain analog of spring coupling with the elasticity coefficients α. Such forces are called *central forces*.

In going over to the long-wave vibrations with the replacement $u_n \to u(x)$ can be effected, the leading term when expanding the difference ξ_n in powers of a/λ (1.2.8), where λ is the characteristic wavelength, is proportional to the first derivative of $u(x)$ with respect to x. The crystal potential energy (1.2.4) then becomes

$$U = \int \varphi\left(au'(x)\right)\frac{dx}{a}, \tag{1.3.1}$$

and according to (1.2.5) the energy density is

$$\varphi = \frac{1}{2}\alpha a^2\left(u'\right)^2. \tag{1.3.2}$$

Equation (1.1.14) with $s^2 = \alpha a^2/m$ is obtained in a standard way from (1.3.1), (1.3.2).

If the atoms in a linear chain are displaced perpendicular to the x-axis, in the harmonic approximation central interaction forces do not arise and the crystal energy depends on the relative rotations of the segments connecting atoms in neighboring pairs rather than on the relative displacement of neighboring atoms. We assume that the transverse displacements of all atoms lie in one plane and denote the transverse displacement in this plane as v_n, the angle of similar rotation by θ (Fig. 1.5). Then, as seen from the figure, for small θ one can write for the nearest neighbors

$$\theta_n = \frac{1}{a}\left(v_{n+1} - v_n\right) - \frac{1}{a}\left(v_n - v_{n-1}\right) = \frac{1}{a}\left(v_{n+1} + v_{n-1} - 2v_n\right).$$

Thus, the crystal energy with noncentral interaction forces of the nearest atom pairs taken into account will have an additional term

$$V = \sum_n \psi(a\theta_n) = \sum \psi\left(v_{n+1} + v_{n-1} - 2v_n\right),$$

and, in the harmonic approximation,

$$\psi(\eta) = \frac{1}{2\beta}\eta^2.$$

In the long-wave limit, when v_n is replaced by a continuous function of the coordinate $v(x)$, the leading term of the θ_n angle expansion in powers of a/λ is proportional to the second derivative of $v(x)$ with respect to x:

$$a\theta_n = v_{n+1} + v_{n-1} - 2v_n = a^2\frac{\partial^2 v}{\partial x^2}. \tag{1.3.3}$$

The crystal energy V takes the form of an integral over the whole length of a linear chain

$$V = \int \psi\left(a^2 v''(x)\right)\frac{dx}{a},$$

where the energy density is

$$\psi = \frac{1}{2}\beta a^4\left(v''\right)^2. \tag{1.3.4}$$

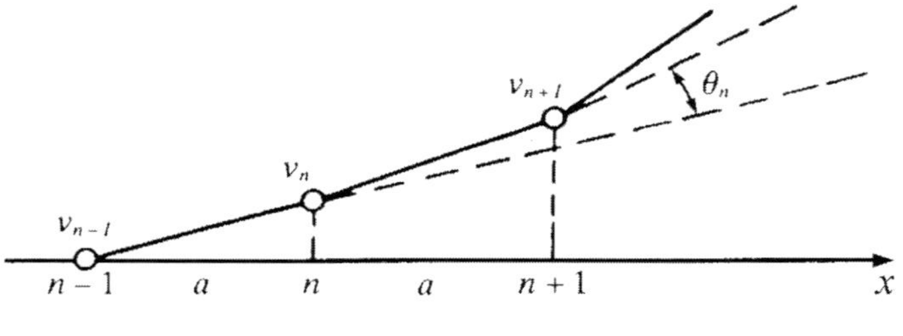

Fig. 1.5 Atom configuration in a 1D chain with transverse (bending) vibrations.

The potential energy density (1.3.4) leads to the following equation

$$m\frac{\partial^2 v}{\partial t^2} + aA^2\frac{\partial^4 v}{\partial x^4} = 0, \tag{1.3.5}$$

where $A^2 = \beta a^2/m$. Equation (1.3.5) describes the so-called *bending vibrations*.

Comparing (1.1.14) and (1.3.5) shows that the term with the fourth-order derivative in (1.3.5) includes a small parameter of the order $(a/\lambda)^2$ unavailable in the term with the second-order derivative in (1.1.14). With such small terms preserved, the linear approximation may be insufficient. The nonlinearity should be taken into account, in particular, in the presence of static stretching forces applied to chain ends. Under the action of such forces there arises a homogeneous longitudinal deformation $\partial u/\partial x = \epsilon_0 = \text{const}$ dependent on stretching load, so that it can be large.

In constructing an elementary nonlinear theory of vibrations of the chain concerned, anharmonicity should be taken into account only in the terms associated with central forces and the potential energy of small noncentral forces V can be calculated in the ordinary harmonic approximation.

The main nonlinearity is generated by the anharmonicities of central forces. If the nonlinearity is small, the crystal energy can be described by (1.2.4), (1.2.5) having determined more exactly the relative distance between neighboring atoms δl_n, with account taken of displacements in a direction perpendicular to the x-axis (Fig. 1.6)

$$\delta l_n = \left\{ \left(a + u_n^z - u_{n-1}^z \right)^2 + (v_n - v_{n-1})^2 \right\}^{\frac{1}{2}} - a = \left\{ (a + \xi_n)^2 + \eta_n^2 \right\}^{\frac{1}{2}} - a,$$

where $v(u_n^y, u_n^z)$ is the 2D transverse displacement vector, $\xi_n = u_n^x - u_{n-1}^x = u_n - u_{n-1}$; $\eta_n = v_n - v_{n-1}$.

To describe the nonlinear bending vibrations of a chain, we use the expression for δl_n, written with sufficient accuracy

$$\delta l_n^2 = \xi_n^2 + \frac{1}{a}\xi_n \eta_n^2 + \frac{1}{4a^2}\eta_n^2 \eta_n^2. \tag{1.3.6}$$

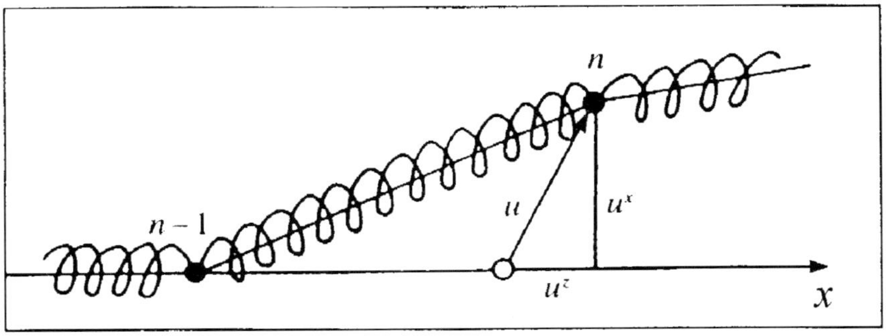

Fig. 1.6 A scheme of displacements at transverse 1D chain vibrations (the displacement of an atom with the number $n - 1$ equals zero).

Comparison of the first two terms on the r.h.s. of (1.3.6) reveals that under bending vibrations, the value of ξ_n is commensurate with $(l/a)\eta_n^2$; therefore, we retain the last term proportional to η_n^4.

Taking into account the above arguments one should write instead of (1.2.4)

$$U_* = \sum_n \varphi(\delta l_n) = \frac{1}{2}\alpha \sum \delta l_n^2,$$

conserving, for the noncentral interaction energy, the expression

$$V = \sum_n \psi \left(v_{n+1} + v_{n-1} - 2v_n \right) = \sum_n \psi \left(\eta_{n+1} + \eta_n \right) = \frac{1}{2}\beta \sum_n \left(\eta_{n+1} + \eta_n \right)^2.$$

It is seen that the total potential energy of the chain W in the nonlinear approximation can be divided into three parts

$$W = U_* + V = U + U_\perp + U_{\text{int}}, \tag{1.3.7}$$

where U is the total energy of longitudinal vibrations given by (1.2.4); $U_\perp$ is the energy of transverse (bending) vibrations

$$U_\perp = \frac{\alpha}{8a^2}\sum_n \eta_n^4 + \frac{1}{2}\beta\sum_n (\eta_{n+1} - \eta_n)^2, \tag{1.3.8}$$

and U_{int} is the energy of the interaction between transverse and longitudinal vibrations

$$U_{\text{int}} = \frac{\alpha}{2a}\sum_n \xi_n\eta_n^2. \tag{1.3.9}$$

The presence of the energy (1.3.9) means that in a nonlinear approximation the independence of longitudinal and transverse vibrations in a chain vanishes.

We compose the nonlinear equations of motion by using

$$m\frac{d^2u_n}{dt^2} = -\frac{\partial W}{\partial u_n}, \qquad m\frac{d^2v_n}{dt^2} = -\frac{\partial W}{\partial v_n}.$$

Simple calculations lead to

$$m\frac{d^2u_n}{dt^2} = \alpha\left(\xi_{n+1} - \xi_n\right) + \frac{\alpha}{2a}\left(\eta_{n+1}^2 - \eta_n^2\right), \tag{1.3.10}$$

$$\begin{aligned} m\frac{d^2v_n}{dt^2} &= \alpha\left(\xi_{n+1}\eta_{n+1} - \xi_n\eta_n\right) + \frac{\alpha}{2a^2}\left(\eta_{n+1}^2\eta_{n+1} - \eta_n^2\eta_n\right) \\ &\quad -\beta\left(\eta_{n+2} - 3\eta_{n+1} + 3\eta_n - \eta_{n-1}\right). \end{aligned} \tag{1.3.11}$$

These equations allow us to describe the bending vibrations, taking into account the influence of longitudinal chain vibrations. In order to derive the nonlinear equations of vibrations in a continuous approximation we make use of the expansions such as (1.2.8) and (1.3.3), retaining the higher space derivatives and the leading nonlinear terms containing the functions $u(x)$ and $v(x)$:

$$\frac{\partial^2 u}{\partial t^2} = s^2\frac{\partial^2 u}{\partial x^2} + \frac{1}{2}s^2\frac{\partial}{\partial x}\left(\frac{\partial v}{\partial x}\right)^2; \tag{1.3.12}$$

$$\frac{\partial^2 v}{\partial t^2} = s^2\frac{\partial}{\partial x}\left(\frac{\partial u}{\partial x}\frac{\partial v}{\partial x}\right) + \frac{1}{2}s^2\frac{\partial}{\partial x}\left[\left(\frac{\partial v}{\partial x}\right)^2\left(\frac{\partial v}{\partial x}\right)\right] - a^2A^2\left(\frac{\partial^4 v}{\partial x^4}\right). \tag{1.3.13}$$

The total potential energy of a linear chain (1.3.7), in the long-wave approximation corresponding to (1.3.12), (1.3.13), is equal to

$$\begin{aligned} W = \frac{1}{2}a^2\int\Bigg\{&\alpha\left(\frac{\partial u}{\partial x}\right)^2 + \alpha\left(\frac{\partial u}{\partial x}\right)\left(\frac{\partial v}{\partial x}\right)^2 \\ &+ \frac{\alpha}{4}\left(\frac{\partial v}{\partial x}\right)^2\left(\frac{\partial v}{\partial x}\right)^2 + a^2\beta\left(\frac{\partial^2 v}{\partial x^2}\right)^2\Bigg\}\frac{dx}{a}. \end{aligned} \tag{1.3.14}$$

Using the rules of functional differentiation it is easy to find from (1.3.14) the forces acting on vibrating atoms and leading to (1.3.12), (1.3.13).

Neglecting anharmonicities of transverse displacements reduces (1.3.12) to (1.1.14) with the dispersion law $\omega = sk$ and simplifies (1.3.13). If the linear chain is free (static stresses are absent), the nonlinear term retained in (1.3.13) that contains $\partial u/\partial x$ is also small. If the action of external forces generates a considerable static homogeneous deformation $\partial u/\partial x = \epsilon_0 =$ constant, (1.3.13) in a linearized form is reduced to

$$v_{tt} = \epsilon_0 s v_{xx} - a^2 A^2 v_{xxxx}. \tag{1.3.15}$$

Equation (1.3.15) is associated with the dispersion relation

$$\omega^2 = \epsilon_0 s^2 k^2 + a^2 A^2 k^4. \tag{1.3.16}$$

It is interesting that the relation (1.3.16) is similar to the dispersion law (1.2.13) but with the opposite sign in front of the term proportional to k^4.

For $ak \ll \sqrt{\epsilon_0 (s/A)}$ the dependence typical for the acoustic branch follows from (1.3.16)

$$\omega = \sqrt{\epsilon_0}(sk) = s_* k. \tag{1.3.17}$$

Its sound velocity $s_* = s\sqrt{\epsilon_0}$ is small. Thus, the long-wave dispersion law of bending vibrations of a linear chain that experienced a static longitudinal stretching does not differ qualitatively from the dispersion law of longitudinal vibrations of this chain.

In the region of wavelengths for which $\sqrt{\epsilon_0}(s/A) \ll ak \ll 1$, we obtain the dispersion law

$$\omega = aAk^2,$$

which is typical for bending waves.

1.4 Solitons of Bending Vibrations of a Linear Chain

We analyze (1.3.12), (1.3.13) to clarify their purely nonlinear properties. Recall that the interest in the above equations is explained by the fact that they model the dynamics of homopolymer molecules.

To simplify the system of nonlinear equations (1.3.12), (1.3.13), we introduce, instead of time, a variable $\tau = st$. The one-dimensional parameter aA/s will then remain in (1.3.12), (1.3.13). Estimating it we assume that with regard to order of magnitude, $A \sim s$. The equations will be written in the form

$$u_{\tau\tau} = \frac{\partial}{\partial x}\left(u_x + \frac{1}{2}v_x^2\right); \tag{1.4.1}$$

$$v_{\tau\tau} = \frac{\partial}{\partial x}\left[\left(u_x + \frac{1}{2}v_x^2\right)v_x - \left(\frac{aA}{s}\right)^2 v_{xxx}\right]. \tag{1.4.2}$$

We remind ourselves of the notations $u_x = \partial u/\partial x$, $v_x = \partial v/\partial x$, $v_{xx} = \partial^2 v/\partial x^2$, etc. It is easy to find the solutions to (1.4.1), (1.4.2) in the form of a stationary profile whose dependence on the coordinate and time is represented through a combination $\zeta = x - V\tau$. The waves of a stationary profile are the solutions to a system of ordinary differential equations

$$(1 - V^2)u_{xx} + \frac{1}{2}\frac{d}{dx}v_x^2 = 0, \tag{1.4.3}$$

$$V^2 v_{xx} + \left(\frac{aA}{s}\right)^2 v_{xxxx} = \frac{d}{dx}\left[\left(u_x + \frac{1}{2}v_x^2\right) v_x\right]. \tag{1.4.4}$$

The equations given admit one trivial integration. To perform this, we assume the linear chain experiences a static longitudinal stretching (longitudinal strain) equal to ε_0 ($\varepsilon_0 \ll 1$). Besides that, as we are interested primarily in the solitary waves, we assume all velocities and all gradients vanish at infinity. Under such boundary conditions we get from (1.4.3)

$$u_x = \varepsilon_0 - \frac{v_x^2}{2(1 - v^2)}; \qquad \varepsilon_0 = \text{constant}. \tag{1.4.5}$$

It follows then from (1.4.4) that

$$(V^2 - \varepsilon_0)v_x + \left(\frac{aA}{s}\right)^2 v_{xxx} = \frac{V^2}{2}\frac{v_x^2 v_x}{V^2 - 1}. \tag{1.4.6}$$

We use the fact that (1.4.6) involves only the derivatives of the vector function $\boldsymbol{v}$ and denote $\boldsymbol{w} = \boldsymbol{v}_x$. Equation (1.4.6) will then take the form

$$\boldsymbol{w}_{xx} + (\gamma + \beta \boldsymbol{w}^2)\boldsymbol{w} = 0, \tag{1.4.7}$$

where

$$\beta = \frac{1}{2}\left(\frac{s}{aA}\right)\frac{V^2}{1 - V^2}, \qquad \gamma = \left(\frac{s}{aA}\right)\left(V^2 - \varepsilon_0\right).$$

We introduce the amplitude and the phase φ of the transverse motion velocity by means of the relation

$$\boldsymbol{w} = w(\boldsymbol{i}_1 \cos\varphi + \boldsymbol{i}_2 \sin\varphi),$$

where $\boldsymbol{i}_1, \boldsymbol{i}_2$ are the unit vectors of two coordinate axes perpendicular to the direction of a nondeformed chain. The vector equation (1.4.7) will then be reduced to the two equations

$$w_{xx} - w\varphi_x^2 + \beta w^3 + \gamma w = 0; \tag{1.4.8}$$

$$\frac{d}{dx}\left(w^2 \varphi_x\right) = 0. \tag{1.4.9}$$

Equation (1.4.9) has the form of the area conservation law (if by the variable x we understand the time) and gives the integral of motion

$$I = w^2 \varphi_x = \text{const}. \tag{1.4.10}$$

Taking into account (1.4.10), (1.4.8) is simplified as

$$w_{xx} + \left(\beta w^2 + \gamma - \frac{I^2}{w^4}\right) w = 0. \tag{1.4.11}$$

It is easy to see that solitary waves cannot have nonzero integral of motion I. Indeed, we set $I \neq 0$ and integrate (1.4.11)

$$w_x + \frac{1}{2}\beta w^4 + \gamma w + \frac{I^2}{w^2} = c, \qquad c = \text{const.} \tag{1.4.12}$$

It is clear that if $I \neq 0$ it is impossible to get a phase trajectory passing through a point $w = w_x = 0$ on the phase plane at any choice of the integration constant c.

Such solutions exist only at $I = 0$. The case $I = 0$ corresponds to $\varphi = \text{const}$ and refers to a plane-polarized bending wave running along the chain. Phase trajectories for the plane-polarized wave follow from (1.4.12)

$$w_x^2 + \gamma w^2 + \frac{1}{2}\beta w^4 = c. \tag{1.4.13}$$

The general behavior of the trajectories (1.4.13) is determined by the ratio of the signs of the parameters β and γ. It follows from the definition of the latter that both of them can be positive or have different signs, but cannot be negative simultaneously. Solitary waves are possible only when $\beta > 0$ and $\gamma < 0$, i. e., for $V^2 < \varepsilon_0$. In this case there exists a typical soliton separatrix (Fig. 1.7), corresponding to $c = 0$, among the phase trajectories. Equation (1.4.13) at $c = 0$ is easily integrated

$$w(\xi) \equiv v_x(\xi) = \frac{\sqrt{2\,|\gamma|\,/\beta}}{\cosh\sqrt{|\gamma|}\xi}. \tag{1.4.14}$$

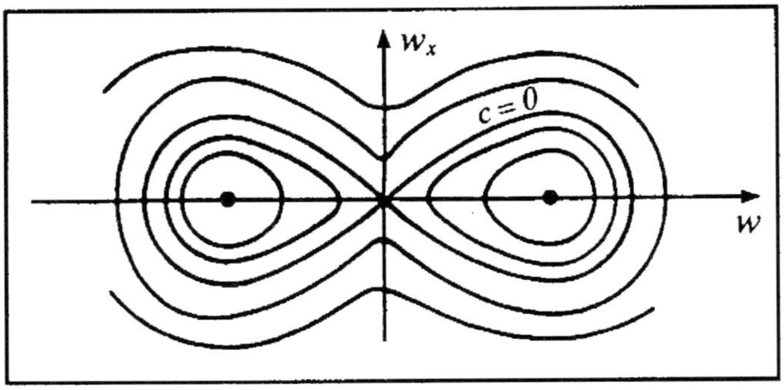

Fig. 1.7 Separatrix ($c = 0$) corresponding to a soliton.

A solitary wave of the transverse velocity field (1.4.14) is a slow soliton with velocity $V < \sqrt{\varepsilon_0}$. The absence of solitons for $V > \sqrt{\varepsilon_0}$ is not surprising and is explained

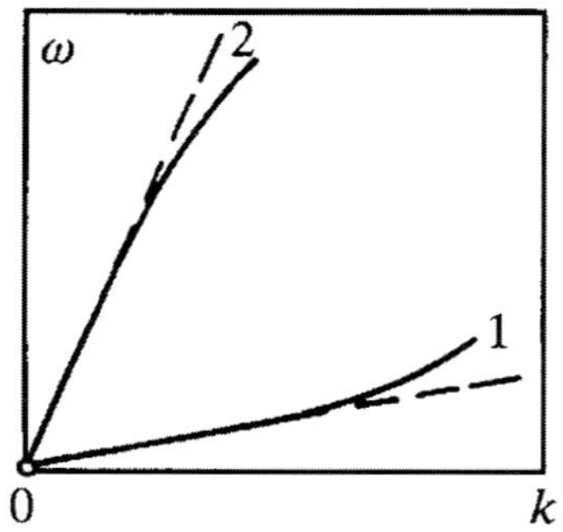

Fig. 1.8 Dispersion laws of a linear chain for the waves:
1 – bending; 2 – longitudinal.

by the form of the dispersion law for harmonic eigenvibrations of the chain. The plots of bending wave laws (Fig. 1.8 curve 1) and the longitudinal chain deformation waves (curve 2) are described by (1.2.13) and (1.3.16).

The condition $V < \sqrt{\varepsilon_0}$ thus means that the soliton velocity V does not get into the interval of phase velocities of the linear chain modes and, under positive dispersion of the latter, is less than the minimum phase velocity of harmonic vibrations.

The region where the slow soliton is localized is inversely proportional to its amplitude: $\Delta x = 1/\sqrt{|\gamma|} = a(A/s)\sqrt{\varepsilon_0 - V^2}$. For $V^2 \to \varepsilon_0$ the soliton amplitude decreases proportionally to $\sqrt{\varepsilon_0 - V^2}$ and the localization region increases as $(\varepsilon_0 - V^2)^{-1/2}$. It can be concluded that the soliton smears out as its velocity approaches the limiting one.

1.5
Dynamics of Biatomic 1D Crystals

A biatomic 1D crystal is different from a monatomic chain in that its unit cell contains two atoms. Let us consider the vibrations of a two-atomic 1D crystal with nearest-neighbor interaction. Let atoms with masses m_1 and m_2 alternate, located at a distance b apart. The translational period of the crystal is $a = 2b$, and its lattice (Fig. 1.9) has an inversion center. Practically this is the simplest example of a superlattice consisting of two sublattices, i. e., of two Bravais lattices. Assuming the lattice point to coincide with the first atom type we obtain the following equation of motion

$$\begin{aligned} m_1 \frac{d^2 u_1(n)}{dt^2} &= -\alpha[2u_1(n) - u_2(n) - u_2(n-1)], \\ m_2 \frac{d^2 u_2(n)}{dt^2} &= -\alpha[2u_2(n) - u_1(n) - u_1(n+1), \end{aligned} \tag{1.5.1}$$

where α is a unique elastic parameter of the model.

It is easily seen that the dispersion relation for stationary vibrations is determined by the solution to

$$\begin{vmatrix} m_1\omega^2 - 2\alpha & \alpha(1+e^{iak}) \\ \alpha(1+e^{-iak}) & m_2\omega^2 - 2\alpha \end{vmatrix} = 0,$$

whose roots are

$$\begin{aligned} \omega_1^2(k) &= \frac{1}{2}\omega_0^2\left\{1 - \sqrt{1-\gamma^2 \sin^2\frac{ak}{2}}\right\}, \\ \omega_2^2(k) &= \frac{1}{2}\omega_0^2\left\{1 + \sqrt{1-\gamma^2 \sin^2\frac{ak}{2}}\right\}, \end{aligned} \tag{1.5.2}$$

where $\omega_0^2 = 2\alpha(m_1+m_2)/m_1m_2, \gamma^2 = 4m_1m_2/(m_1+m_2)^2$.

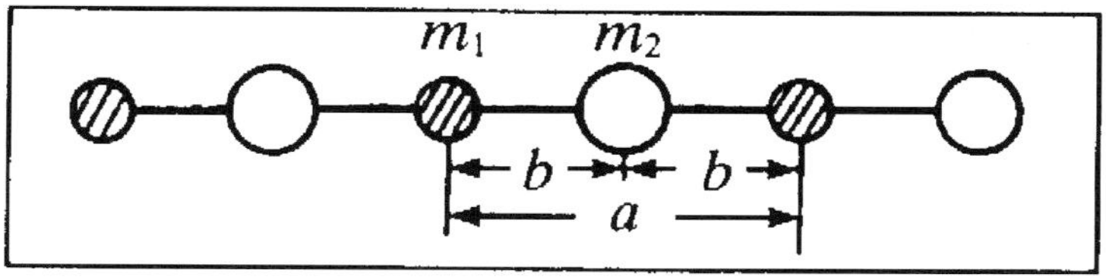

Fig. 1.9 Diatomic one-dimensional crystal.

Therefore, every value of the quasi-wave number k meets two frequencies in the spectrum of the biatomic chain $\omega = \omega_\beta(k)$, $\beta = 1, 2$, i.e., the dispersion relation possesses two branches of the dependence of ω on k. Two branches appear in the spectrum due to the presence of two degrees of freedom in the unit cell.

The two branches of vibrations are different. The most essential differences of the branches take place at small $k(ak \ll 1)$. At $k = 0$ we have $\omega_1(0) = 0$, $\omega_2(0) = \omega_0$ and we have in the long-wave limit $(ak \ll 1)$

$$\omega_1(k) = sk, \qquad s = \frac{1}{4}\omega_0\gamma a, \qquad \omega_2(k) = \omega_0\left(1 - \frac{\gamma^2 a^2}{32}k^2\right). \tag{1.5.3}$$

In order to formulate a difference of two types of vibrations, it is convenient to introduce the displacement of the center of mass of a pair of atoms, $u(n)$, i.e., the center of mass of a unit cell, and the relative displacement of atoms in a pair, $\xi(n)$:

$$u(n) = \frac{1}{m}(m_1u_1(n) + m_2u_2(n)), \qquad \xi(n) = u_1(n) - u_1(n). \tag{1.5.4}$$

Under long-wave vibrations with the dispersion law $\omega(k) = sk$, the unit cell centers of mass vibrate with the relative position of atoms in a pair remaining unchanged. Therefore, we have $u(n) = u_0 e^{i\omega t}$, $\xi(n) = 0$.

A feature of the second dispersion law in (1.5.3) is that the corresponding vibrations with an infinitely large wavelength have the finite frequency ω_0. This vibration at $k = 0$ is

$$u(n) = 0, \quad \xi(n) = \xi_0 e^{-i\omega t}.$$

Under such crystal vibrations the centers of mass of the unit cells are at rest and the motion in the lattice is reduced to relative vibrations inside the unit cells. The presence of such vibrations distinguishes a diatomic crystal lattice from a monatomic one.

The low-frequency branch of the dispersion law ($\omega < \omega_m$) describes the *acoustic* vibrations, and the high-frequency one ($\omega_1 < \omega < \omega_2$) the *optical* vibrations of a crystal. Thus, the biatomic 1D crystal lattice, apart from acoustic vibrations (A) also has optical vibrations (O). The optical branch of the dispersion law (1.5.2) is separated from the acoustic branch by the gap $\delta\omega$ on the Brillouin zone boundary (Fig 1.10a), and this separation at $m_1 > m_2$ is equal:

$$\delta\omega = \omega_2 - \omega_1 = \sqrt{2\alpha/m_2} - \sqrt{2\alpha/m_1}. \qquad (1.5.5)$$

Formation of a gap in 1D vibration spectrum is an inevitable consequence of the appearance of a superlattice.

The frequency $\omega_1 = \sqrt{2\alpha/m_1}$ is a frequency of homogeneous vibrations of the superlattice 1 relative to the resting superlattice 2, and the frequency $\omega_2 = \sqrt{2\alpha/m_2}$ is the frequency of homogeneous vibrations of the superlattice 2 relative the motionless superlattice 1.

However, for $m_1 = m_2 = m$ the parameter $\gamma^2 = 1$, the gap $\delta\omega$ vanishes and the dispersion laws (1.5.2) degenerate:

$$\omega_1(k) = 2\sqrt{\frac{\alpha}{m}}\,\left|\,\sin\frac{ak}{4}\,\right|, \quad \omega_2(k) = 2\sqrt{\frac{\alpha}{m}}\,\left|\,\cos\frac{ak}{4}\,\right|. \qquad (1.5.6)$$

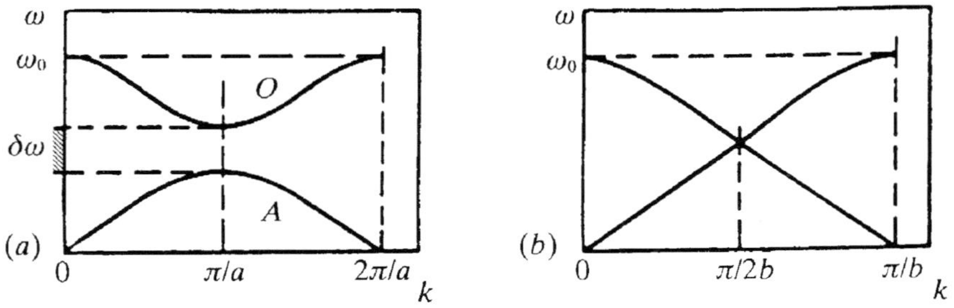

Fig. 1.10 Transformation of the dispersion law when the crystal period reduces twice: (*a*) dispersion branches at $m_1 \neq m_2$, (*b*) junction of branches at $m_1 = m_2$.

The degeneracy corresponds to transformation of a biatomic lattice into a monatomic one with the period $b = a/2$. Since (1.5.6) implies that $\omega_1(k) = \omega_2(k + (2\pi/a))$, both equations (1.5.6) describe in fact the same acoustic dispersion law of dispersion of this 1D lattice (Fig. 1.10b).

1.6
Frenkel–Kontorova Model and sine-Gordon Equation

In Section 1.2 we analyzed a deformation soliton that describes a specific elastic perturbation of the 1D crystal that belongs to the acoustic type of collective excitations, but solitary deformation waves may also arise due to the nonlinearity of optical crystal excitations.

We assume a 1D crystal to be in a given external periodic field whose period coincides with the 1D lattice constant a. The crystal energy will then be determined not only by a relative displacement of neighboring atoms, but also by an absolute displacement of separate atoms in an external potential field. We write this additional crystal energy as

$$W = \sum_n F(u_n), \quad F(u+a) = F(u). \tag{1.6.1}$$

This situation may arise in the case when the atomic chain is on the ideally smooth surface of a 3D crystal that serves as substrate and determines a periodic potential (1.6.1).

The availability of this potential essentially affects the dynamics of a 1D system, since the equation of crystal motion now takes the form

$$m\frac{d^2u_n}{dt^2} = \sum_{n'} \alpha(n-n')[u(n) - u(n')] - \frac{dF(u_n)}{du_n}. \tag{1.6.2}$$

If the displacements $u(t)$ are small then

$$F(u) = \frac{1}{2}K^2u^2, \quad K^2 = F''(0) > 0, \tag{1.6.3}$$

and, thus, taking (1.6.1), (1.6.3) into account in studying small crystal vibrations would result in the appearance in the dispersion law $\omega = \omega(k)$ of a nonzero frequency of extremely long-wave vibrations: $\omega(0) = \omega_0 = K/\sqrt{m}$.

We shall be interested in the solutions to (1.6.2) whose behavior corresponds to the plot in Fig. 1.11 and that satisfy the boundary conditions

$$u(-\infty) = a, \quad u(\infty) = 0.$$

These solutions describe such a deformation of a 1D crystal under which the atoms of the left-hand end ($x = -\infty$) are displaced into their neighboring equilibrium positions. At $x = \infty$, the crystal remains undeformed $u = 0$, i. e., all atoms are in their positions. As the number of atoms in the crystal is fixed, the number of atoms in the vicinity of a certain point $x = x_0$ is larger by unity than the number of the initial static equilibrium positions (Fig. 1.12).

The problem of such an aggregate of atoms near some point (a *crowdion*) was first considered and solved by Frenkel and Kontorova (1938), thus, the corresponding model is named after them. In *the Frenkel–Kontorova model* the additional assumption

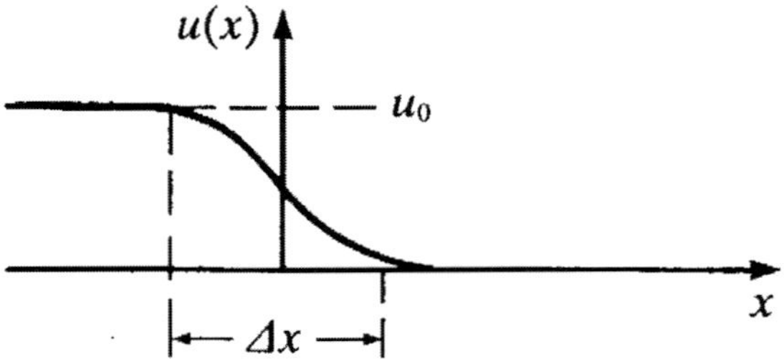

Fig. 1.11 Step perturbation ("overfall") in a 1D crystal.

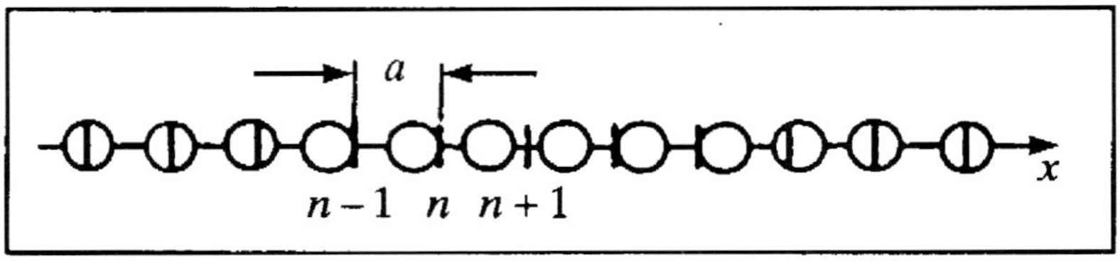

Fig. 1.12 Atomic distribution in a crowdion.

of a simple form of the function $F(u)$ is made

$$F(u) = \frac{1}{2}K^2\tau^2 \sin^2\frac{\pi u}{a}, \tag{1.6.4}$$

but the latter seems to be unimportant, although it greatly simplifies the solution to the nonlinear problem that we encounter here.

The nonlinearity of the function $F(u)$ such as (1.6.4) is basic in the Frenkel–Kontorova model. Thus, the interatomic interaction of the nearest neighbors along a 1D crystal is sufficient to be taken into account within the harmonic approximation. In other words, the equation of crystal motion should be taken in the form

$$m\frac{d^2u_n}{dt^2} = \alpha_0\left[u_{n+1} + u_{n-1} - 2u_n\right] - \frac{aK^2}{2\pi}\sin\frac{2\pi u_n}{a}, \tag{1.6.5}$$

where $K = \pi\tau/a = \sqrt{m}\omega_0$.

Considering only the case of long waves ($\lambda \gg a$), we go over to a continuum treatment by replacing (1.6.5) with the equation in partial derivatives

$$\frac{\partial^2 u}{\partial t^2} = s_0^2\frac{\partial^2 u}{\partial x^2} - \frac{a}{2\pi}\omega_0^2 \sin\frac{2\pi u}{a}, \qquad s_0^2 = \frac{a^2\alpha_0}{m}. \tag{1.6.6}$$

Equation (1.6.6) can be associated with the Lagrange function whose density is

$$\mathcal{L} = \frac{m}{2a}\left\{\left(\frac{\partial u}{\partial t}\right)^2 - s_0^2\left(\frac{\partial u}{\partial x}\right)^2 - \omega_0^2\left(\frac{a}{\pi}\right)^2\sin^2\frac{\pi u}{a}\right\}. \tag{1.6.7}$$

In the harmonic approximation ($2\pi u \ll a$), (1.6.6) describes optical vibrations with the dispersion law:

$$\omega^2 = \omega_0^2 + s_0^2 k^2. \tag{1.6.8}$$

The dispersion law plot (1.6.8) is characterized by two parameters (Fig. 1.13): the limiting frequency ω_0 and the minimum phase velocity of optical vibrations s_0.

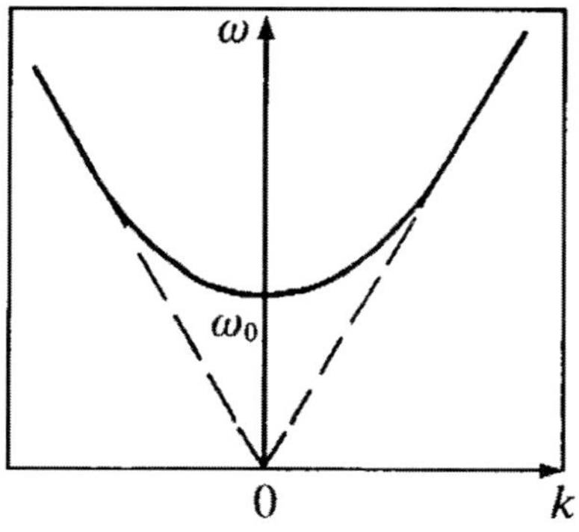

Fig. 1.13 The dispersion law of harmonic vibrations.

To simplify the equations further, we introduce the dimensionless displacement $2\pi u/a$ and denote it by the same letter u. Equation (1.6.6) will then change slightly:

$$\frac{\partial^2 u}{\partial t^2} = s_0^2 \frac{\partial^2 u}{\partial x^2} - \omega_0^2 \sin u. \tag{1.6.9}$$

This equation is called the *sine-Gordon equation*. All its physically meaningful solutions are now systematized and studied.

Using the results of the previous sections, we look for a solution to (1.6.9) in the form

$$u = u(x - Vt), \quad V = \text{const},$$

where the function $u(x)$ satisfies the boundary conditions and has a plot such as in Fig. 1.11. The equation for the function $u(x)$ follows from (1.6.9)

$$(s_0^2 - V^2)u_{xx} - \omega_0^2 \sin u = 0. \tag{1.6.10}$$

This is actually the equation of a mathematic pendulum and can easily be integrated. The first integral of (1.6.10) corresponding to the desired function is

$$\frac{d^2 u}{dt^2} = \pm \frac{2\omega_0}{\sqrt{s_0^2 - V^2}} \sin \frac{u}{2}. \tag{1.6.11}$$

It is also easy to perform a second integration that leads us to the following result

$$u(x,t) = 4 \arctan \left[\exp\left(\frac{x_0 - x}{l} \right) \right], \tag{1.6.12}$$

where l is a parameter determining the crowdion width

$$l = \frac{\sqrt{s_0^2 - V^2}}{\omega_0} = \frac{s_0}{\omega_0}\sqrt{1 - \left(\frac{V}{s_0}\right)^2}. \tag{1.6.13}$$

When V approaches the limiting velocity s_0, the crowdion width experiences a relativistic reduction. In the limit $V = s_0$ there arises a step with $\delta u = a$. At $V = 0$ a fixed inflection of width $l_0 = s_0/\omega_0$ remains in the crystal.

Since a homogeneous translation onto the crystal period is a symmetry transformation, the physical state of a 1D crystal far from the point $x = x_0$ ($|x - x_0| \gg x_0$) is similar to that in the absence of a crowdion. The derivative du/dx coinciding with a linear chain deformation is evidence of the change in the state of a one-dimensional system in the presence of a crowdion. It follows from (1.6.12) (in initial dimensional quantities)

$$\frac{du}{dx} = -\frac{a}{\pi l}\frac{1}{\cosh\left(\dfrac{x - x_0}{l}\right)}. \tag{1.6.14}$$

The localized deformation (1.6.14) moving in the crystal has the form of a solitary wave and is another example of a strain soliton in a 1D crystal. Finally, coming back to a dimensional displacement and to a complete dependence on the coordinates and time in (1.6.12), we get the final expression for the displacements

$$u(x,t) = \frac{2a}{\pi}\arctan\left\{\exp\left[-\omega_0\frac{x - Vt}{\sqrt{s_0^2 - V^2}}\right]\right\}. \tag{1.6.15}$$

The perturbation described by this formula moves with velocity V (necessarily less than that of a sound s_0). This makes it different from the shock-wave perturbation. The velocity is determined by the total energy associated with this perturbation.

The last observation will be supported with a certain qualitative argument of general character. It follows from a comparison of the properties of solitons of the two types considered by us. Irrespective of the character of nonlinearity that generates a soliton, the value of its limiting velocity is completely determined by the dispersion law of harmonic vibrations that can exist in the system under study.

If a plot of the dispersion law of linear vibrations is convex upwards similar to the plot of the dispersion law (1.2.13), it is characterized by some maximum phase velocity s_0. The velocity of a soliton (if it arises) will exceed s. If a plot of the dispersion law shows convexity downwards, as in the case of a sine-Gordon equation, it is characterized by the minimum phase velocity s_0, and the velocity of the existing soliton should be less than s_0. This conclusion will be proved in the following.

1.7 Soliton as a Particle in 1D Crystals

The local deformation associated with a crowdion (1.6.15) moved along a 1D crystal, remaining undeformed and losing no velocity, i. e., moves "by inertia" like a particle. However, the similarity between the soliton of the sine-Gordon equation and a particle is much stronger. It can be observed in the soliton dynamics in an external field.

In the long-wave approximation, the soliton energy is

$$\begin{aligned} E &= \int_{-\infty}^{\infty} \left\{ \frac{m}{2} \left[\left(\frac{\partial u}{\partial t} \right)^2 + s_0^2 \left(\frac{\partial u}{\partial x} \right)^2 \right] + F(u) \right\} \frac{dx}{a} \\ &= \frac{E_0}{\sqrt{1 - \left(\frac{V}{s_0} \right)^2}}, \qquad E_0 = \frac{2}{\pi} a \tau \sqrt{\alpha_0}. \end{aligned} \tag{1.7.1}$$

For small velocities ($V \ll s_0$) we have

$$E = m^* s_0^2 + \frac{1}{2} m^* V^2, \qquad m^* = \frac{E_0}{s_0^2} = \frac{2m\tau}{\pi a \sqrt{\alpha_0}}, \tag{1.7.2}$$

where m^* is the effective soliton (crowdion) mass. The effective mass is less than that of an individual atom m as measured relative to the ratio of a lattice period to the rest soliton width l_0:

$$m^* = m \frac{2a}{l_0} \ll m. \tag{1.7.3}$$

The energy (1.7.1) can be calculated in a standard way by means of (1.6.7) as the field energy of elastic displacement. It is natural to assume that the soliton momentum may also be determined as the momentum of the field of displacements. Then, by the definition of the field momentum,

$$P = -\int \frac{\partial \mathcal{L}}{\partial u_t} \frac{\partial u}{\partial x} dx = -\rho \int \frac{\partial u}{\partial t} \frac{\partial u}{\partial x} dx, \tag{1.7.4}$$

where $\rho = m/a$.

If a soliton moves with constant velocity, i. e., $u(x,t)$ is a function of the form (1.6.15), then $\partial u/\partial t = -V \partial u/\partial x$, and from (1.7.4) and (1.6.14) it follows[1] that

$$P = \rho V \int \left(\frac{\partial u}{\partial x} \right)^2 dx = -\rho \left(\frac{a}{\pi l} \right)^2 V \int \frac{dx}{\cosh^2 \left(\frac{x}{l} \right)} = \frac{2ma}{\pi^2 l} V.$$

1) We note that the momentum of atoms P_{at} is determined otherwise:

$$P_{at} = \frac{m}{a} \int \frac{\partial u}{\partial t} dx = -\frac{m}{a} V \int_{-\infty}^{\infty} \frac{\partial u}{\partial x} dx = \frac{mV}{a} [u(-\infty) - u(\infty)] = mV$$

and equals the product of the mass of one "extra" atom m and the crowdion velocity V.

Using (1.6.13) and (1.7.3), we obtain a relativistic relation between the momentum and the soliton velocity

$$P = \frac{m^* V}{\sqrt{1 - \left(\frac{V}{s_0}\right)^2}}. \tag{1.7.5}$$

Thus, the soliton behaved as a classical particle with the Hamiltonian $\mathcal{H}$, where $\mathcal{H}^2 = m^* s_0^2 + s_0^2 P^2$. For small momenta the Hamiltonian function agrees with (1.7.2)

$$\mathcal{H} = E_0 + \frac{P^2}{2m^*}. \tag{1.7.6}$$

We suppose now that a 1D crystal experiences a weak external influence and denote by $f(x)$ the density of the force acting on a 1D crystal. Equation (1.6.6) should then be replaced by

$$\rho \frac{\partial^2 u}{\partial t^2} = \rho s_0^2 \frac{\partial^2 u}{\partial x^2} - \frac{m\omega_0^2}{2\pi} \sin \frac{2\pi u}{a} + f(x). \tag{1.7.7}$$

The long-wave approximation assumes that $f(x)$ is a smooth function of the coordinate x varying considerably at distances large as compared to a. If the value of $f(x)$ is small, the presence of such a perturbation in (1.7.7) cannot change considerably the soliton form, affecting mainly its dynamics. This allows us to seek a solution to (1.7.7) satisfying the boundary conditions formulated before as a soliton (1.6.15)

$$u(x,t) = u^s(x - \zeta) \equiv \frac{2a}{\pi} \arctan \left\{ \exp \left[-\frac{x - \zeta}{l_0 \sqrt{1 - (V/s_0)^2}} \right] \right\}, \tag{1.7.8}$$

whose parameters ζ and V change slowly with the soliton moving along the crystal. Since the point $x = \zeta$ is the "center of mass" for the deformation $\varepsilon(x) = \partial u / \partial x$, the soliton velocity may be assumed to be $V = d\zeta/dt$.

To deduce the equation determining the change in the velocity V with time, we use the following procedure. We multiply (1.7.7) by $\partial u/\partial x$ and regroup the terms in the relation obtained

$$-\rho \frac{\partial}{\partial t}(u_t u_x) + \rho \frac{\partial}{\partial x} \left[\frac{1}{2} u_t^2 + \frac{1}{2} s_0^2 u_x^2 + \left(\frac{a\omega_0}{2\pi} \right)^2 \left(1 - \cos \frac{2\pi u}{a} \right) \right] = -f(x) u_x. \tag{1.7.9}$$

We now integrate (1.7.9) over x from $-\infty$ to ∞ and use the definition (1.7.4) as well as the boundary conditions at infinity

$$\frac{dP}{dt} = -\int_{-\infty}^{\infty} f(x) \frac{\partial u}{\partial x} \, dx. \tag{1.7.10}$$

Within the main approximation for small perturbations, we can substitute $u = u^s(x - \zeta)$ on the r.h.s. of (1.7.10). $\partial u/\partial x = -\partial u/\partial \zeta$, and we obtain

$$\frac{dP}{dt} = -\frac{\partial W}{\partial \zeta}, \qquad W(\zeta) = -\int_{-\infty}^{\infty} f(x) u^s(x - \zeta)\, dx. \tag{1.7.11}$$

Equation (1.7.11) has the form of one of the Hamilton equations and $W(\zeta)$ plays the role of the soliton potential energy in an external field. If the force density $f(x)$ is of purely elastic origin, it can be written as $f(x) = \partial \sigma^e/\partial x$, where σ^e are the elastic stresses created in a 1D crystal by external loads. In this case

$$W(\zeta) = -\int_{-\infty}^{\infty} \frac{\partial \sigma^e}{\partial x} u^s\, dx = a\sigma^e(-\infty) + \int_{-\infty}^{\infty} \sigma^e(x) u^s_x(x - \zeta)\, dx. \tag{1.7.12}$$

The first term on the r.h.s. of (1.7.12) is a constant that may not be taken into account in writing the potential energy of a soliton $W(\zeta)$.

We assume further that the field of stresses $\sigma^e(x)$ varies smoothly in space and the characteristic distance of its variation is much larger than l (soliton width). Then we may replace (1.7.12) by

$$W(\zeta) = \sigma^e(\zeta) \int_{-\infty}^{\infty} \frac{\partial u^s}{\partial x}\, dx = a\sigma^e(\zeta). \tag{1.7.13}$$

Comparing (1.7.11), (1.7.13) we conclude that the force acting on a soliton is determined by the gradient (space derivative) of elastic stresses created by external loads at the point where the soliton center of mass is located.

On the other hand, the relation (1.7.13) makes it possible to generalize the expression for the Hamilton function that, for small soliton momenta, takes the form

$$\mathcal{H}(\zeta, P) = E_0 + \frac{P^2}{2m^*} + a\sigma^e(\zeta). \tag{1.7.14}$$

Using (1.7.14), a standard pair of Hamilton equations is constructed

$$\frac{dP}{dt} = -\frac{\partial \mathcal{H}}{\partial \zeta}, \qquad \frac{d\zeta}{dt} = \frac{\partial \mathcal{H}}{\partial P}. \tag{1.7.15}$$

The equations for soliton motion thus take the form of equations of the motion of a particle with mass m^* and effective "charge" a, reflecting its interaction with the elastic potential field, where $\sigma^e(x)$ is the field potential.

The soliton, a collective excitation of a 1D crystal, has very different properties from the collective small harmonic vibrations of a crystal lattice. The primary difference is that the soliton is generated by the nonlinear dynamics of a 1D crystal. Thus, the usual superposition principle is inapplicable to solitons. One may expect that this restricts the possibility of using solitons to describe the excited states of a crystal. Problems may arise when we try to consider many solitons in the crystal or the interaction between soliton perturbation and small harmonic vibrations of the crystal.

The remarkable properties of (1.6.9) eliminate these problems. First, the asymptotic superposition principle is applicable to solitons. If at time $t = 0$ two solitons with velocities V_1 and V_2 (assume them moving in the opposite directions) exist in a 1D crystal, then at $t \to \infty$ the same two solitons with velocities V_1 and V_2 will remain in the crystal. In other words, the asymptotic behavior of soliton solutions to the nonlinear equation (1.6.9) is analogous to the independent behavior of the eigensolutions to a linear equation.

Then, the asymptotic superposition principle is valid also for the interaction of collective excitations of different types: solitons and small harmonic vibrations. Without going into details of the nonlinear mechanics to justify the first remark, we illustrate the validity of the second one by studying the properties of small harmonic vibrations in a 1D crystal containing one crowdion (soliton).

1.8
Harmonic Vibrations in a 1D Crystal Containing a Crowdion (Kink)

We use the simplest (linear) perturbation theory to describe how a harmonic vibration moves through a soliton (1.6.12). We assume a soliton to be at rest (results for a moving soliton can be obtained by means of the Lorentz transformation) and represent the soliton to (1.6.9) in the form

$$u = u^s + u^1, \qquad u^s = 4\arctan\left[\exp\left(-\frac{x}{l_0}\right)\right]. \tag{1.8.1}$$

We substitute (1.8.1) into (1.6.9) and linearize the equation obtained in u^1. Then

$$u^1_{tt} - s_0^2 u^1_{xx} + \omega_0^2\left[1 - \operatorname{sech}^2\left(\frac{x}{l_0}\right)\right] u^1 = 0.$$

We seek for the solution to this linear equation in a standard form $u^1 = \psi(x)e^{-i\omega t}$. For the function $\psi(x)$ we have

$$\omega^2\psi = -s_0^2\frac{d^2\psi}{dx^2} + \omega_0^2\left[1 - \operatorname{sech}^2\left(\frac{x}{l_0}\right)\right]\psi. \tag{1.8.2}$$

Equation (1.8.2) is a Schrödinger stationary equation with a reflectionless potential. One localized eigenstate is among its solutions

$$\psi_{loc}(x) = \frac{2}{l_0}\operatorname{sech}\left(\frac{x}{l_0}\right), \tag{1.8.3}$$

corresponding to a zero frequency ($\omega = 0$), and a set of harmonic vibrations of the continuum spectrum

$$\psi_k(x) = \frac{1}{\sqrt{2\pi}} \frac{\omega_0}{\omega(k)} \left[kl_0 + i \tanh\left(\frac{x}{l_0}\right)\right] e^{ikx}, \tag{1.8.4}$$

with the dispersion law (1.6.8).

Formula (1.8.3) gives a translational mode in the linear approximation

$$\psi_{loc}(x) = \frac{du^s(x)}{dx}.$$

It reflects the homogeneity of a 1D crystal and the possibility to choose arbitrarily the position of the center of gravity of a soliton. Indeed, the linear approximation gives

$$u^s(x) + \delta x \psi_{loc}(x) = u^s(x) + \delta x \frac{du^s(x)}{dx} \cong u^s(x + \delta x).$$

The solutions (1.8.4) describe the harmonic vibrations with the background of a crowdion at rest. Their form confirms the asymptotic superposition principle according to which the independent eigenvibrations are only slightly modulated near the soliton center and change insignificantly. Each eigenvibration is still characterized by the wave number k and the dispersion law (the vibration frequency dependence on k) does not change.

A set of functions (1.8.3), (1.8.4), as a set of eigenfunctions of the self-adjoint operator (1.8.2), forms a total basis in the space of functions of the variable x. This is the most natural basis for the representation of perturbations of a soliton solution, as it allows one to give a clear physical interpretation. A translational mode describes the motion of a soliton mass center, and the continuum spectrum modes refer to the change in its form and the resulting "radiation" of small vibrations.

It follows from (1.8.4) that on passing through a soliton, the eigenvibration reproduces its standard coordinate dependence $\sim e^{ikx}$, but the vibration phase $\eta(x, k)$ is shifted by

$$\eta_k = \eta(+\infty, k) - \eta(-\infty, k) = \pi - 2\arctan(kl_0). \tag{1.8.5}$$

The phase shift (1.8.5) affects the vibration density in a 1D crystal. In the absence of a soliton, the expression for the density of states in the specimen of the finite dimension L follows from the requirement $kL = 2\pi n$, $n = 1, 2, 3 \ldots$, that is a consequence of cyclicity conditions (1.1.8) for the eigenvibrations. In the presence of a soliton an additional phase shift (1.8.5) results in a change in the allowed wave-vector values: $kL + \eta(k) = 2\pi n$, $n = 0, 1, 2, \ldots$. In the limit $L \to \infty$, the spectrum of the values of k becomes continuous. The vibration density (the distribution function for the wave vector k) then equals

$$\nu(k) = \frac{dn}{dk} = \frac{L}{2\pi} + \frac{1}{2\pi}\frac{d\eta(k)}{dk} = \frac{L}{2\pi} - \frac{1}{\pi}\frac{l_0}{1 + (l_0 k)^2}, \tag{1.8.6}$$

where $L/2\pi$ is the vibration density in the absence of a soliton. It is easily seen that

$$\int_{-\infty}^{\infty} \left[\nu(k) - \frac{L}{2\pi} \right] dk = -1.$$

The number of plane-wave vibrations (with a continuous frequency spectrum) scattered by the soliton is reduced by unity because of the presence of one local state with a discrete frequency $\omega = 0$ (translational mode).

Analyzing the changes in the vibration distribution, it is of interest to find the frequency spectrum function $\nu(\omega)$ in a crystal with a soliton. Since the dispersion law (1.6.8) does not change in the presence of a soliton, the function $\nu(\omega)$ is found by multiplying (1.8.6) by $dk/d\omega$:

$$\nu(\omega) = \nu_0(\omega) - \frac{1}{\pi}\frac{\omega_0}{\omega}\frac{1}{\sqrt{\omega^2 - \omega_0^2}} = \nu_0(\omega)\left\{1 - \frac{2l_0}{L}\left(\frac{\omega_0}{\omega}\right)^2\right\}, \qquad (1.8.7)$$

where $\nu_0(\omega)$ is the frequency spectrum in the absence of a soliton

$$nu_0(\omega) = \frac{L}{2\pi s_0}\frac{\omega}{\sqrt{\omega^2 - \omega_0^2}}, \qquad \omega > \omega_0.$$

If a soliton moves with velocity V, the formula for the phase shift is obtained from (1.8.5) by replacing $l_0 \to l(V) = l_0\sqrt{1 - (V/s_0)^2}$:

$$\eta_k = \pi - 2\arctan\left[l_0 k \sqrt{1 - \left(\frac{V}{s_0}\right)^2} \right]. \qquad (1.8.8)$$

Finally, if there are several solitons (crowdions with different velocities) in the system, the total phase shift of an eigenvibration is equal to the sum of the phase shifts for each soliton

$$\eta_k = N\pi - 2\sum_{\alpha=1}^{N} \arctan\left[l_0 k \sqrt{1 - \left(\frac{V_\alpha}{s_0}\right)^2} \right], \qquad (1.8.9)$$

where V_α is the soliton velocity with number α, N is the number of solitons. It becomes clear that a set of kinks and small harmonic vibrations can be considered as "nonlinear normal modes". It follows from the "asymptotic independence" of these modes that the energy is the sum of soliton energies (1.7.1) and the energy of small vibrations with the density of states determined by (1.8.6), (1.8.9).

1.9 Motion of the Crowdion in a Discrete Chain

According to the classical equations of crowdion motion, the crowdion Hamiltonian function (1.7.6) is independent of the position of its center and this is a result of the continuum approximation in which this function is derived.

The simplest way to take into account the discreteness of the system concerned is the following. We use the solution (1.6.12) to the continuous differential equation (1.6.9) and calculate the static energy of a discrete atomic chain with a crowdion at rest by using the formula

$$E = \sum_{n=-\infty}^{\infty} \left\{ \frac{1}{2}\alpha_0 \left(u_{n+1} - u_n \right)^2 + f(u_n) \right\}, \tag{1.9.1}$$

where $u_n = u(x_n) \equiv u(an)$.

It is easily seen that the static crowdion energy in the continuum approximation is equally divided between the interatomic interaction energy and the atomic energy in an external field. It can be assumed that the same equal energy distribution remains in a discrete chain, and instead of (1.9.1) we then write

$$E = 2 \sum_{n=-\infty}^{\infty} F(u_n) = \tau^2 \sum_{n=-\infty}^{\infty} \sin^2 \left(\frac{\pi u_n}{a} \right).$$

We substitute here (1.6.12), assuming the point $x = x_0$ to be a crowdion center:

$$E = \tau^2 \sum_{n=-\infty}^{\infty} \operatorname{sech}^2 \left(\frac{an - x_0}{l_0} \right), \tag{1.9.2}$$

where $l_0 = as\sqrt{m}/(\pi\tau) = a^2\sqrt{\alpha_0}/(\pi\tau)$.

Using the Poisson summation formula,

$$\sum_{n=-\infty}^{\infty} f(n) = \sum_{m=-\infty}^{\infty} \int_{-\infty}^{+\infty} f(k) e^{2\pi i m k}\, dk \,, \tag{1.9.3}$$

we find

$$E = \frac{\tau^2}{a} \int_{-\infty}^{+\infty} \frac{dx}{\cosh^2\left(\frac{x}{l_0}\right)} + \frac{2\tau^2}{a} \sum_{m=1}^{\infty} e^{2\pi i m (x_0/a)} \int_{-\infty}^{+\infty} \frac{e^{2\pi i m (x/a)}}{\cosh^2\left(\frac{x}{l_0}\right)}\, dx. \tag{1.9.4}$$

The first term in (1.9.4) coincides with the energy E_0 of a crowdion at rest found in the continuum approximation (1.7.1), and the second term is a periodic function of the coordinate x_0 with period a: $E = E_0 + U(x_0)$, where

$$U(x) = \frac{2l_0\tau^2}{a}\sum_{m=1}^{\infty} e^{2\pi i m(x/a)} \int_{-\infty}^{+\infty} \frac{\cos\left(\frac{2\pi m l_0 \xi}{a}\right)}{\cosh^2 \xi} d\xi$$
$$= 4\alpha_0 a \sum_{m=1}^{\infty} \frac{m}{\sinh\left(\frac{\pi^2 m l_0}{a}\right)} \cos\left(2\pi m \frac{x}{a}\right). \tag{1.9.5}$$

Since we have assumed that $l_0 \gg a$, it then suffices to keep one term with $m = 1$

$$U(x_0) = U_1 \cos\left(2\pi \frac{x_0}{a}\right), \qquad U_1 = 8\alpha_0 a^2 e^{-\pi^2(l_0/a)}. \tag{1.9.6}$$

It is clear that the crowdion energy is periodically dependent on the coordinate of its center x_0 that may be regarded as a quasi-particle coordinate. We set $x_0 = Vt$ in (1.6.15) using the ordinary relation between the coordinate and the velocity at $V = \text{const} \neq 0$. The part of the energy (1.9.6) that is dependent on the coordinate plays the role of the crowdion potential energy. Minima of the potential energy determine possible equilibrium states of the crowdion ($x_0 = a/2 \pm na$, where $n = 0, \pm 1, \pm 2, \ldots$), and the crowdion can oscillate relative to these equilibrium positions with the frequency

$$\Omega_0^2 = \left(\frac{2\pi}{a}\right)^2 \frac{U_0}{m^*}.$$

Vibration motion with such a frequency can really be a free harmonic oscillation if the quantum energy of the ground (zero) state of the oscillator $\hbar\Omega_0$ is much smaller than U_0:

$$\frac{\hbar^2}{m^* a^2} \ll U_1. \tag{1.9.7}$$

Under such a condition the crowdion in a discrete structure possesses an internal vibration degree of freedom with the frequency Ω_0.

However, in the case $l_0 \gg a$, the potential energy curve (1.9.6) creates very weak potential barriers between the neighboring energy minima, and the crowdion may overcome them through quantum tunneling. Thus, the crowdion migrates really in a discrete atomic chain overcoming the potential relief (1.9.6).

The Hamiltonian

$$\mathcal{H} = E_0 + \frac{p^2}{2m^*} + U_1 \cos\left(2\pi \frac{x}{a}\right) \tag{1.9.8}$$

is used for the quantum description of crowdion motion.

As both m^* and U_1 decrease with increasing parameter $l_0/a \sim a\sqrt{\alpha_0}/\tau$ the physical situations, where

$$\frac{\hbar^2}{m^* a^2} \gg U_1, \tag{1.9.9}$$

are quite reasonable. The inequality (1.9.9) means that the potential energy contribution that is dependent on the coordinate is a weak perturbation of the kinetic energy of free crowdion motion. In other words, the amplitude of zero crowdion vibrations in one of the potential minima (1.9.9) greatly exceeds much the one-dimensional crystal period and the crowdion transforms into a *crowdion wave*.

The energy spectrum of a crowdion wave with the Hamiltonian (1.9.7) is rather complicated and consists of many bands in each of which the energy is a periodic function of the quasi-wave number k with period $2\pi/a$. However, small crowdion wave energies for $k \ll 2\pi/a$ are not practically distinguished from the free particle energy with the Hamiltonian (1.9.8) under the condition (1.9.9). Indeed, if we calculate quantum-mechanical corrections to the free particle energy in the second order of perturbation theory in the potential (1.9.6), then

$$\varepsilon(k) = E_0 - U_1 \frac{m^* a^2 U_1}{(2\pi\hbar)^2} + \frac{\hbar^2 k^2}{2M},$$

$$M = m^* \left\{ 1 + \frac{1}{2} \left[\frac{m^* a^2 U_1}{(\pi\hbar)^2} \right]^2 \right\}.$$

Thus, in spite of the presence of a potential energy curve (1.9.6), the crowdion wave moves through a crystal as a free particle with a mass close to the crowdion effective mass.

1.10 Point Defect in the 1D Crystal

Any distortion of regularity in the crystal atom arrangement is regarded as a crystal lattice defect. A *point* (from the macroscopic point of view) defect is a lattice distortion concentrated in the volume of the order of magnitude of the atomic volume. The typical point defects in a 1D crystal are as follows: an *interstitial atom* is an atom occupying position between the equilibrium positions of ideal lattice atoms (a crowdion can be considered as the extended interstitial atom); a *substitutional impurity* is a "strange" atom that replaces the host atom in a lattice site (Fig. 1.14).

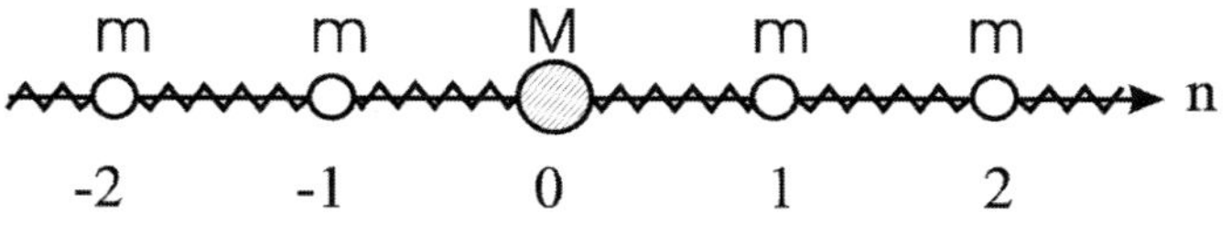

Fig. 1.14 Substitutional impurity in 1D crystal.

The simplest point defect arises in a monatomic species when one of the lattice sites is occupied by an isotope of the atom making up the crystal. Since the isotope atom differs from the host atom in mass only, it is natural to assume that the crystal perturbation does not change the elastic bond parameters. Let the isotope be situated at the origin ($n = 0$) and have a mass M different from the mass of the host atom m. With such a defect we get in the case of stationary vibrations the following set of equations

$$\begin{aligned} &\omega^2 M u(0) - \alpha\left[u(+1) - 2u(0) + u(-1)\right] = 0, \qquad n = 0; \\ &\omega^2 m u(n) - \alpha\left[u(n+1) - 2u(n) + u(n-1)\right] = 0, \quad n \neq 0. \end{aligned} \tag{1.10.1}$$

Equations (1.10.1) can be written more compactly as

$$m\omega^2 u(n) - \alpha[u(n+1) - 2u(n) + u(n-1)] = (m - M)\omega^2 \delta_{n0}, \tag{1.10.2}$$

by introducing the Kronecker delta $\delta_{nn'}$.

We denote

$$\Delta m = M - m, \qquad U_0 = -\frac{\Delta m}{m}\omega^2, \tag{1.10.3}$$

and rewrite (1.10.2) in a form typical for such problems

$$\omega^2 u(n) - \frac{\alpha}{m}\left[u(n+1) - 2u(n) + u(n-1)\right] = U_0 u(0)\delta_{n0}. \tag{1.10.4}$$

We write a formal solution to (1.10.4) as

$$u(n) = U_0 G_\varepsilon^0(n) u(0), \tag{1.10.5}$$

where G_ε^0 is the Green function for ideal lattice vibrations, $\varepsilon = \omega^2$; $u(0)$ is a constant multiplier still to be defined.

Setting $n = 0$ in (1.10.5), we find that (1.10.5) is consistent only when

$$1 - U_0 G_\varepsilon^0(0) = 0. \tag{1.10.6}$$

Equation (1.10.6) is an equation to determine the squares of frequencies ε at which the atomic displacements around an isotope have the form of (1.10.5). In the theory of crystal vibrations with a point defect, an equation such as (1.10.6) was first obtained by Lifshits, 1947.

The expression (1.1.22) for the Green function for $\varepsilon > \omega_m^2$ at $n = 0$ is substituted into (1.10.6):

$$1 - \frac{U_0}{\sqrt{\varepsilon(\varepsilon - \omega_m^2)}} = 0. \tag{1.10.7}$$

Since $\varepsilon > 0$, a solution to (1.10.7) exists only at $U_0 > 0$. It is not difficult to find this solution:

$$\varepsilon = \frac{\omega_m^2}{1 - (\Delta m/m)^2}. \tag{1.10.8}$$

Thus we obtain a discrete frequency corresponding to vibrations of the crystal with the single point defect. For $|\Delta m| \ll m$ this frequency is slightly shifted relative to the upper edge of the frequency spectrum:

$$\frac{\omega - \omega_m}{\omega_m} = \frac{1}{2}\left(\frac{\Delta m}{m}\right)^2. \tag{1.10.9}$$

Crystal vibrations with the frequencies described are called *local vibrations*, and the discrete frequencies are called *local frequencies* ω_d. These names are attributed to the fact that the amplitude of the corresponding vibration is only nonzero in a small vicinity near the point defect. The local vibration amplitude is given by (1.10.6), implying its coordinate dependence is completely determined by the behavior of the ideal crystal Green function. In order to obtain the Green function using (1.1.22) for $\varepsilon > \omega_m^2$ it is convenient to take the quasi-wave number in the complex form (1.1.24):

$$G_\varepsilon^0(n) = \frac{(-1)^n e^{-\kappa n a}}{\sqrt{\varepsilon(\varepsilon - \omega_m^2)}}, \tag{1.10.10}$$

and take into account the connection of the frequency with the parameter κ (1.1.25) for the discrete local frequency ($\varepsilon_d = \omega_d^2$):

$$\varepsilon_d = \omega_{\mathrm{m}}^2 \cosh^2 \frac{a\kappa}{2}. \tag{1.10.11}$$

Combining Eqs (1.10.8) and (1.10.11) one can get

$$\kappa a = \log \frac{2m - M}{M}. \tag{1.10.12}$$

At $m - M \ll m$ (1.10.12) can be simplified:

$$\kappa a = 2\frac{|\Delta m|}{m}. \tag{1.10.13}$$

We substitute the result (1.10.12) in (1.10.10) and rewrite (1.10.5)

$$u(n) = u(0)(-1)^n \left(\frac{M}{2m - M}\right)^n. \tag{1.10.14}$$

Thus, the local vibration amplitude decreases if the distance na increases and this decay of the amplitude confirms the fact of the vibration localization.

Let us introduce a length of the localization region of vibrations $l = 1/\kappa$. As it results from (1.10.12), under the condition $|\Delta m| \ll m$ the length of localization is very large ($l = am/\Delta m \gg a$). In this connection it is interesting to consider a long-wave description of problems concerning with the localization of crystal vibrations near a point defect. Returning to (1.10.4) and using (1.1.15) in the long-wave approximation the Kronecker delta δ_{n0} in the r.h.p. of (1.10.4) can be substituted by the Dirac delta-function

$$\delta_{n0} = a\delta(na) = a\delta(x), \qquad x = na,$$

and the finite differences in the l.h.p. can be substituted with the partial derivatives of the function $v(x)$ (see (1.1.16) written in the same approximation):

$$(\omega^2 - \omega_m^2)v(x) - s^2\frac{\partial^2 v(x)}{\partial x^2} = aU_0 v(0)\delta(x), \qquad s = \frac{1}{2}a\omega_m. \tag{1.10.15}$$

The presence of the delta-function in the r.h.p. of (1.10.15) is equivalent to the following boundary condition for (1.1.16)

$$-s^2\left[\frac{\partial u(x)}{\partial x}\right]_{-0}^{+0} = aU_0 u(0), \tag{1.10.16}$$

which determines a jump of the first spatial derivative of the continuous function $v(x)$ at the point $x = 0$ (on the isotopic defect).

Take a solution to (1.10.15) in the form

$$u(x) = u(0)\exp(-\kappa\,|x|), \qquad s^2\kappa^2 = \omega^2 - \omega_m^2,$$

and find the parameter κ from (1.10.16):

$$\kappa a = 2\frac{|\Delta m|}{m}\left(\frac{\omega}{\omega_m}\right)^2. \tag{1.10.17}$$

In the long-wave approximation $a\kappa \ll 1$ and $\omega - \omega_m \ll \omega_m$, and then the simplification of (1.10.17) is possible:

$$\kappa a = 2\frac{|\Delta m|}{m}.$$

The result obtained coincides with (1.10.13).

Therefore, the long-wave approximation allows us to solve problems associated with local vibrations in the frequency interval $\omega - \omega_m \ll \omega_m$.

1.11 Heavy Defects and 1D Superlattice

In the previous section we analyzed the local vibrations and found that only a light isotope defect could produce such a vibration with a frequency higher than all frequencies of a defectless chain. Under the conditions $m - M \ll m$ and $\omega - \omega_m \ll \omega_m$ description of the problems under consideration could be performed in the long-wave approximation. It was explained why a heavy isotope defect could not produce a local vibration. Nevertheless, the heavy defect influences the continuous vibration spectrum. Obviously a heavy defect can influence low vibrational frequencies, because heavier masses are associated with lower frequencies. One can expect to find essential effects at very low frequencies $\omega \ll \omega_m$ at $M - m \gg m$.

Consider a periodical array of isotope defects with $M > m$ separated by the distance $d = N_0 a$ in the linear chain. Suppose $d \gg a$ ($N_0 \gg 1$) and $M - m \gg m$;

such conditions allow us to use a long-wave approximation. The continuous analog of (1.10.4) for one isolated isotope defect at the frequencies $\omega \ll \omega_m$ is the following

$$\frac{\partial^2 u(x)}{\partial t^2} - s^2 \frac{\partial^2 u(x)}{\partial x^2} = aU_0 u(0)\delta(x - x_0), \qquad (1.11.1)$$

where x_0 is the defect coordinate. See (1.1.4) and compare (1.11.1) with (1.10.15) for $\omega - \omega_m \ll \omega_m$.

The presence of the delta-function in the r.h.p. of (1.11.1) determines a jump of the first spatial derivative of the continuous function $u(x)$ at the point $x = x_0$ and is equivalent to the following boundary condition for (1.1.16)

$$-s^2 \left[\frac{\partial u(x)}{\partial x}\right]_{x_0-0}^{x_0+0} = aU_0 u(0). \qquad (1.11.2)$$

From the macroscopic point of view the model proposed for consideration is a 1D acoustic superlattice consisting of elastic elements of the length d separated by set of joints. Excitations inside the elastic elements are described by the wave equation (1.1.4) with the boundary conditions (1.11.2) on the all joints. From the microscopic point of view such a system is a polyatomic 1D crystal with the unit cell including a very large number N_0 of atoms. Taking into account this fact, let the large unit cells (elements of the superlattice) be numbered by n.

Then the eigenvibration in the n-th element can be written in the form of the Bloch wave

$$u_n(x) = v_k(x)e^{ikx}, \qquad v_k(x + d) = v_k(x), \qquad (1.11.3)$$

where the Bloch amplitude $v(x)$ is a solution to the wave equation (1.1.4) and the function $u(x)$ obeys the boundary conditions (1.11.2).

Let the origin of coordinates coincide with one of the joints (it will have the number $n = 0$). Then (1.11.3) can be written in the form

$$u_n(x) = v(x - nd)e^{ikx}.$$

Take the function $v(x)$ as a general solution to (1.1.4)

$$v(x) = Ae^{iqx} + Be^{-iqx}, \qquad q = \frac{\omega}{s}, \qquad (1.11.4)$$

and consider the solution inside the interval $0 < x < d$.

Writing the boundary conditions such that at the left joint

$$u_{-1}(-0) = u_0(d - 0)e^{-ikd} = v(d)e^{-ikd},$$

and at the right joint

$$u_{+1}(d + 0) = u_0(+0)e^{ikd} = v(0)e^{ikd}.$$

The boundary conditions (1.11.2) and continuity of the function $u(x)$ at the point $x = 0$ or $x = d$ turn into the following set of linear algebraic equations for the coefficients A and B

$$\begin{aligned} A - B - i\left(\frac{aM\omega}{ms}\right)(A+B) &= (Ae^{iqd} - Be^{-iqd})e^{-ikd}, \\ A + B &= (Ae^{iqd} - Be^{-iqd})e^{-1kd}. \end{aligned} \tag{1.11.5}$$

The determinant of (1.11.5) can be easy calculated and equality of the determinant to zero leads to the following expression (the calculation can be considered as a problem for readers)

$$\cos kd = \cos qd - \frac{1}{2}\left(\frac{aM\omega}{ms}\right) q \sin qd. \tag{1.11.6}$$

Introduce a new variable $z = qd$ and rewrite (1.11.6)

$$\cos kd = \cos z - Qz \sin z, \qquad Q = \frac{aM}{2md} = \frac{M}{2N_0 m}. \tag{1.11.7}$$

It is assumed that $M \gg m$ and $N_0 \gg 1$. Since the long-wave approximation was used the variable z can not be very large ($z \ll N_0$).

Equation (1.11.7) gives the dispersion law for the superlattice and corresponds to a band structure of the spectrum. The allowed vibrational frequencies of a continuous spectrum of the system under study can be qualitatively found by analyzing graphically (1.11.7) as shown in Fig. 1.15: if the expression $\cos z - Qz \sin z$ has values between ± 1, the roots of the equation have values in the intervals shown on the abscissa. As a result the vibration spectrum consists of bands of two types: allowed frequency bands and forbidden bands (gaps).

Note that, as z increases, the allowed frequencies are localized within the narrowing intervals near the values $k_1 d = \pm m\pi$ where m is a large integer. For the condition $m^2 Q \gg 1$ the dispersion law for the m-th band can be found. Indeed, near odd $m = 2p + 1$ (see the vicinity of $z = 3\pi$ in Fig. 1.15) we can write with sufficient accuracy

$$\cos kd = -1 + Qm\pi(z - m\pi) = -1 + \frac{m\pi d}{s} Q\left(\omega - \frac{m\pi s}{d}\right),$$

which yields

$$\omega = m\Omega_0 + \frac{Q}{m}(1 + \cos kd), \tag{1.11.8}$$

where $\omega_0 = \pi s/d$ and $\Omega = s/(\pi Q d)$. Similarly, near even $m = 2p$ (see the vicinity of $z = 4\pi$ in Fig. 1.15) we can write

$$\cos kd = 1 - Qm\pi(z - m\pi) = 1 + \frac{m\pi d}{s} Q\left(\omega - \frac{m\pi s}{d}\right),$$

which gives

$$\omega = m\Omega_0 + \frac{Q}{m}(1 - \cos kd). \tag{1.11.9}$$

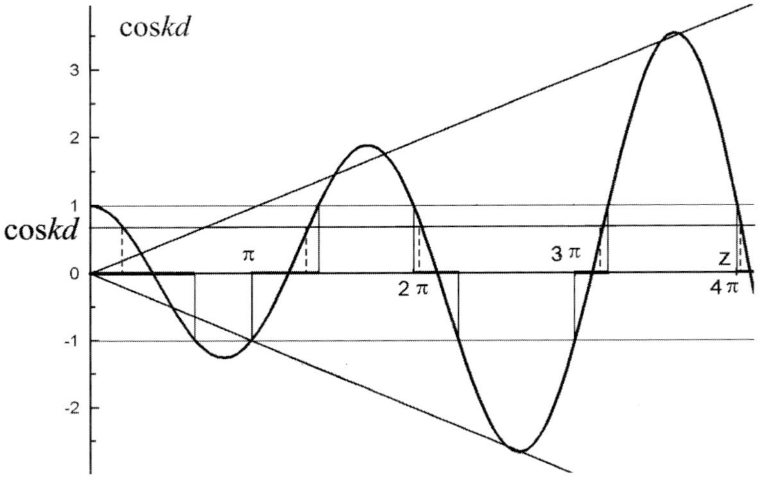

Fig. 1.15 Graphical solution of (1.11.7). If $\cos z - Qz \sin z$ has values between ± 1, the roots of the equation have values in the intervals indicated on the abscissa.

By combining (1.11.8) and (1.11.9), we obtain the dispersion relations for the m-th band:

$$\omega = m\omega_0 + \frac{2Q}{m} \begin{cases} \sin^2(kd/2), & m = 2p\,, \\ \cos^2(kd/2), & m = 2p+1. \end{cases} \tag{1.11.10}$$

One can easily see that expressions (1.11.10) represent the size-quantization spectrum of vibrations in a 1D elastic element of length d with levels (frequencies) split up into minibands due to a low "transparency" of the joints dividing the 1D crystal into elements.

Thus studying a biatomic 1D crystal in Section 1.5 and a simple continuous model of the superlattice in this Section we consider two limiting cases of polyatomic lattices: a lattice with 2 atoms in the unit cell and a lattice with $N_0 \gg 1$ atoms in the unit cell. And the number of gaps in the continuous vibrational spectrum is determined by the number of atoms in the unit cell.

2
General Analysis of Vibrations of Monatomic Lattices

2.1
Equation of Small Vibrations of 3D Lattice

According to the definition of a crystal structure all spatial lattice points can be reproduced by the integral vector $\boldsymbol{n} = (n_1, n_2, n_3)$ where n_α ($\alpha = 1, 2, 3$) are integers. If we choose the coordinate origin at one of the sites, the position vector of an arbitrary lattice site with "number" $\boldsymbol{n}$ will have the form[1]

$$\boldsymbol{r}_n \equiv \boldsymbol{r}(\boldsymbol{n}) = \sum_{\alpha=1}^{3} n_\alpha \boldsymbol{a}_\alpha. \tag{2.1.1}$$

In a monatomic lattice the sites are numbered in the same way as the atoms, so that the integral vector $\boldsymbol{n}$ is at the same time "the number" or "the number vector" of a corresponding atom. However, a lattice site coordinate (2.1.1) differs from the coordinate of the corresponding atom when atoms are displaced relative to their equilibrium positions. If the atom equilibrium positions coincide with the lattice sites (2.1.1), to define the coordinates of an atom, it is necessary, apart from its "number", to indicate its displacement with respect to its own site.

We denote the displacement of an atom with "number" $\boldsymbol{n}$ from its equilibrium position by $\boldsymbol{u}(\boldsymbol{n})$.

The existence of the crystal state means that over a wide temperature range the relative atomic displacements are small compared with the lattice constant a (by the lattice constant we mean the value whose order coincides with the values of fundamental translation vectors $\boldsymbol{a}_\alpha$, in a cubic lattice. Therefore, we shall begin studying crystal lattice vibrations with the case of small harmonic oscillations.

It is easy to derive and to write the equations for small lattice vibrations. We consider the potential energy of a crystal whose atoms are displaced from their equilibrium positions and express it through the displacement $\boldsymbol{u}(\boldsymbol{n})$. The crystal energy is

1) In the following, for the coordinates of the crystal lattice sites, we shall use the notation $\boldsymbol{r}(\boldsymbol{n})$ instead of $\boldsymbol{R}(\boldsymbol{n})$ used in the Introduction.

The Crystal Lattice: Phonons, Solitons, Dislocations, Superlattices, Second Edition. Arnold M. Kosevich

ISBN: 3-527-40508-9

dependent on the coordinates of all particles making up the crystal, that is, the atomic nuclei and electrons. The latter, however, are so mobile that they manage to adapt to nuclear motion. Thus, at every given moment the electronic state is described by a function dependent on the positions of the nuclei. Excluding the electronic coordinates from the crystal energy is the essence of the so-called *adiabatic approximation* that is justified both in studying harmonic vibrations and in the case of small anharmonicity.

Regarding the potential energy of a crystal U as a function only of the coordinates of atomic nuclei that coincide with the coordinates of atomic centers of mass we expand it in powers of $\boldsymbol{u}(\boldsymbol{n})$ and consider the first nonvanishing expansion terms. Assuming the crystal to be in equilibrium at $\boldsymbol{u}(\boldsymbol{n}) = 0$ we write the potential energy in the form

$$U = U_0 + \frac{1}{2} \sum_{\boldsymbol{n},\boldsymbol{n}'} \beta^{ik}(\boldsymbol{n},\boldsymbol{n}') u^i(\boldsymbol{n}) u^k(\boldsymbol{n}'), \tag{2.1.2}$$

where U_0 = constant and the summation is over all crystal sites. The Latin letters i, k (also $j, l, m, \ldots$) are the coordinate indices. The summation over doubly repeated coordinate indices from 1 to 3 is assumed. To simplify the notation of the sums over the numbers $\boldsymbol{n}, \boldsymbol{n}'$, etc., we do not indicate the summation indices under the summation sign if the summation is over all indices involved under the summation sign.

Since we consider that the crystal at $\boldsymbol{u}(\boldsymbol{n}) = 0$ is in mechanical equilibrium, the matrix $\beta^{ik}(\boldsymbol{n},\boldsymbol{n}')$ should be defined positively; in particular, the following inequalities must be satisfied

$$\beta_{11}(\boldsymbol{n},\boldsymbol{n}) > 0, \quad \beta_{22}(\boldsymbol{n},\boldsymbol{n}) > 0, \quad \beta_{33}(\boldsymbol{n},\boldsymbol{n}) > 0.$$

Other important properties of the matrix will be discussed below.

The positively defined matrix with elements $\beta^{ik}(\boldsymbol{n},\boldsymbol{n}')$ is generally called the *matrix of atomic force constants* or *dynamical matrix* of a crystal.

With a general expression for the potential energy (2.1.2) we can easily write down the equation of motion of every atom in a crystal:

$$m\frac{d^2u^i}{dt^2} = -\frac{\partial U}{\partial u^i(\boldsymbol{n})} = -\sum_{\boldsymbol{n}'} \beta^{ik}(\boldsymbol{n},\boldsymbol{n}') u^k(\boldsymbol{n}'), \tag{2.1.3}$$

where m is the atomic mass.

To simplify formulae and equations such as (2.1.2), (2.1.3) containing the $u^i(\boldsymbol{n})$ displacement vectors and the $\beta^{ik}(\boldsymbol{n},\boldsymbol{n}')$ dynamical force matrix or similar matrices, we omit the coordinate indices $i, k, l, \ldots$ for the matrix and vector values, and the square (3×3) matrices are denoted by the corresponding capital letters, e. g., the matrix of force constants is denoted by the letter $\boldsymbol{B}$:

$$\boldsymbol{B} = \beta^{ik}, \quad i, k = 1, 2, 3.$$

By $\boldsymbol{u}(\boldsymbol{n})$ we mean the displacement vector that is in the form of a column consisting of three elements: $u = col(u_1, u_2, u_3)$. Such a notation is only "decoded" when possible misunderstanding may appear.

Thus, the quadratic expression for the potential energy (2.1.2) is

$$U = U_0 + \frac{1}{2}\sum B(\boldsymbol{n},\boldsymbol{n}')u(\boldsymbol{n})u(\boldsymbol{n}'). \tag{2.1.4}$$

In the new notation we write the equations of motion in the form

$$m\frac{d^2u^k}{dt^2} = -\sum_{\boldsymbol{n}'} B(\boldsymbol{n},\boldsymbol{n}')u(\boldsymbol{n}'). \tag{2.1.5}$$

Here we easily see one important property of the coefficients $B(\boldsymbol{n},\boldsymbol{n}')$. Assume the crystal to be displaced as a whole: $\boldsymbol{u}(\boldsymbol{n}) = u_0 =$ constant. Then the internal crystal state cannot change (absence of external forces). Thus, the atoms are not affected by additional forces. Consequently,

$$\sum_{\boldsymbol{n}'} B(\boldsymbol{n},\boldsymbol{n}') \equiv \sum_{\boldsymbol{n}'} \beta^{ik}(\boldsymbol{n},\boldsymbol{n}') = \sum_{\boldsymbol{n}'} \beta^{ik}(\boldsymbol{n}',\boldsymbol{n}) = 0. \tag{2.1.6}$$

The requirement (2.1.6) follows from the conservation of total momentum in the crystal. Another obvious requirement is that the total angular momentum conservation law be automatically obeyed in the absence of external forces. We assume rigid rotation of the crystal effected and described by the axial vector $\boldsymbol{\Omega}$:

$$u^i(\boldsymbol{n}) = \epsilon_{ikl}\Omega_k x^l(\boldsymbol{n}), \tag{2.1.7}$$

where ϵ_{ikl} is a rank three unit antisymmetric tensor and $x^l(\boldsymbol{n})$ is determined by (2.1.1). The above rotation causes no change in the internal lattice state if external forces are absent. Hence, with (2.1.7) substituted into (2.1.3), (2.1.5) we have

$$\epsilon_{klm}\Omega_l \sum_{\boldsymbol{n}'} \beta^{ik}(\boldsymbol{n},\boldsymbol{n}')x^m(\boldsymbol{n}') = 0. \tag{2.1.8}$$

Since the vector $\boldsymbol{\Omega}$ is arbitrary

$$\sum_{\boldsymbol{n}'} \beta^{ik}(\boldsymbol{n},\boldsymbol{n}')x^l(\boldsymbol{n}') = \sum_{\boldsymbol{n}'} \beta^{il}(\boldsymbol{n},\boldsymbol{n}')x^k(\boldsymbol{n}'). \tag{2.1.9}$$

The conditions (2.1.6), (2.1.9) should be satisfied for any atom in the lattice (for any vector $\boldsymbol{n}$). The last point is especially important when not all crystal sites are equivalent. For example, in describing the vibrations of a finite crystal specimen (the atoms near the surface are under conditions differing from those in the crystal interior) or a real crystal with broken translational symmetry (a crystal with defects).

A detailed discussion of constraints imposed on the elements of the matrix $\boldsymbol{B}$ is due to a usual simplification of crystal models when the minimum possible number of elements of this matrix is nonzero, thus describing the interaction of only the nearest-neighbor atoms in the lattice. But it is questionable whether such a choice of model agrees with general physical properties of a crystal, i. e., with mechanical conservation laws.

We now turn to (2.1.4) and discuss firstly the bulk properties of a crystal assuming the crystal lattice to be unbounded. For an unbounded homogeneous lattice, due to its homogeneity the matrix $\boldsymbol{B}$ has the form

$$\boldsymbol{B}(\boldsymbol{n},\boldsymbol{n}') = \boldsymbol{A}(\boldsymbol{n}-\boldsymbol{n}'),$$

where we introduced $\boldsymbol{A} = \alpha_{ik}, \quad i,k = 1,2,3$ whose elements are satisfied by the conditions

$$\sum_{n'} \boldsymbol{A}(\boldsymbol{n}-\boldsymbol{n}') \equiv \sum_{n'} \alpha^{ik}(\boldsymbol{n}-\boldsymbol{n}') = \sum_{n'} \alpha^{ik}(\boldsymbol{n}) = 0. \qquad (2.1.10)$$

Then (2.1.4) is written as

$$U = U_0 + \frac{1}{2}\sum \boldsymbol{A}(\boldsymbol{n}-\boldsymbol{n}')u(\boldsymbol{n})u(\boldsymbol{n}'), \qquad (2.1.11)$$

and the equations of motion (2.1.3) take the form

$$m\frac{d^2u(\boldsymbol{n})}{dt^2} = -\sum_{\boldsymbol{n}'} \boldsymbol{A}(\boldsymbol{n}-\boldsymbol{n}')\boldsymbol{u}(\boldsymbol{n}'). \qquad (2.1.12)$$

We substitute the condition (1.1.10) into the equation of motion (1.1.12) to obtain

$$m\frac{d^2u(\boldsymbol{n})}{dt^2} = \sum_{\boldsymbol{n}'} \boldsymbol{A}(\boldsymbol{n}-\boldsymbol{n}')\left[\boldsymbol{u}(\boldsymbol{n}) - \boldsymbol{u}(\boldsymbol{n}')\right]. \qquad (2.1.13)$$

The equation of motion (2.1.13) illustrates the invariance of a crystal state with respect to its displacement as a whole.

Because (2.1.4) and (2.1.11)–(2.1.13) are equations for three-dimensional displacement vectors we shall use the simplest model to describe the vibration of a crystal lattice where all atoms are displaced in one direction. Then the atom displacement from its equilibrium position is determined by a scalar rather than a vector one. This model allows us to describe the main properties of a vibrating crystal using simple formulae and to obtain the correct quantitative estimates. This model, here called conventionally *scalar*, is used to illustrate the calculation scheme or the ideology of mathematical methods that produce, in a real three-dimensional crystal, the same results as those obtained by employing sophisticated and cumbersome calculations.

In the scalar model the quadratic expressions for the potential energy (1.1.2), (1.1.11) preserve their form; however, we see here no coordinate indices, and the dynamical matrix of an unbounded crystal is reduced to a scalar function of the vector argument $\alpha(\boldsymbol{n})$:

$$U = U_0 + \frac{1}{2}\sum \alpha(\boldsymbol{n}-\boldsymbol{n}')u(\boldsymbol{n})u(\boldsymbol{n}'). \qquad (2.1.14)$$

We include in the properties of a scalar model the energy invariance (1.1.14) with respect to the displacement of a crystal as a whole assuming the function $\alpha(\boldsymbol{n})$ to be satisfied by the condition (1.1.10), i. e.,

$$\sum \alpha(\boldsymbol{n}) = 0. \qquad (2.1.15)$$

Replacing (2.1.11) by (2.1.14) in the analysis of crystal lattice vibrations will in some cases allow us to avoid cumbersome calculations.

2.2 The Dispersion Law of Stationary Vibrations

The stationary crystal vibrations for which the displacement of all atoms are time dependent only by the factor $e^{-i\omega t}$ are of special interest. For such vibrations we obtain instead of (2.1.12)

$$m\omega^2 u(\boldsymbol{n}) = \sum_{\boldsymbol{n}'} A(\boldsymbol{n} - \boldsymbol{n}') u(\boldsymbol{n}'). \tag{2.2.1}$$

If we introduce the notation $\epsilon = \omega^2$ we can rewrite (1.2.1) in a canonical form:

$$\frac{1}{m} \sum_{\boldsymbol{n}'} A(\boldsymbol{n} - \boldsymbol{n}') u(\boldsymbol{n}') - \epsilon u(\boldsymbol{n}) = 0. \tag{2.2.2}$$

Equation (2.2.2) together with certain boundary conditions for the function u is the eigenvalue problem for the linear Hermitian operator $(1/m)A$.

Let the solution to (2.2.1), (2.2.2) has the form

$$u(\boldsymbol{n}) = u e^{i\boldsymbol{k}\boldsymbol{r}(\boldsymbol{n})}. \tag{2.2.3}$$

The vector $\boldsymbol{k}$ is analogous to a wave vector of crystal vibrations and is regarded as a *quasi-wave vector*. It is now a free parameter that determines the solution. Substituting (2.2.3) into (2.2.1), we obtain for the displacement a system of linear equations

$$m\omega^2 u = \sum_{\boldsymbol{n}'} A(\boldsymbol{n} - \boldsymbol{n}') u e^{i\boldsymbol{k}[\boldsymbol{r}(\boldsymbol{n}') - \boldsymbol{r}(\boldsymbol{n})]}.$$

From the definition (1.1.1) it follows that

$$\boldsymbol{r}(\boldsymbol{n}') - \boldsymbol{r}(\boldsymbol{n}) = \sum_{\alpha=1}^{3} (n_\alpha - n'_\alpha)\boldsymbol{a}_\alpha = \boldsymbol{r}(\boldsymbol{n} - \boldsymbol{n}'),$$

hence

$$m\omega^2 u - A(\boldsymbol{k}) u = 0, \tag{2.2.4}$$

where the matrix $A(\boldsymbol{k})$ is given by

$$A(\boldsymbol{k}) = \sum_{\boldsymbol{n}} A(\boldsymbol{n}) e^{-i\boldsymbol{k}\boldsymbol{r}(\boldsymbol{n})}. \tag{2.2.5}$$

Here, and in what follows, we denote with the same capital letter the matrix in various "representations" and distinguish the form of representation by the argument only. Since throughout the book the vectors $\boldsymbol{n}$ or $\boldsymbol{r}$ refer to the real space of sites or the coordinates in a crystal and the vectors $\boldsymbol{k}$ to reciprocal space, this generally accepted system of notations will be quite clear. Thus, in the given case, $\boldsymbol{A}(\boldsymbol{n})$ is the force dynamical matrix in the site representation, $\boldsymbol{A}(\boldsymbol{k})$ is the same matrix in $\boldsymbol{k}$-representation.

The condition for the system (2.2.4) to be compatible has the form

$$\text{Det}\,\|m\omega \boldsymbol{I} - \boldsymbol{A}(\boldsymbol{k})\| = 0, \tag{2.2.6}$$

where $\boldsymbol{I}$ is the unit matrix; $\boldsymbol{I} \equiv \delta_{ik}, \quad i,j = 1,2,3.$

Express now (2.2.6) as

$$D(\omega^2) = 0, \tag{2.2.7}$$

introducing the new notation

$$D(\epsilon) = \text{Det}\left\|\epsilon \boldsymbol{I} - \frac{1}{m}\boldsymbol{A}(\boldsymbol{k})\right\|. \tag{2.2.8}$$

In mechanics, the relation (2.2.6) or (2.2.8) is called the characteristic equation for eigenfrequencies and its solution relates the frequency of possible crystal vibrations to a quasi-wave vector $\boldsymbol{k}$. The wave-vector dependence of frequency is called the *dispersion law* or *dispersion relation* and the equation is referred to as a dispersion equation. Thus, solving the dispersion equation we obtain the dispersion law

$$\epsilon = \omega^2(\boldsymbol{k})$$

for crystal lattice vibrations.

In a monatomic lattice each atom is an inversion center, therefore,

$$\boldsymbol{A}(\boldsymbol{n}) = \boldsymbol{A}(-\boldsymbol{n}),$$

and (2.2.5) is reduced to the following expression, where it is obvious that the matrix $\boldsymbol{A}(\boldsymbol{k})$ is real:

$$\boldsymbol{A}(\boldsymbol{k}) = \sum \boldsymbol{A}(\boldsymbol{n}) \cos \boldsymbol{kr}(\boldsymbol{n}) = \sum \boldsymbol{A}(\boldsymbol{n}) \left[\cos \boldsymbol{kr}(\boldsymbol{n}) - 1\right]. \tag{2.2.9}$$

The last part of (2.2.9) follows from the force matrix property (2.1.10).

Thus, the plane-wave vibration (2.2.3) can be traveling in a crystal if its frequency ω is connected with the quasi-wave vector $\boldsymbol{k}$ by the dispersion law (2.2.6) or (2.2.7).

In a scalar model instead of the system of (2.2.4) there is only one equation of vibrations, and the dispersion law can be written explicitly

$$\omega^2(k) = \frac{1}{m}\sum \alpha(\boldsymbol{n}) e^{-ikr(n)} = \frac{1}{m}\sum \alpha(\boldsymbol{n}) \left[\cos \boldsymbol{kr}(\boldsymbol{n}) - 1\right]. \tag{2.2.10}$$

From (2.2.6) or (2.2.10) we note that the dispersion law determines the frequency as a periodic function of the quasi-wave vector with a period of a reciprocal lattice

$$\omega(\boldsymbol{k}) = \omega(\boldsymbol{k} + \boldsymbol{G}).$$

This proves to be the basic distinction between the dispersion law of crystal vibrations and that of continuous medium vibrations, since the monotonic wave-vector dependence of the frequency is typical for the latter. The difference between the quasi-wave vector $\boldsymbol{k}$ and the ordinary wave vector is also observed here. Only vector $\boldsymbol{k}$ values lying inside one unit cell of a reciprocal lattice correspond to physically nonequivalent states of a crystal.

Recall that free space is homogeneous, i. e., invariant with respect to a translation along an arbitrary vector including an infinitely small one. A set of all similar displacements forms a continuous group of translations. The operator of a translation onto an infinitely small vector is the momentum operator (the momentum operator is said to be the *generator* of a continuous group of translations).

The invariance of mechanical equations with respect to a continuous group of translations generates the momentum vector $\boldsymbol{p}$ as the main characteristic of a free particle state or the wave vector $\boldsymbol{k}$ as the main characteristic of the wave process in a vacuum. The vectors $\boldsymbol{p}$ and $\boldsymbol{k}$ in free space are not restricted in value by any conditions and are related by

$$\boldsymbol{p} = \hbar \boldsymbol{k}. \tag{2.2.11}$$

A crystal lattice, unlike a vacuum, has no homogeneity, but is spatially periodic. We see that the quasi-wave vector is the result of translational symmetry of the periodic structure to the same extent as the wave vector is the result of free-space homogeneity. Thus, in an unbounded crystal the wave processes can be described using the concept of a quasi-wave vector $\boldsymbol{k}$, and the motion of particles using the concept of quasi-momentum related to the vector $\boldsymbol{k}$ via (2.2.11). The wave function corresponding to a quasi-momentum (or quasi-wave vector) represents a plane wave modulated with a lattice period.

When the minimum space dimension (lattice period) tends to zero, the Brillouin zone dimensions become unbounded and we go over to a homogeneous space and return to the concept of momentum and its eigenfunctions in the form of plane waves.

Returning to the dispersion law as a solution of the dispersion equation (2.2.6), we take into account that it is a cubic algebraic equation with respect to ω^2:

$$\text{Det} \left\| m\omega^2 \delta_{ij} - A_{ij}(\boldsymbol{k}) \right\| = 0, \quad i, j = 1, 2, 3. \tag{2.2.12}$$

The roots of this equation determine the three branches for monatomic crystal lattice vibrations specified by the dispersion law: $\epsilon = \omega_\alpha^2(\boldsymbol{k}), \quad \alpha = 1, 2, 3$, where α is the number of a branch of vibrations.

But the characteristic equation (2.2.12) only determines the squared frequency ω^2. Thus, the α-branch dispersion law actualizes each value of the vector $\boldsymbol{k}$ with two frequencies: $\omega = \pm\omega_\alpha(\boldsymbol{k})$. Hence the spectrum of squared frequencies of a vibrating crystal seems to be doubly degenerate. However, as follows from (1.2.9), (1.2.12) the dispersion law is invariant relative to the change in sign of the quasi-wave vector: $\omega^2(\boldsymbol{k}) = \omega^2(-\boldsymbol{k})$. Therefore, the wave with a quasi-wave vector $\boldsymbol{k}$ and frequency

$\omega = -|\omega_\alpha(\boldsymbol{k})|$ describes the same crystal vibrational state as the wave with vector $-\boldsymbol{k}$ and frequency $\omega = -|\omega_\alpha(-\boldsymbol{k})|$. Consequently, in order to describe independent crystal states it suffices to consider the frequency of one sign that corresponds to all possible $\boldsymbol{k}$ vectors inside a single unit cell of the reciprocal lattice. This allows us in what follows to discuss vibrations with positive frequencies only.

2.3
Normal Modes of Vibrations

We have seen that crystal eigenvibrations can be represented in the form of plane waves (2.2.3) whose frequencies are connected with a quasi-wave $\boldsymbol{k}$ by the dispersion law $\omega = \omega_\alpha(\boldsymbol{k})$, $\alpha = 1, 2, 3$. To distinguish between the displacements of different branches of the vibrations we explicitly write (2.2.3)

$$\boldsymbol{u}(\boldsymbol{n}, t) = \boldsymbol{e}(\boldsymbol{k}, \alpha) e^{i[\boldsymbol{k}\boldsymbol{r}(\boldsymbol{n}) - \omega t]}. \tag{2.3.1}$$

Since the equations of motion (2.1.12) or (2.2.1) are homogeneous, their solutions are found up to a constant factor. With this in mind we determine $\boldsymbol{e}(\boldsymbol{k}, \alpha)$, as the unit vector called the *polarization vector*. The dependence of the vector $\boldsymbol{e}$ on $\boldsymbol{k}$ and α follows from equation like (2.2.4), which makes it possible to choose the vectors $\boldsymbol{e}$ real[2] and possessing the property $\boldsymbol{e}(\boldsymbol{k}, \alpha) = \boldsymbol{e}(-\boldsymbol{k}, \alpha)$. Various branches of vibrations correspond to different solutions of some eigenvalue problem, which is why the linear dependence of the eigenfunctions (2.3.1) requires the polarization vectors of vibrations of different branches to be orthogonal:

$$\boldsymbol{e}(\boldsymbol{k}, \alpha)\boldsymbol{e}(\boldsymbol{k}, \alpha') = \delta_{\alpha,\alpha'}. \tag{2.3.2}$$

If the vector $\boldsymbol{k}$ is directed along a highly symmetrical direction (e. g., along a four-fold symmetry axis) there is one longitudinal vibration whose vector $\boldsymbol{e}$ is a simple classification of the possible types of wave polarization breaks down, only three polarization vectors remain mutually orthogonal (2.3.2). For some highly symmetrical directions in a crystal the vibration of the same branch corresponding to a certain value of the index α can be either transverse or longitudinal, depending on the direction of the vector $\boldsymbol{k}$.

We choose the time dependence as $e^{-i\omega t}$ and consider the normalized solutions to (2.2.1) or (2.2.2) in the form

$$\boldsymbol{\phi}_{\boldsymbol{k}\alpha}(\boldsymbol{n}) = \frac{1}{\sqrt{N}} \boldsymbol{e}(\boldsymbol{k}, \alpha) e^{i\boldsymbol{k}\boldsymbol{r}(\boldsymbol{n})}. \tag{2.3.3}$$

According to the properties of a quasi-wave vector we assume the vector $\boldsymbol{k}$ to be in one unit cell of a reciprocal lattice (or in the first Brillouin zone). Under this condition

2) The polarization vector can be chosen as real only in a monatomic lattice.

the eigenfunctions (1.3.3) possess the natural orthogonality properties

$$\sum \boldsymbol{\phi}^*_{k\alpha}(\boldsymbol{n})\boldsymbol{\phi}_{k'\alpha'}(\boldsymbol{n}) = \delta_{kk'}\delta_{\alpha,\alpha'}, \tag{2.3.4}$$

where the asterisk denotes complex conjugation; $\delta_{kk'}$ is the three-dimensional Kronecker symbol.

Thus, the crystal eigenvibrations (2.3.3) are numbered by $(\boldsymbol{k}, \alpha)$. The eigenfunctions (2.3.3) are often called the *normal modes* of the vibrations.

There is no concept of polarization in a scalar model and the coordinate dependence of normal vibrations is written in the form

$$\phi_k(\boldsymbol{n}) = \frac{1}{\sqrt{N}} e^{i\boldsymbol{k}\boldsymbol{r}(\boldsymbol{n})}, \tag{2.3.5}$$

which provides the normalization condition

$$\sum \phi^*_k(\boldsymbol{n})\phi_{k'}(n) = \delta_{kk'}. \tag{2.3.6}$$

2.4 Analysis of the Dispersion Law

To analyze the dispersion law we write (2.2.10) for a scalar model. Consider the vibrations with small $\boldsymbol{k}$, i. e., those for which $ak \ll 1$. We expand the cosine on the r.h.s. of (2.2.11) in powers of its argument and use the fact that the function $\alpha(\boldsymbol{n})$ decreases rapidly with increasing $\boldsymbol{n}$. Then, in the main approximation

$$\omega^2(\boldsymbol{k}) = -\frac{1}{2m} \sum_{\alpha,\beta=1}^{3} (\boldsymbol{k}\boldsymbol{a}_\alpha)(\boldsymbol{k}\boldsymbol{a}_\beta) \sum_n \alpha(\boldsymbol{n}) n_\alpha n_\beta. \tag{2.4.1}$$

We denote $\boldsymbol{k} = k\boldsymbol{\kappa}$, by introducing the unit vector in reciprocal space $\boldsymbol{\kappa}$ and represent (2.4.1) as $\omega^2 = s^2(\boldsymbol{\kappa})k^2$, where

$$s^2(\kappa) = -\frac{1}{2m} \sum_{\alpha,\beta=1}^{3} (\boldsymbol{\kappa}\boldsymbol{a}_\alpha)(\boldsymbol{\kappa}\boldsymbol{a}_\beta) \sum_n \alpha(\boldsymbol{n}) n_\alpha n_\beta. \tag{2.4.2}$$

Thus, for small k we get the linear dispersion law of sound vibrations that is typical for an anisotropic continuum

$$\omega = s(\boldsymbol{\kappa})k, \tag{2.4.3}$$

here s is the phase velocity of an acoustic wave. This result seems to be quite natural, so long as small k correspond to large wavelengths λ, and the condition $ak \ll 1$ determines the requirement $\lambda \gg a$ determining the possibility of passing over from crystal-lattice mechanics to that of a continuum.

According to (2.2.11), for arbitrary $\boldsymbol{k}$ values, in particular for $ak \sim 1$, the character of the dispersion law is determined mainly by the specific form of the matrix $\alpha(\boldsymbol{n})$. In the general case one can assert that for the coefficients $\alpha(\boldsymbol{n})$ decreasing fast enough with increasing number n the function $\omega(\boldsymbol{k})$ is continuous, differentiable, and always bounded.

Thus, the following is typical of the dispersion law: the possible frequencies of crystal vibrations fill the band of a finite width $(0, \omega_m)$ beyond which there are no vibrational frequencies. It is easy to evaluate the order of the maximum frequency value ω_m, which is of the order of magnitude $\omega \sim s/a \sim 10^{13}\ \mathrm{s}^{-1}$ (the typical sound velocity in a crystal $s \sim 10^5$ cm/s).

There exists a very simple model of the spectrum of crystal vibrations that takes into account the availability of a maximum frequency and permits one easily to perform a lot of calculations explicit using the dispersion law. This is the so-called *Debye model* based on the assumption that the dispersion law is linear for all k, but is restricted in frequencies: $\omega = sk, \omega < \omega_D$. The frequency $\omega_D \sim \omega_m \sim 10^{13}\ \mathrm{s}^{-1}$ is called the *Debye frequency*.

In the real situation near the upper edge of the band of possible frequencies, i. e., when $\omega_m - \omega \ll \omega_m$ the frequency and quasi-wave vector are described by the following quadratic dependence

$$\omega = \omega_m - \frac{1}{2}\gamma_{ij}(k_i - k_i^{\mathrm{m}})(k_j - k_j^{\mathrm{m}}), \tag{2.4.4}$$

or by

$$\omega^2 = \omega_{\mathrm{m}}^2 - \omega_{\mathrm{m}}\gamma_{ij}(k_i - k_i^{\mathrm{m}})(k_j - k_j^{\mathrm{m}}). \tag{2.4.5}$$

Here the vector $\boldsymbol{k}_{\mathrm{m}}$ is determined by the condition $\omega(\boldsymbol{k}_{\mathrm{m}}) = \omega_{\mathrm{m}}$ and the matrix of constant coefficients γ_{ik} is defined positively. The terms linear in $\boldsymbol{k} - \boldsymbol{k}_m$ do not enter in (2.4.4) as the frequency ω_m is maximum by definition.

A dispersion law of the type (2.4.4) or (2.4.5) is known as a *quadratic dispersion law*.

Bearing in mind the results of the dispersion law of a scalar model, we go over to considering the general case when the frequency dependence on a quasi-wave vector is obtained by solving (2.2.12).

We first note that from (2.2.9) there follows the property of the dynamical matrix in $\boldsymbol{k}$-representation: $\boldsymbol{A}(\boldsymbol{k}) = \boldsymbol{A}(-\boldsymbol{k})$.

Thus, the solution to (2.2.12) has the same property, namely, the dispersion law is described by the function invariant relative to an inversion in reciprocal space

$$\omega(\boldsymbol{k}) = \omega(-\boldsymbol{k}). \tag{2.4.6}$$

For a scalar model this property follows directly from (2.2.10).

In the limiting case of long waves $(ak \ll 1)$ for the matrix $\boldsymbol{A}(\boldsymbol{k})$ there holds an expansion of the type (2.4.1) that follows directly from (2.2.9):

$$\begin{aligned} \boldsymbol{A}(\boldsymbol{k}) &= -\frac{1}{2}\sum_{\alpha,\beta=1}^{3}(\boldsymbol{k}\boldsymbol{a}_\alpha)(\boldsymbol{k}\boldsymbol{a}_\beta)\sum_{\boldsymbol{n}}\boldsymbol{A}(\boldsymbol{n})n_\alpha n_\beta \\ &= -\frac{k^2}{2}\sum_{\alpha,\beta=1}^{3}(\boldsymbol{\kappa}\boldsymbol{a}_\alpha)(\boldsymbol{\kappa}\boldsymbol{a}_\beta)\sum_{\boldsymbol{n}}\boldsymbol{A}(\boldsymbol{n})n_\alpha n_\beta . \end{aligned} \tag{2.4.7}$$

All elements of the matrix $\boldsymbol{A}(\boldsymbol{k})$ are thus proportional to the square of the wave vector k^2. Therefore, the squares of frequencies, being the solution to (2.2.12), are also proportional to k^2

$$\omega^2 = s_\alpha^2(\boldsymbol{\kappa})k^2, \quad \alpha = 1,2,3, \tag{2.4.8}$$

and in the long-wave limit we get three sound dispersion laws

$$\omega = s_\alpha(\boldsymbol{\kappa})k, \quad \alpha = 1,2,3. \tag{2.4.9}$$

Three branches of vibrations for which (2.4.9) generalize the relation (2.4.3) correspond to the three different sound velocities $s_\alpha(\boldsymbol{\kappa})$.

Consequently, at the point $\boldsymbol{k} = 0$ there is a degeneration, i. e., several branches of vibrations coincide. Due to unambiguity of ω^2 as the wave-vector function at the point $\boldsymbol{k} = 0$, its expansion as a power series in k_i is impossible. The relation (2.4.8) cannot generally be considered as an expansion of the function ω^2 in powers of the wave-vector components. This is just the point in which the long-wave dispersion law of a three-dimensional crystal differs from the dispersion law (1.4.1) for a scalar model.

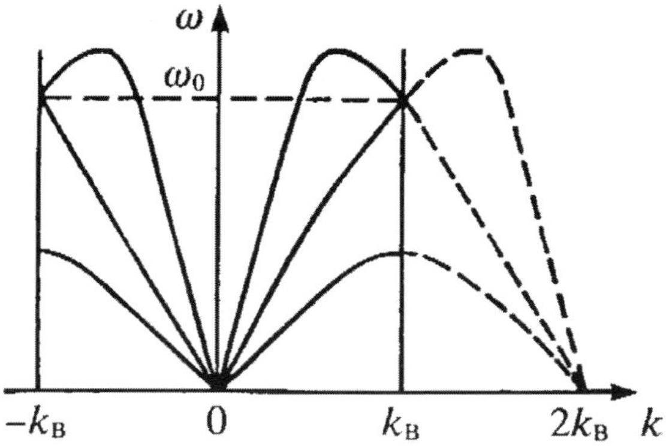

Fig. 2.1 Dispersion diagram with a point of degeneracy (k_B, ω_0) on the Brillouin zone boundary.

The form of the dispersion law at $ak \sim 1$ reflects the specific properties of a real crystal. Thus, we can make some general remarks concerning the dispersion law on Brillouin zone boundaries.

The normal component of the gradient in the $\boldsymbol{k}$-space, $\nabla_k \omega$, vanishes on the Brillouin zone boundary, if at the corresponding point there is no degeneracy. This property of the dispersion law has a simple physical meaning. The gradient determines the group velocity of the wave (2.3.1):

$$v = \frac{\partial \omega}{\partial \boldsymbol{k}} = \nabla_k \omega. \tag{2.4.10}$$

When the vector $\boldsymbol{k}$ ends on the Brillouin zone boundary the group-velocity component normal to it vanishes and the vibrational motion (2.3.1) acquires the character of a stationary wave with respect to this direction.

If at the point considered the degeneration occurs on the zone boundary, the dispersion law plots may approach the Brillouin zone boundary at an arbitrary angle (Fig. 2.1), the points $\boldsymbol{k} = \pm \boldsymbol{k}_B$ give the zone boundary positions. The degeneracy point on the zone boundary corresponds to the frequency $\omega = \omega_0$.

Finally, near the upper edge of frequencies for each branch of vibrations one can expect a quadratic dispersion law of the type (2.4.4) or (2.4.5).

2.5 Spectrum of Quasi-Wave Vector Values

Some vibrations (2.2.3), being independent states of motion of the whole crystal lattice, are characterized by different quasi-wave $\boldsymbol{k}$ values.

It is known that for physically nonequivalent crystal vibrations it suffices to consider the $\boldsymbol{k}$ values lying inside one unit cell of the reciprocal lattice. However, not all points inside the unit cell in $\boldsymbol{k}$-space may correspond to independent crystal states. This follows from the fact that a set of points inside a unit cell composes a continuum but the set of independent vibrations coinciding with the set of degrees of freedom of the crystal lattice turns out to be countable even in the case of an infinite crystal.

For a crystal of finite dimensions the above-mentioned fact is obvious. Therefore, the general qualitative study of crystal vibrations should not be regarded as complete until the spectrum of possible $\boldsymbol{k}$ values has been determined.

When the form of the equations of motion is given, i. e., with the given force matrix $\boldsymbol{A}(\boldsymbol{k})$ in (2.2.2), certain boundary conditions should be formulated to define the spectrum of eigenvalues. However, it seems that a specific form of reasonable boundary conditions has little influence on the spectrum of $\boldsymbol{k}$ values in a crystal consisting of a great number of atoms. Proceeding from this assumption we choose the boundary condition so that it simplifies the solution of the problem as much as possible. Such a condition is the cyclicity requirement according to which

$$u(\boldsymbol{r_n}) = u(r_n + N_1 \boldsymbol{a}_1) = u(r_n + N_2 \boldsymbol{a}_2) = u(r_n + N_3 \boldsymbol{a}_3). \tag{2.5.1}$$

In formulating (2.5.1) it is assumed that a crystal has a form of a parallelepiped with edges $N_\alpha a_\alpha$, $\alpha = 1, 2, 3$, i. e., it contains $N = N_1 N_2 N_3$ atoms.

The cyclicity conditions (2.5.1) are called the *Born–Karman conditions*. In a one-dimensional case the Born–Karman conditions admit a very simple interpretation. We can close up a linear periodic chain of N_1 points into a ring, after which the $N_1 + n$ atom actually coincides with the n-th atom. But in a three-dimensional case a similar attempt to interpret the conditions (2.5.1) does not produce a clear representation.

In studying the bulk dynamical properties of the crystal, we will always proceed from the boundary conditions (2.5.1).

Imposing the requirement (2.5.1) on (2.2.3), we obtain

$$\mathbf{k}\mathbf{a}_\alpha = \frac{2\pi}{N_\alpha} p_\alpha, \quad \alpha = 1, 2, 3, \tag{2.5.2}$$

where p_α are integers. To consider the $\mathbf{k}$ values lying in one unit cell of the reciprocal lattice, we assume p_α to belong to a set $p_\alpha = 0, 1, 2, \ldots, N_\alpha$. In a cubic lattice the formula (1.5.2) will be simplified if we put $L_\alpha = N_\alpha a$ and direct the coordinate axes along the four-fold symmetry axes:

$$k_x = \frac{2\pi}{L_1} p_1, k_y = \frac{2\pi}{L_2} p_2, k_z = \frac{2\pi}{L_3} p_3. \tag{2.5.3}$$

Finally, the set p_α is generally taken to be symmetrical with respect to the numbering axis

$$p_\alpha = 0, \pm 1, \pm 2, \ldots, \pm \frac{N_\alpha + 1}{2}, \quad \alpha = 1, 2, 3.$$

It follows from (2.5.2) or (2.5.3) that the discrete values of the $\mathbf{k}$ vector components are divided by the intervals $\Delta k \sim 1/L$ that decrease with increasing the linear crystal dimensions. Therefore, when all linear crystal dimensions are macroscopic the spectrum of $\mathbf{k}$ values can be regarded as quasi-continuous. The last property of the $\mathbf{k}$ spectrum was used to analyze the dispersion law considering the frequency as a continuous function of quasi-wave vector.

Proceeding further from the quasi-continuity of the spectrum of $\mathbf{k}$ values we change the summation over the discrete values of a quasi-wave vector for the integration. Taking (2.5.2), (2.5.3) into account it is easy to obtain the rule governing this transition to the integration

$$\sum_{\mathbf{k}} f(\mathbf{k}) = \frac{V}{(2\pi)^3} \int f(\mathbf{k})\, d^3k, \tag{2.5.4}$$

where the integration is carried out over the volume of a single unit cell in $\mathbf{k}$-space (or the Brillouin zone).

We note that if we put $f(\mathbf{k}) \equiv 1$ in (1.5.4), we obtain the simple relation

$$\sum_{\mathbf{k}} = \frac{V}{(2\pi)^3} \int d^3\mathbf{k} = \frac{V}{V_0} = N,$$

implying that the number of independent $\mathbf{k}$ vector values in one unit cell equals the number of unit cells (the number of atoms in a monatomic lattice).

2.6
Normal Coordinates of Crystal Vibrations

We have seen that the crystal eigenvibrations can be represented in the form of plane monochromatic waves (2.3.1) where the frequency ω is related to the quasi-wave vector $\boldsymbol{k}$ by the dispersion law $\omega = \omega(\boldsymbol{k})$.

It is clear that the harmonic waves (2.3.1) do not describe the most general motion of atoms in a crystal. But the general solution of the equations of motion (2.1.12) can certainly be expressed through a sum of all possible waves such as (2.3.1). In particular, an arbitrary coordinate dependence of the displacement of a vibrating crystal can be realized by an appropriate set of normal modes (2.3.3).

We shall now expand the crystal vibrations into normal modes for a scalar model, disregarding the polarization vectors and the presence of several branches of the dispersion law. The generalization to a real scheme of three-dimensional lattice vibrations involves no difficulties. It will be carried out after performing all the necessary calculations.

Thus, we represent an arbitrary motion of atoms of the crystal lattice as a superposition of normal vibrations (2.3.5):

$$u(\boldsymbol{n},t) = \sum_{k} Q_{\boldsymbol{k}}(\boldsymbol{n})\psi_{\boldsymbol{k}}(\boldsymbol{n}) = \frac{1}{\sqrt{m}}\phi_{\boldsymbol{k}}(\boldsymbol{n}). \qquad (2.6.1)$$

The quantities $Q_{\boldsymbol{k}}(t) \equiv Q_(\boldsymbol{k})$ are called complex normal coordinates of lattice vibrations[3]. Since the atom displacements (2.6.1) are described by a real function the normal coordinates should have an obvious property

$$Q^*_{\boldsymbol{k}}(t) = Q_{-\boldsymbol{k}}(t). \qquad (2.6.2)$$

Therefore, (2.6.1) is equivalent to

$$u(\boldsymbol{n}) = \frac{1}{2\sqrt{mN}}\sum_{k}\left[Q_{\boldsymbol{k}}e^{i\boldsymbol{k}\boldsymbol{r}(\boldsymbol{n})} + Q^*_{\boldsymbol{k}}e^{-i\boldsymbol{k}\boldsymbol{r}(\boldsymbol{n})}\right], \qquad (2.6.3)$$

showing that the displacements $u(\boldsymbol{n})$ are real.

Now express the mechanical energy of a vibration crystal through normal coordinates. For the kinetic energy K we have

$$\begin{aligned} K = & \frac{m}{2}\sum_{n}\left[\frac{du(\boldsymbol{n})}{dt}\right]^2 = \sum_{k,k'}\dot{Q}_{\boldsymbol{k}}\dot{Q}_{\boldsymbol{k}'}\sum_{n}\phi_{\boldsymbol{k}}(\boldsymbol{n})\phi_{\boldsymbol{k}'}(\boldsymbol{n}) \\ = & \frac{1}{2}\sum_{k}\dot{Q}_{\boldsymbol{k}}\dot{Q}_{-\boldsymbol{k}} = \frac{1}{2}\sum_{k}|\dot{Q}(\boldsymbol{k})|^2. \end{aligned} \qquad (2.6.4)$$

3) The factor $\sqrt{m}$ in the definition of normal coordinates reflects the specific feature of (2.2.1) and is introduced to describe the dynamics of a polyatomic crystal lattice (Section 3.2) in a convenient way.

Performing transformations in (2.6.4) we used the definition (2.3.4) as well as the properties (2.3.6) and (2.6.2).

Let us transform the potential energy of small crystal vibrations, depending on squared displacement, as follows in terms of normal coordinates

$$\begin{aligned} U &= \frac{1}{2}\sum_{n,n'} \alpha(\boldsymbol{n}-\boldsymbol{n}')u(\boldsymbol{n})u(\boldsymbol{n}') \\ &= \frac{1}{2m}\sum_{k,k'} Q(\boldsymbol{k})Q(\boldsymbol{k}')\sum_{n,n'}\alpha(\boldsymbol{n}-\boldsymbol{n}')\phi_{k}(\boldsymbol{n})\phi_{k'}(\boldsymbol{n}) \\ &= \frac{1}{2m}\sum_{k,k'} Q(\boldsymbol{k})Q(-\boldsymbol{k})\sum_{n}\alpha(\boldsymbol{n})e^{-i\boldsymbol{k}\boldsymbol{r}(\boldsymbol{n})} = \frac{1}{2}\sum_{k}\omega^2(\boldsymbol{k})\left|Q_k\right|^2 . \end{aligned}$$

The last in the chain of transformations was performed by making use of (2.2.10) that determines the dispersion law.

Thus, the energy and, hence, the Lagrangian function of crystal vibrations are reduced to a sum of terms that refer to separate normal coordinates. In particular, the Lagrangian function has the form

$$L = K - U = \frac{1}{2}\sum_{k}\left[\dot{Q}_k^*\dot{Q}_k - \omega^2(k)Q_k^*Q_k\right] . \tag{2.6.5}$$

The equation of motion for every normal coordinate follows from (2.6.5):

$$\frac{d^2Q(\boldsymbol{k})}{dt^2} + \omega^2(\boldsymbol{k})Q(\boldsymbol{k}) = 0. \tag{2.6.6}$$

We introduce the generalized momentum conjugate to Q_k

$$P_k = \frac{\partial L}{\partial \dot{Q}(\boldsymbol{k})} = \dot{Q}^*(\boldsymbol{k}),$$

and obtain the Hamiltonian function for crystal vibrations:

$$H = \frac{1}{2}\sum_{k}\left[\left|P(\boldsymbol{k})\right|^2 + \omega^2(\boldsymbol{k})\left|Q(\boldsymbol{k})\right|^2\right] . \tag{2.6.7}$$

Changing to a final formula of the type of (2.6.7) is trivial in general: it suffices to take into account that the normal modes (2.3.3) refer to certain branches of vibrations and, therefore, the coordinates Q and moments P

$$Q = Q_\alpha(\boldsymbol{k}), \quad P = P_\alpha(\boldsymbol{k}).$$

Then,

$$H = \frac{1}{2}\sum_{k,\alpha}\left[\left|P_\alpha(\boldsymbol{k})\right|^2 + \omega_\alpha^2(\boldsymbol{k})\left|Q_\alpha(\boldsymbol{k})\right|^2\right] . \tag{2.6.8}$$

Thus, the independent oscillations are numbered by a pair of indices $(\boldsymbol{k}, \alpha)$ and their number equals that of the vibrational degrees of freedom of a monatomic lattice, i. e., $3N$ (three branches of vibrations and N physically inequivalent $\boldsymbol{k}$ values for each branch).

2.7 The Crystal as a Violation of Space Symmetry

The motion of a crystal lattice in which each atom vibrates around its equilibrium position can be expanded in terms of motions of independent oscillators, i. e., normal vibrations. The crystal energy (or its Hamiltonian function) is separated into the terms corresponding to individual normal modes.

Separation of independent motions that may be superpositioned to compose any complex motion of a system of many particles (atoms) is known as the procedure of introducing the *collective excitations* and relevant collective coordinates (or variables). For small crystal vibrations, i. e., mechanically weakly excited states of a crystal body, the collective excitations are represented by normal modes and collective coordinates by normal coordinates.

The dispersion law of collective vibrations of a monatomic lattice has the universal property: the frequencies of all three branches of vibrations vanish at $k \to 0$. The extremely long-wave vibrations ($k = 0, \lambda = \infty$) are equivalent to the displacement of a lattice as a whole and this property is a direct result of the crystal-energy invariance with respect to its translational motion as a whole. In proving the relation (2.1.6) we proceeded from the fact that due to space homogeneity the internal state of a body is independent of the position of its center of masses.

However, this property (i. e., the condition $\omega(\boldsymbol{k}) \to 0$ as $k \to 0$) of the frequency spectrum of crystal eigenvibrations can be explained in another way. Since the space where a crystal exists is homogeneous, the movement from one point of free space to another by an arbitrary vector, including an infinitely small one, is equal to the transformation into an equivalent state. For this reason the energy of a system of interacting atoms does not change for arbitrary translations of the whole system. The symmetry connected with the Lagrangian function (or Hamiltonian function) invariance relative to transformations of a continuous group of translations is inherent to any system of particles.

However, in the crystal ground state the atoms form a space lattice whose symmetry is lower than the initial one: the physical characteristics of an equilibrium crystal are invariant under a discrete group of translations, since they are described by some periodic functions reflecting the lattice periodicity.

When the symmetry of the ground state of a system is lower than that of the corresponding Lagrangian function, the initial *symmetry* is *broken spontaneously*.

If the properties of the ground state of a system with a large number of degrees of freedom break its symmetry with respect to transformations of a certain continuous group then the collective excitations whose frequencies tend to zero at $k \to 0$ arise in the system (Goldstone, 1961). These excitations seem to strive to re-establish the broken symmetry of the system. The number of branches of such *Goldstone excitations* is determined by the number of broken independent elements of a continuous symmetry group of the Lagrangian function of the system (by the number of "disappeared" generators of the initial symmetry group).

The ground state in a crystal breaks the symmetry relative to continuous translations in three independent directions, which is "generated" by three components of the momentum. The role of the three branches of the collective excitations generated by a spontaneous symmetry breaking is played by the three branches of harmonic crystal vibrations. Thus, the crystal eigenvibrations are Goldstone excitations and, for this reason, their dispersion laws should possess the properties discussed ($\omega(\boldsymbol{k}) \to 0$ as $\boldsymbol{k} \to 0$).

2.8
Long-Wave Approximation and Macroscopic Equations for the Displacements Field

We know that the dispersion law for long-wave vibrations ($ak \ll 1$) coincides with the dispersion law of sound vibrations in a continuous medium. Now we show how the equations of motion of a crystal are simplified for long-wave vibrations, i. e., in what manner the limiting transition from the equations of crystal lattice mechanics to those of a continuous solid is made. It is clear that as one of the results of such a limiting transition we should obtain the known equations of elasticity theory.

Generally, the equation of motion of a homogeneous three-dimensional crystal lattice may be written as

$$m\frac{d^2u_i(\boldsymbol{n})}{dt^2} = -\sum_{\boldsymbol{n}'} \alpha^{ik}(\boldsymbol{n}-\boldsymbol{n}')u^k(\boldsymbol{n}'), \tag{2.8.1}$$

and the matrix $\boldsymbol{A}(\boldsymbol{n})$ obeys the requirement

$$\boldsymbol{A}(\boldsymbol{n}) = \boldsymbol{A}(-\boldsymbol{n}), \quad \sum \boldsymbol{A}(\boldsymbol{n}) = 0.$$

Consider those solutions to (2.8.1) that describe displacement fields weakly varying in space. Let λ be the characteristic wavelength of the relevant displacement with $\lambda \gg a$. The difference in atom displacements in neighboring unit cells $\delta u_{\boldsymbol{n}}$ is then very small compared to the displacement $\delta u(\boldsymbol{n}) \ll u$. Since the natural discrete step in the lattice is small due to the condition $a \ll \lambda$, to analyze the displacements weakly varying in space we use some simplifications. First, we assume that the atom coordinate $\boldsymbol{r} = \boldsymbol{x}(\boldsymbol{n})$ takes a continuous series of values and then the displacement $\boldsymbol{u}(\boldsymbol{n})$ is a continuous function of $\boldsymbol{r}$. We next denote the new function by the same letter and write (2.8.1) as

$$m\frac{d^2u_i(\boldsymbol{r})}{dt^2} = -\sum_{\boldsymbol{n}'} \alpha^{ik}(\boldsymbol{n}-\boldsymbol{n}')u^k(\boldsymbol{r}'), \tag{2.8.2}$$

where $\boldsymbol{r} = \boldsymbol{x}(\boldsymbol{n}); \boldsymbol{r}' = \boldsymbol{x}(\boldsymbol{n}')$.

Expanding the function $\boldsymbol{u}(\boldsymbol{r}')$ as a series near the point $\boldsymbol{r}(x_1, x_2, x_3)$ retaining only the second-order terms in $|x_k(\boldsymbol{n}) - x_k(\boldsymbol{n}')|$:

$$u^k(\boldsymbol{r}') = u^k(\boldsymbol{r}) + (x'_l - x_l)\nabla_l u^k(\boldsymbol{r}) + \frac{1}{2}(x'_l - x_l)(x'_m - x_m)\nabla_l \nabla_m u^k(\boldsymbol{r}). \quad (2.8.3)$$

Here we introduced the notation $\nabla_{i\phi} \equiv \partial\phi/\partial x_i$; it will often be used later.

We now make use of the fast decay of coefficients $\alpha(\boldsymbol{n})$ with increasing n and substitute the expansion (2.8.3) into (1.8.2), taking no account of the terms with higher space derivatives of the displacements:

$$m\frac{\partial^2 u^i}{\partial t^2} = c^{ik}u^k + c^{ikl}\nabla_l u^k + c^{iklm}\nabla_l \nabla_m u^k. \quad (2.8.4)$$

The constant coefficients on the r.h.s. of (1.8.4) are defined by the force matrix elements as:

$$\begin{aligned} c^{ik} = \quad & c^{ki} = -\sum_n \alpha^{ik}(\boldsymbol{n}) = 0; \\ c^{ikl} = \quad & \sum_{n'} \alpha^{ik}(\boldsymbol{n} - \boldsymbol{n}')[x^l(\boldsymbol{n}) - x^l(\boldsymbol{n}')] = -\sum_n \alpha^{ik}(\boldsymbol{n})x^l(\boldsymbol{n}) = 0; \\ c^{iklm} = \quad & -\frac{1}{2}\sum_{n'} \alpha^{ik}(\boldsymbol{n} - \boldsymbol{n}')[x^l(\boldsymbol{n}) - x^l(\boldsymbol{n}')][x^m(\boldsymbol{n}) - x^m(\boldsymbol{n}')] \\ = \quad & -\frac{1}{2}\sum_n \alpha^{ik}(\boldsymbol{n})x^l(\boldsymbol{n})x^m(\boldsymbol{n}). \end{aligned} \quad (2.8.5)$$

Thus, to describe the long-wave (slowly varying in space) crystal displacements we have the following system of second-order differential equations in partial derivatives

$$m\frac{\partial^2 u^i}{\partial t^2} = c^{iklm}\nabla_l \nabla_m u^k. \quad (2.8.6)$$

Equation (2.8.6) coincides in its notation with the dynamical equation of elasticity theory

$$\rho\frac{\partial^2 u^i}{\partial t^2} = \lambda_{iklm}\nabla_k \nabla_l u^m, \quad (2.8.7)$$

where $\rho = m/V_0$ is the mean mass density in a crystal (V_0 is the unit cell volume).

The coefficients on the r.h.s. of (2.8.7) that give the crystal elastic moduli tensor have, however, known symmetry with respect to a permutation of the first and second pairs of indices.

Comparing (2.8.6), (2.8.7) we see that for these equations to be the same, the following equality should hold:

$$\frac{1}{2}(\lambda_{iklm} + \lambda_{ilkm}) = \frac{1}{V_0}c^{imkl}. \quad (2.8.8)$$

The relation (2.8.8) will not be inconsistent only in the presence of the symmetry

$$c^{iklm} = c^{lmik}. \tag{2.8.9}$$

The property (2.8.9) does not follow immediately from the definition (2.8.5) and imposes additional constraints on the force matrix elements of a crystal. We make use of (2.8.5) and write these constraints as

$$\sum \alpha^{ik}(\boldsymbol{n})x^l(\boldsymbol{n})x^m(\boldsymbol{n}) = \sum \alpha^{lm}(\boldsymbol{n})x^i(\boldsymbol{n})x^k(\boldsymbol{n}). \tag{2.8.10}$$

It is clear that the conditions (2.8.10) are actually the result of the invariance of crystal energy relative to a hard rotation of the type (2.1.7).

By imposing the constraints (2.8.10) on the matrix of atomic force constants we provide for the symmetry (2.8.9). This allows us to establish a relation between the tensors λ_{iklm} and c^{iklm}, by solving the relation (2.8.8) for the elastic modulus tensor:

$$\lambda_{iklm} = \frac{1}{V_0}\left(c^{imkl} + c^{kmil} - c^{lmki}\right). \tag{2.8.11}$$

The equality (2.8.11) determines the crystal moduli through the atom force constants, i. e., gives an exact relationship between macroscopic mechanical monocrystal characteristics and microscopic crystal lattice properties.

2.9 The Theory of Elasticity

The transformation from crystal lattice equations (2.8.1) to those of elasticity theory (2.8.7) is accomplished by changing the model of the substance construction. We go over from a discrete structure to a continuum, i. e., the lattice is replaced by a continuous medium. This radical change from a microscopic description of a crystal to a macroscopic one entails new concepts, terms and relations.

For a macroscopic (or continuum) description of crystal deformation, the concept of a displacement vector $\boldsymbol{u}$ as a function of coordinates $\boldsymbol{r}(x, y, z)$ and time t: $\boldsymbol{u} = \boldsymbol{u}(\boldsymbol{r}, t)$ is normally used. Using space derivatives of the displacement vector, the *strain tensor* $(i, k = 1, 2, 3)$ is written as

$$\epsilon_{ik} = \frac{1}{2}\left(\frac{\partial u_i}{\partial x_k} + \frac{\partial u_k}{\partial x_i} + \frac{\partial u_l}{\partial x_i}\frac{\partial u_l}{\partial x_k}\right). \tag{2.9.1}$$

The latter is the main geometrical characteristic of the deformed state of a medium.

The tensor ϵ_{ik} defined by (2.9.1) is sometimes called the *finite strain tensor*, and the part that is linear in displacements,

$$\epsilon_{ik} = \frac{1}{2}\left(\frac{\partial u_i}{\partial x_k} + \frac{\partial u_k}{\partial x_i}\right) \equiv \frac{1}{2}\left(\nabla_i u_k + \nabla_k u_i\right), \tag{2.9.2}$$

is the *small strain tensor*. The linear elasticity theory that we will be concerned with is based on the definition of the strain tensor (2.9.2).

Six different elements of the strain tensor (2.9.2) cannot be absolutely independent since all of them are generated by differentiating three components of displacement vectors. Indeed the strain tensor components ϵ_{ik} are related by differential relations known as the *Saint Venant compatibility conditions*:

$$e_{ilm}e_{kpn}\nabla_l\nabla_p\epsilon_{mn} = 0. \tag{2.9.3}$$

All the components of the tensor of homogeneous deformations (independent of the coordinates) can, however, be arbitrary.

In addition to the strain tensor (2.9.2), often the *distortion tensor* $u_{ik} = \nabla_i u_k$ whose symmetrical part determines the tensor ϵ_{ik} introduced. The antisymmetric part of the distortion tensor gives the vector $\boldsymbol{\omega}$ of the local crystal lattice rotation due to deformation:

$$\boldsymbol{\omega} = \frac{1}{2}\,\mathrm{curl}\,\boldsymbol{u}. \tag{2.9.4}$$

The sum of diagonal elements of the tensor ϵ_{ik}, i. e., the value of e_{kk} equals the relative increase in the volume element under deformation. Consequently, the total change in the volume as a result of deformation ΔV can be written as

$$\Delta V = \int \epsilon_{kk}\,dV. \tag{2.9.5}$$

The time derivative of the vector $\boldsymbol{u}$ determines the velocity of displacements $\boldsymbol{v} = \partial\boldsymbol{u}/\partial t$. If a crystal is deformed but retains its continuity (no breaks, cracks, cavities, etc.) the displacement velocity satisfies the continuity equation

$$\frac{\partial\rho}{\partial t} + \mathrm{div}\,\rho\boldsymbol{v} = 0, \tag{2.9.6}$$

where ρ is the crystal density (the mass of a unit volume).

The forces of internal stresses arising under crystal deformation are characterized by the symmetric *stress tensor* σ_{ik}; the force that acts on unit area is

$$F_i = \sigma_{ik}n_k^0,$$

where $\boldsymbol{n}^0$ is the unit vector normal to the area. If the area concerned is chosen on an external body surface then $\boldsymbol{F}$ equals the force created by external loads.

In the case of hydrostatic crystal compression under the pressure p the tensor σ_{ik} reads

$$\sigma_{ik} = -p\delta_{ik}. \tag{2.9.7}$$

On the basis of (2.9.7)

$$p_0 = -\frac{1}{3}\sigma_{kk} \tag{2.9.8}$$

is called the *mean hydrostatic pressure* when the stress tensor does not coincide with (2.9.7) and describes a more complex crystal state. When the tensor σ_{ik} is different from (2.9.7) displacement stresses are present, which are usually characterized by the *deviator tensor*:

$$\sigma'_{ik} = \sigma_{ik} - \frac{1}{3}\delta_{ki}\sigma_{ll} = \sigma_{ik} + \delta_{ik}p_0. \qquad (2.9.9)$$

If the crystal deformation is purely elastic, the stresses are related linearly to strains ϵ_{ik} by the generalized Hooke's law:

$$\sigma_{ik} = \lambda_{iklm}\epsilon_{lm}, \qquad (2.9.10)$$

where λ_{iklm} is the tensor of crystal elasticity moduli. For a cubic crystal, there are three independent elastic moduli (or stiffness constants):

$$\lambda_1 = \lambda_{1111} = \lambda_{2222} = \lambda_{3333}, \quad \lambda_2 = \lambda_{1122} = \lambda_{1133} = \lambda_{2233};$$

$$G = \lambda_{1212} = \lambda_{1313} = \lambda_{2323}. \qquad (2.9.11)$$

In an isotropic approximation these three moduli are related through $\lambda_1 - \lambda_2 - 2G = 0$. Therefore, the tensor λ_{iklm} for an isotropic medium reduces to two independent moduli that can be represented, for example, by the *Lamé* coefficients $\lambda = \lambda_2$ and G:

$$\lambda_{iklm} = \lambda\delta_{ik}\delta_{lm} + G(\delta_{il}\delta_{km} + \delta_{im}\delta_{kl}). \qquad (2.9.12)$$

The coefficient G in (2.9.11), (2.9.12) often denoted by μ is called the shear modulus and relates the nondiagonal ("oblique") elements of the σ_{ik} and ϵ_{ik} tensors in an isotropic medium and in a cubic crystal

$$\sigma_{ik} = 2G\epsilon_{ik}, \qquad i \neq k. \qquad (2.9.13)$$

Note that there is an obvious relation between the mean hydrostatic pressure p_0 and the relative compression of an isotropic medium or a cubic crystal. From (2.9.10)–(2.9.12), we have

$$\sigma_{ll} = 3K\epsilon_{ll}, \qquad (2.9.14)$$

where K is the modulus of hydrostatic compression that, in a cubic crystal, is found from

$$3K = \lambda_1 + 2\lambda_2, \qquad (2.9.15)$$

and, in an isotropic medium, from

$$K = \lambda + \frac{2}{3}G. \qquad (2.9.16)$$

To determine the deformed and stressed crystal states in the presence of bulk forces, it is necessary to solve the following equation for an elastic medium

$$\rho\frac{\partial^2 u_i}{\partial t^2} = \nabla_k\sigma_{ki} + f_i, \qquad (2.9.17)$$

where the vector f describes the density of bulk forces acting on a crystal (the mean force applied to a crystal unit volume), and the tensor σ_{ik} is related to the strains through Hooke's law (2.9.10).

The dynamics of a free elastic field ($f = 0$) is described by

$$\rho \frac{\partial^2 u_i}{\partial t^2} - \lambda_{iklm} \nabla_k \nabla_l u_m = 0. \qquad (2.9.18)$$

Equation (2.9.18) corresponds to the Lagrangian function

$$L = \int \left\{ \frac{1}{2} \rho \left(\frac{\partial \boldsymbol{u}}{\partial t} \right)^2 - \frac{1}{2} \frac{\partial u_k}{\partial x_i} \frac{\partial u_m}{\partial x_l} \right\} dV. \qquad (2.9.19)$$

Equation (2.9.18) and the corresponding Lagrangian function, even in an isotropic case, are rather complicated. One of the difficulties in solving (2.9.18) for the three components of the displacement vector $\boldsymbol{u}$ is the following. Equation (2.9.18) is similar to a wave equation. In transforming to normal vibrations we can reduce it to three wave equations. But the latter describe the waves propagating with different velocities. Even in an isotropic approximation the elastic field has two different characteristic wave velocities (the velocities of longitudinal and transverse waves). This very much complicates the solution of dynamic problems.

To simplify the equations reflecting the main physical properties of an elastic medium, we formulate the analog of the scalar model for an elastic continuum, i. e., we introduce a *scalar elastic field*. We take as a generalized field coordinate the scalar value $u(\boldsymbol{r}, t)$ and assume the Lagrangian function of this field to be

$$L = \int \left(\frac{1}{2} \rho \left(\frac{\partial u}{\partial t} \right)^2 - \frac{1}{2} G \left(\frac{\partial u}{\partial x_i} \right)^2 \right) dV. \qquad (2.9.20)$$

The equation for the field motion stemming from (2.9.20) is

$$\rho \frac{\partial^2 u}{\partial t^2} - G \Delta u = 0, \qquad (2.9.21)$$

where Δ is the Laplace operator ($\Delta = \nabla_k^2$). This is an ordinary equation of the waves propagation with the acoustic dispersion law:

$$\omega^2 = s^2 k^2, \qquad s = G/\rho. \qquad (2.9.22)$$

The main disadvantage of (2.9.21) as a model equation for crystal dynamics is its scalar character, which does not allow one to describe transverse elastic vibrations.

2.10 Vibrations of a Strongly Anisotropic Crystal (Scalar Model)

We write the dispersion law for monatomic lattice vibrations with interatomic nearest-neighbor interactions in an explicit form, using a scalar model that enables us to find the dependence of vibration frequencies on the quasi-wave vector through the simplest elementary functions. It turns out that for crystal directions where it is possible to distinguish longitudinal and transverse vibrations a scalar model describes the crystal longitudinal vibrations well. This can be explained as follows. There is no polarization of displacements in a scalar model and the only vector characteristic of a normal vibration is vector $\boldsymbol{k}$. Therefore, the atomic displacements described by such a model can be associated only with the quasi-wave vector direction.

Going over to the formulation of a concrete problem, we simplify a model to describe in detail some interesting physical properties of a vibrating crystal, in particular, the vibrations of strongly anisotropic crystal lattices.

As an example of such an anisotropic model we consider a tetragonal lattice with different interactions of the nearest atoms in the basal plane (xOy) and along the four-fold axis (z). Choosing naturally the translation vectors, we denote $|\boldsymbol{a}_1| = |\boldsymbol{a}_2| = a$, $|\boldsymbol{a}_3| = b$. The neighboring atom interaction in the basal plane will be described by the force matrix element α_1 and the interaction along the axis z by the element α_2. On the basis of the relation (2.1.10), we have

$$\alpha_0 + 4\alpha_1 + 2\alpha_2 = 0. \tag{2.10.1}$$

Since $\alpha_0 = \alpha(0) > 0$, it follows from (2.10.1) that $2\alpha_1 + \alpha_2 < 0$. We assume that $\alpha_1 < 0$ and $\alpha_2 < 0$.

The dispersion law (2.2.10) for the lattice concerned is written as

$$\omega^2(\boldsymbol{k}) = \omega_1^2 \left(\sin^2 \frac{ak_x}{2} + \sin^2 \frac{ak_y}{2} \right) + \omega_2^2 \sin^2 \frac{ak_z}{2}, \tag{2.10.2}$$

where $\omega_1^2 = -4\alpha_1/m$ and $\omega_2^2 = -4\alpha_2/m$.

We assume the atomic interaction in the basal plane to be much stronger than that along the four-fold axis:

$$\omega_1 \gg \omega_2. \tag{2.10.3}$$

This assumption transforms a tetragonal lattice into a crystal lattice with a layered structure, whose separate atom layers are interrelated weakly. The formula (2.10.2) for such a crystal determines an extremely anisotropic dispersion relation, which is well shown in the low-frequency part of the vibration spectrum.

Consider the frequencies $\omega \ll \omega_1$ (e. g., $\omega \leq \omega_2$) for which the formula (2.10.2) is much simplified

$$\omega^2 = s_1^2 k_\perp^2 + \omega_2^2 \sin^2 \frac{ak_z}{2}; \qquad k_\perp^2 = k_x^2 + k_y^2, \qquad s_1^2 = \frac{1}{4} a^2 \omega_1^2, \tag{2.10.4}$$

where s_1 has a meaning of a sound velocity in the basal plane.

In view of the condition (2.10.3), we keep the second term in the r.h.s. of (2.10.4) unchanged, in so far as the assumption $\omega \ll \omega_1$ does not imply that bk_z is small. Thus, we take into account that at comparatively low frequencies the quasi-wave vector component along the z-axis may be large.

Within the long-wave limit when $bk_z \ll 1$, (2.10.4) gives the dispersion law of sound vibrations in an anisotropic medium

$$\omega^2 = s_1^2 k_\perp^2 + s_2^2 k_z^2; \qquad s_2^2 = \frac{1}{2} b^2 \omega_2^2, \tag{2.10.5}$$

where s_2 is the sound velocity along the z-axis.

If the lattice parameters a, b are little different (have the same order of magnitude), the sound velocity in the basal plane of a "layered" crystal will be much larger than that in a perpendicular direction ($s_2 \ll s_1$).

Equation (2.10.4) is also simplified in the case when the vibration frequencies have the range

$$\omega_2 \ll \omega \ll \omega_1. \tag{2.10.6}$$

For such frequencies the second term in (2.10.4) should not be taken into account, and the dispersion relation reduces to

$$\omega = s_1 k_\perp \equiv s_1 \sqrt{k_x^2 + k_y^2}, \tag{2.10.7}$$

which coincides with the dispersion relation for sound vibrations in a two-dimensional elastic medium. Hence, under the conditions (2.10.6) the frequency of vibrations of a three-dimensional "layered" crystal is independent of the wave-vector component along the direction perpendicular to its "layers".

Along with a "layered" crystal, one can consider a crystal model with a "chain" structure where one-dimensional chains of atoms weakly interact one with another. In order to obtain this model it suffices to assume

$$\omega_1 \ll \omega_2. \tag{2.10.8}$$

Then the results stated above are easily transformed by changing the numbers 1 and 2 and also the components k_i and k_z.

In particular, at frequencies $\omega \leq \omega_1$ the dispersion relation (2.10.2) reduces to

$$\omega^2 = \omega^1 \left(\sin^2 \frac{ak_x}{2} + \sin^2 \frac{ak_y}{2} \right) + s_2^2 k_z^2.$$

In the long-wave limit ($ak_\perp \ll 1$) we come again to (2.10.5), and in the frequency range $\omega_1 \ll \omega_2$ the dispersion law of vibrations of such a crystal coincides with the dispersion law of elastic vibrations of a one-dimensional system

$$\omega = s_2 k_z,$$

i. e., the vibration frequency is independent of the wave-vector projection onto the plane perpendicular to the direction of a strong interaction between atoms.

2.11 "Bending" Waves in a Strongly Anisotropic Crystal

We consider a crystal with a simple hexagonal lattice in which the atoms interact in different ways in the basal plane xOy and along the six-fold axis Oz. We assume the crystal structure to be layered and the atom interaction in the plane xOy to be much larger than the atom interaction in neighboring basal planes. In describing the vibration of such "layered" crystal one can proceed from the model that takes exact account of the strong interaction between all atoms lying in the basal plane, and the weak interaction of neighboring atomic layers is taken into account in the nearest-neighbor approximation along the six-fold axis.

A crystal with a chain structure may be considered simultaneously. A crystal with such a structure consists of weakly interacting parallel linear chains. In the model proposed, this corresponds to the fact that the atomic interaction along the six-fold axis is much stronger that the interaction between neighboring chains (or the nearest neighbors in the plane xOy).

An example of a chemical element that has three possible crystalline forms (approximately isotropic, layered and chain) is carbon. It exists in the form of diamond (an extremely hard crystal with a three-dimensional lattice), in the form of graphite (layered crystal) and in the form of carbene (a synthetic polymer chain structure).

For definiteness the following arguments are given for a layered crystal and intended for the model formulated above. The latter makes it possible to qualitatively describe the acoustic vibrations in graphite – a layered hexagonal crystal with very weak interactions between the layers[4]. The atomic forces between the neighboring layers in graphite are almost two orders less that the nearest-neighbor interaction forces within the layer.

Let a and b be interatomic distances in the xOy plane and along the Oz-axis, respectively. The vector $\boldsymbol{n}_1$ represents a set of two-dimensional number vectors connecting any one of the atoms with all remaining atoms in the same basal plane, $\boldsymbol{n}_3$ is the unit vector of the Oz-axis. Then nonzero elements of the matrix $\alpha^{ik}(\boldsymbol{n})$ in our model are represented by $\alpha^{ik}(\boldsymbol{n}_1)$ and $\alpha^{ik}(\boldsymbol{n}_2)$. Making use of the obvious force matrix symmetry in a hexagonal crystal, we write the elements $\alpha^{ik}(\boldsymbol{n}_3)$ responsible for the weak atomic layer interaction as follows

$$\begin{aligned} \alpha^{ik}(\boldsymbol{n}_3) &= \alpha_1\delta_{ik}, \quad i,k = 1,2, \\ \alpha^{zz}(\boldsymbol{n}_3) &= \alpha_2, \qquad \alpha^{xz}(\boldsymbol{n}_3) = \alpha^{yz}(\boldsymbol{n}_3) = 0. \end{aligned} \tag{2.11.1}$$

Concerning the spectrum of acoustic vibrations of graphite we note that the parameter α_2 is generally larger than α_1, and α_2 is determined mainly by central forces

$$|\alpha_1| \ll |\alpha_2|, \tag{2.11.2}$$

(for graphite $\alpha_2 \approx 10\alpha_1 \approx 0.610^4$ dyn/cm).

4) Graphite has a complex lattice with atoms positioned in a separate basal plane as shown in Fig. 2.2.

To characterize the strong interaction in the basal plane, we introduce the notation $\alpha_3 = \alpha^{zz}(\boldsymbol{n}_0)$, where $\boldsymbol{n}_0$ is the unit vector directed from an atom to any one of its six nearest neighbors in the plane xOy (Fig. 2.2). It may also be assumed that

$$\alpha^{ik}(\boldsymbol{n}_0) \sim \alpha^{zz}(0) \sim \alpha_3, \qquad i,k = 1,2. \tag{2.11.3}$$

For graphite $\alpha_3 \sim 10^5$ dyn/cm.

Now the assumption of a layered crystal structure can be formulated in the form of a quantitative ratio establishing a hierarchy of interatomic interactions

$$|\alpha_1| \ll |\alpha_2| \ll |\alpha_3|. \tag{2.11.4}$$

We note an important property of anisotropic crystal vibrations whose displacement vector $\boldsymbol{u}$ is perpendicular to the strong interaction layers (perpendicular to the xOy plane). For a very weak layer interaction, these vibrations should resemble the bending waves in the noninteracting layers[5], so that they may tentatively be referred to as "bending" vibrations. Simultaneously, assuming strong anisotropy of interatomic interactions (2.9.4) and the same order of the lattice constant values ($a \sim b$), it is impossible in describing the "bending" vibrations to include only the nearest-neighbor interaction in the basal plane. Noting that the character of the bending vibrations is primarily determined by the force matrix elements $\alpha^{zz}(\boldsymbol{n}_1)$, we take into account the conditions (2.8.10) imposed on the $\boldsymbol{A}(\boldsymbol{n})$ matrix elements, putting $i,k = x$ or $i,k = y$ and $l,m = z$:

$$2\alpha_1 b^2 = \sum_{\boldsymbol{n}_1} \alpha^{zz}(\boldsymbol{n}_1)x^2(\boldsymbol{n}_1) = \sum_{\boldsymbol{n}_1} \alpha^{zz}(\boldsymbol{n}_1)y^2(\boldsymbol{n}_1). \tag{2.11.5}$$

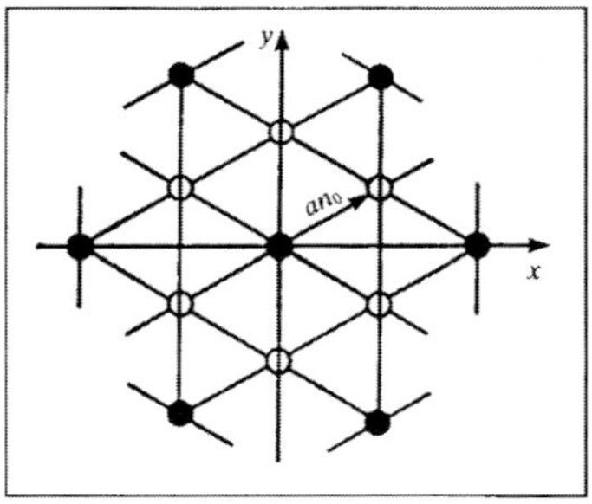

Fig. 2.2 Choice of the nearest neighbors in the basis plane of an hexagonal crystal

By keeping in (2.11.5) the summation over the vectors $\boldsymbol{n}_0$ only, we get the equality

$$2\alpha_1 b^2 = 3\alpha_3 a^2, \tag{2.11.6}$$

5) The need to take into account the bending wave type of vibrations in layered crystals with a weak interlayer interaction was first indicated by Lifshits (1952).

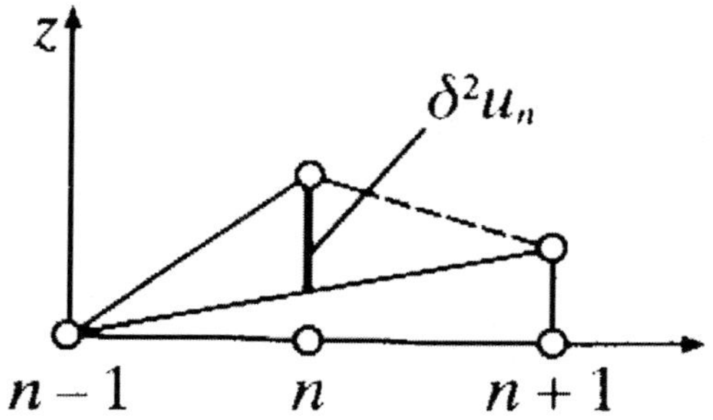

Fig. 2.3 Second difference in atom displacements that determines the bend energy.

which is impossible when the requirements $a \sim b$ and $|\alpha_1| I \ll |\alpha_3|$ are satisfied simultaneously.

Thus, such a model for a layered crystal with interaction between nearest neighbors only is in fact intrinsically inconsistent. To describe a crystal lattice with a characteristic layered structure having "bending" waves it is necessary, while keeping the relations such as (2.11.4), to take into account more distant interatomic interactions in the basal plane. The physical meaning of this assertion is easily understood if we consider the limiting case of noninteracting layers possessing bend rigidity. Analyzing one atomic layer allows one to conclude that the interaction of atoms displaced along the Oz-axis (Fig. 2.3) is determined by the difference of relative pair displacements of at least three atoms rather than by the relative displacement of two neighboring atoms. The atomic interaction energy under bending vibrations of the plane layer depends on $\delta u_n = (1/2)[(u_n - n_{n-1}) - (u_{n+1} - u_n)] = u_n - (1/2)(u_{n-1} + u_{n+1})$.

We now turn to (2.11.5) and note that it does not contradict the assumption of a strong anisotropy of a crystal. This is due to the fact that the elements of the matrix $\alpha^{zz}(\boldsymbol{n}_1)$ describing the interaction not only between nearest neighbors in the plane xOy can be quantities of the same order of magnitude and have opposite signs. The signs of each of them satisfy the condition

$$\sum_n \alpha^{zz}(\boldsymbol{n}) \equiv \alpha^{zz}(0) + \sum_{n \neq 0} \alpha^{zz}(\boldsymbol{n}_1) + 2\alpha^{zz}(\boldsymbol{n}_3) = 0. \tag{2.11.7}$$

Taking into account the inequality (2.11.4), we conclude from (2.11.7) that

$$\sum_{n \neq 0} \alpha^{zz}(\boldsymbol{n}_1) < 0. \tag{2.11.8}$$

To find the dispersion law of the vibrations, we calculate the tensor functions

$$A^{ij}(\boldsymbol{k}) = \sum_n \alpha^{ij}(\boldsymbol{n}) e^{-i\boldsymbol{kr}(\boldsymbol{n})} \equiv \sum_{n \neq 0} \alpha^{ij}(\boldsymbol{n}) \left[\cos \boldsymbol{kr}(\boldsymbol{n}) - 1\right]. \tag{2.11.9}$$

Assume that

$$\alpha^{zz}(\boldsymbol{n}_1) = \alpha^{yz}(\boldsymbol{n}_1) = 0. \tag{2.11.10}$$

It follows from (2.11.10) and also from (2.11.1) that $A^{xz}(\boldsymbol{k}) = A^{yz}(\boldsymbol{k}) = 0$, and in this case the vibrations with the displacement vector $\boldsymbol{u}$ in the xOy plane are independent of the vibrations with the displacement vector $\boldsymbol{u}$, parallel to the Oz-axis, i. e., perpendicular to the layers.

The vibration of the vector $\boldsymbol{u}$ in the plane of the layer has two branches in the strongly anisotropic dispersion law. Since these vibrations are not of main interest the long-wavelength ($ak_x \ll 1, ak_y \ll 1$) dispersion law for one of these branches will be written in a simpler form, analogous to (2.10.4)

$$\omega^2 = \epsilon_1(\boldsymbol{k}) \equiv s_0^2\left(k_x^2 + k_y^2\right) + \omega_1^2 \sin^2\frac{bk_z}{2}, \tag{2.11.11}$$

where $s_0 \sim |\alpha_3|\, a^2/m$, $\omega_1^2 \sim |\alpha_1|\,/m \ll (s_0/b)^2$.

Consider the third type of vibrations, bending. The dispersion law for such vibrations is written as $m\omega^2 = A^{zz}(\boldsymbol{k})$.

Assuming in (2.11.9) $i, j = z$ and performing simple calculations, we have

$$m\omega^2 = \sum_{\boldsymbol{n}_1 \neq 0} \alpha^{zz}(\boldsymbol{n}_1)\left[\cos \boldsymbol{kr}(\boldsymbol{n}) - 1\right] - 4\alpha_2 \sin^2\frac{bk_z}{2}. \tag{2.11.12}$$

From the assumption $|\alpha_2| \ll |\alpha_3| \sim |\alpha^{zz}(\boldsymbol{n}_1)|$, hence (2.11.12) gives a strongly anisotropic dispersion law with two characteristic frequencies ω_0 and ω_1 determined by the relations[6]

$$\omega_0^2 = -\frac{1}{m}\sum_{\boldsymbol{n}_1 \neq 0} \alpha^{zz}(\boldsymbol{n}_1) \approx \frac{1}{m}\alpha^{zz}(0), \quad \omega_2^2 = -\frac{4}{m}\alpha_2, \quad \omega_2 \ll \omega_0.$$

The relation $\eta^2 = (\omega_2/\omega_0)^2 \sim |\alpha_2/\alpha_3| \ll 1$ is the main small parameter that allows one to separate out the bending-type vibrations. For graphite η^2 is equal to several per cent.

We derive from (2.11.12) the dispersion law for the bending waves at small frequencies when $\omega \ll \omega_0$. At such frequencies the cosines in the first sum of the r.h.s. of the formula (2.11.12) can be expanded in powers of ak_x and ak_y, using only a few terms of the expansion. Two characteristic frequencies of different order of magnitude available in the dispersion law require that terms up to the fourth power of ak_x and ak_y be retained in such an expansion:

$$m\omega^2 = -\frac{1}{2}\sum_{\boldsymbol{n}_1} \alpha^{zz}(\boldsymbol{n}_1)\left[\boldsymbol{kr}(\boldsymbol{n})\right]^2 + \frac{1}{4!}\sum_{\boldsymbol{n}_1} \alpha^{zz}(\boldsymbol{n}_1)\left[\boldsymbol{kr}(\boldsymbol{n})\right]^4 - 4\alpha_2 \sin^2\frac{bk_z}{2}.$$

Taking into account the symmetry of the second- and fourth-rank tensors in the presence of the six-fold symmetry axis, we write down the dispersion law for long-

6) It is clear that $\alpha_2 < 0$, otherwise the vibration frequency would not remain real for $k_x = k_y = 0$ and $k_z \to 0$.

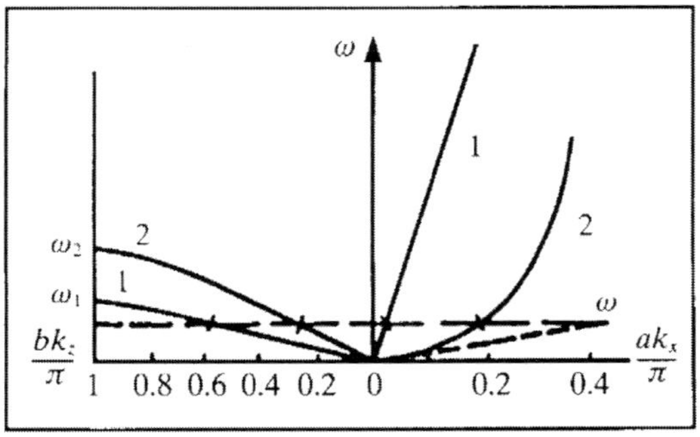

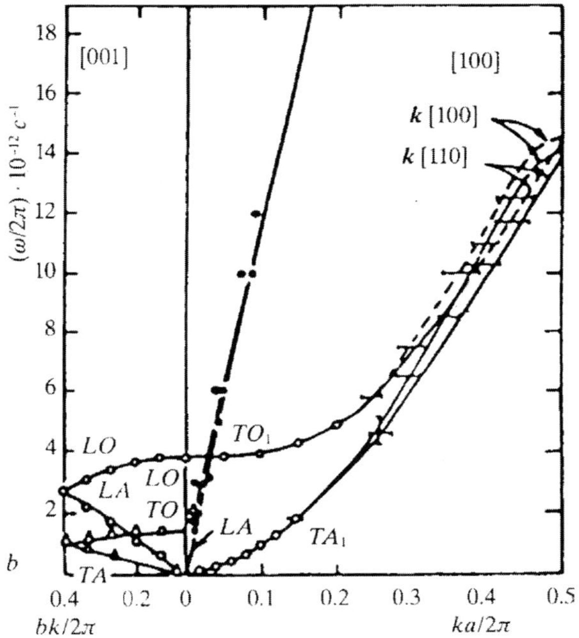

Fig. 2.4 Dispersion laws of a strongly anisotropic layered crystal: (*a*) general scheme; (*b*) experimental curves for graphite (Nicklow et al. 1972). In the upper part of the curves, the dependence on the direction of k in the basal plane is indicated.

wavelength vibrations in the form

$$m\omega^2 = \frac{a^2}{2m}\left(\sum_{n_1}\alpha^{zz}(\boldsymbol{n}_1)n_x^2\right)k_\perp^2 + \frac{a^4}{24}\left(\sum_{n_1}\alpha^{zz}(\boldsymbol{n}_1)n_x^4\right)k_\perp^4 + \omega_2^2\sin^2\frac{bk_z}{2},$$

where $k_\perp^2 = k_x^2 + k_y^2$, and the first term allows us to take the relation (2.11.5) directly into account.

Let us use (2.11.5) and simplify the formula for the dispersion law by introducing new notations

$$\begin{aligned} \omega^2 &= \epsilon_1(\boldsymbol{k}) \equiv s_\perp^2 k_\perp^2 + A^2 a^2 k_\perp^4 + \omega_2^2 \sin^2 \frac{bk_z}{2}; \\ s_\perp^2 &= -\frac{\alpha_1}{m} b^2 \sim \omega_1 b^2, \quad A^2 = \frac{a^4}{24} \sum_{n_1} \alpha^{zz}(\boldsymbol{n}_1) n_x^4 \sim \omega_0 a^2 \gg s_\perp^2 . \end{aligned} \tag{2.11.13}$$

The dispersion law graphs (2.11.13), (2.11.11) for the two main directions (Ok_x and Ok_z, Fig. 2.4a) have different peaks in reciprocal space, so that the scales of units of the quasi-wave vector components k_x and k_z are not the same. The curves 1 refer to the $\boldsymbol{u}$-vector vibrations along the layer and the curves 2 to the $\boldsymbol{u}$-vector vibrations transverse to the layers.

To illustrate the genuine dispersion laws of strongly anisotropic crystals, in Fig. 2.4b we show the graphs of the calculated and experimentally established dependencies $\omega = \omega_z(\boldsymbol{k})$ for graphite. Since graphite has a diatomic lattice, Fig. 2.4b gives not only acoustic (A), but also the optical (O) vibration branches (Chapter 3).

2.11.1 Problem

Find the canonical transformations that allow a transformation from complex normal coordinates to real ones and to reduce (2.6.8) to the form

$$H = \frac{1}{2} \sum_{k\alpha} \left[Y_\alpha^2(\boldsymbol{k}) + \omega_\alpha^2(\boldsymbol{k}) X_\alpha(\boldsymbol{k}) \right] \tag{2.11.14}$$

where $X_\alpha(\boldsymbol{k})$ and $Y_\alpha(\boldsymbol{k})$ are the real normal coordinates and the generalized momenta $Y_\alpha(\boldsymbol{k}) = X_\alpha(\boldsymbol{k})$ conjugated to them.

Solution.

$$\begin{aligned} Q_\alpha(\boldsymbol{k}) &= \frac{1}{2} \left\{ X_\alpha(\boldsymbol{k}) + X_\alpha(-\boldsymbol{k}) + \frac{i}{\omega_\alpha(\boldsymbol{k})} \left[Y_\alpha(\boldsymbol{k}) - Y_\alpha(-\boldsymbol{k}) \right] \right\}; \\ P_\alpha(\boldsymbol{k}) &= \frac{1}{2} \left\{ Y_\alpha(\boldsymbol{k}) + Y_\alpha(-\boldsymbol{k}) - i\omega_\alpha(\boldsymbol{k}) \left[X_\alpha(\boldsymbol{k}) - X_\alpha(-\boldsymbol{k}) \right] \right\} . \end{aligned} \tag{2.11.15}$$

3
Vibrations of Polyatomic Lattices

3.1
Optical Vibrations

A polyatomic crystal lattice is different from a monatomic one in that its unit cell contains more than one atom. In other words, the number of mechanical degrees of freedom per unit cell of a polyatomic lattice is necessarily more than three.

The last point introduces not only quantitative but also qualitative changes in the spectrum of crystal vibrations. The specific features of polyatomic lattice vibrations will first be studied for a very simple model.

Let us consider the vibrations of an atomic crystal with a densely packed hexagonal lattice or a crystal with the NaCl structure (Fig. 0.2). The unit cell of the lattice of such crystals has two atoms, i. e., six degrees of freedom. It is convenient to introduce the vector of displacement of the center of mass of a pair of atoms, $\boldsymbol{u}(\boldsymbol{n})$, i. e., the center of mass of a unit cell (three degrees of freedom), and the vector of a relative displacement of atoms in a pair, $\xi(\boldsymbol{n})$ (three degrees of freedom).

In terms of these displacements, we can write explicitly the energy (or the Lagrangian function) of a vibrating crystal. However, we shall not write this value directly in the general form, but confine ourselves to a scalar model. Now this model cannot be called "scalar", since we have to operate with two scalar functions $u(\boldsymbol{n})$ and $\xi(\boldsymbol{n})$, where $\xi(\boldsymbol{n})$ mean, e. g., changes in the distance between the atoms of a chosen pair. Thus we should actually transform a scalar model to a two-component one.

Assuming that the displacements of the centers of mass of the unit cells $u(\boldsymbol{n})$ and the relative displacements of atom pairs $\xi(\boldsymbol{n})$ are small, we represent the potential energy of crystal vibrations in the form corresponding to a harmonic approximation

$$U = \frac{1}{2}\sum \alpha_1(\boldsymbol{n}-\boldsymbol{n}')u(\boldsymbol{n})u(\boldsymbol{n}') + \frac{1}{2}\sum \alpha_2(\boldsymbol{n}-\boldsymbol{n}')\xi(\boldsymbol{n})\xi(\boldsymbol{n}') + \sum \beta(\boldsymbol{n}-\boldsymbol{n}')u(\boldsymbol{n})\xi(\boldsymbol{n}'), \tag{3.1.1}$$

The Crystal Lattice: Phonons, Solitons, Dislocations, Superlattices, Second Edition. Arnold M. Kosevich

ISBN: 3-527-40508-9

where the new matrices of atomic force constants $\alpha_1(\boldsymbol{n}), \alpha_2(\boldsymbol{n})$ and $\beta(\boldsymbol{n})$ are introduced. Evidently, the matrices $\alpha(\boldsymbol{n})$ may be chosen to be symmetric $\alpha(\boldsymbol{n}) = \alpha(-\boldsymbol{n})$ but the latter is invalid for the matrix $\beta(\boldsymbol{n})$. One can easily construct a model in which the matrix $\beta(\boldsymbol{n})$ responsible for the interaction of displacements of two types, also has a similar symmetry.

The kinetic energy of a diatomic lattice has the standard form

$$K = \frac{1}{2}\sum M\left(\frac{du_n}{dt}\right)^2 + \frac{1}{2}\sum \mu\left(\frac{d\xi_n}{dt}\right)^2, \tag{3.1.2}$$

where M is the unit cell mass; μ is the reduced mass of a pair of atoms. Using (3.1.1), (3.1.2) it is easy to obtain the equations of motion of the crystal

$$\begin{aligned} M\frac{d^2u(\boldsymbol{n})}{dt^2} &= -\sum_{n'}[\alpha_1(\boldsymbol{n}-\boldsymbol{n}')u(\boldsymbol{n}') + \beta(\boldsymbol{n}-\boldsymbol{n}')\xi(\boldsymbol{n}')], \\ \mu\frac{d^2\xi(\boldsymbol{n})}{dt^2} &= -\sum_{n'}[\alpha_2(\boldsymbol{n}-\boldsymbol{n}')\xi(\boldsymbol{n}') + \beta(\boldsymbol{n}-\boldsymbol{n}')u(\boldsymbol{n}')]. \end{aligned} \tag{3.1.3}$$

From (3.1.3), we can immediately derive the constraints imposed on the $\alpha_1(\boldsymbol{n})$ and $\beta(\boldsymbol{n})$ matrix elements that result from the invariance of the energy of a crystal with respect to its motion as a single whole. Indeed, assuming $\xi(\boldsymbol{n}) = 0$ and $u(\boldsymbol{n}) = u^0 =$ const we get

$$\sum \alpha_1(\boldsymbol{n}) = 0, \qquad \sum \beta_1(\boldsymbol{n}) = 0. \tag{3.1.4}$$

We note that the matrix elements of α_2 are not related by a condition such as (3.1.4) since even the homogeneous relative displacement of atoms in all pairs increases the crystal energy. However, the matrix elements of α_2 as well as the elements of α_1 and β obey a set of inequalities providing the positiveness of the energy (3.1.1) for arbitrary $u(\boldsymbol{n})$ and $\xi(\boldsymbol{n})$. In particular,

$$\alpha_2(0) > 0, \qquad \alpha_1(0)\alpha_2(0) > \beta^2(0). \tag{3.1.5}$$

Just as in the case of a monatomic lattice, it is natural to seek a solution to (3.1.3) in the form of vibrations running through a crystal with the given frequency

$$u(\boldsymbol{n}) = u_0 e^{i(\boldsymbol{k}\boldsymbol{r}(\boldsymbol{n})-\omega t)}, \qquad \xi(\boldsymbol{n}) = \xi_0 e^{i(\boldsymbol{k}\boldsymbol{r}(\boldsymbol{n})-\omega t)}. \tag{3.1.6}$$

Substituting (3.1.6) into (3.1.3) and performing simple transformations, we get

$$\begin{aligned} \left[A_1(\boldsymbol{k}) - M\omega^2\right]u_0 + B(\boldsymbol{k})\xi_0 &= 0, \\ B^*(\boldsymbol{k})u_0 + \left[A_2(\boldsymbol{k}) - \mu\omega^2\right]\xi_0 &= 0, \end{aligned} \tag{3.1.7}$$

where

$$A(\boldsymbol{k}) = \sum_n \alpha(\boldsymbol{n})e^{-i\boldsymbol{k}\boldsymbol{r}(\boldsymbol{n})}, \quad B(\boldsymbol{k}) = \sum_n \beta(\boldsymbol{n})e^{-i\boldsymbol{k}\boldsymbol{r}(\boldsymbol{n})}. \tag{3.1.8}$$

The compatibility condition homogeneous equations (3.1.7) for the two unknowns u_0 and ξ_0

$$\begin{vmatrix} A_1(\boldsymbol{k}) - M\omega^2 & B(\boldsymbol{k}) \\ B(\boldsymbol{k}) & A_2(\boldsymbol{k}) - \mu\omega^2 \end{vmatrix} = 0,$$

considered as an equation for finding the squares of the possible vibration frequencies at fixed $\boldsymbol{k}$ transforms to the second-degree algebraic equation, then becomes

$$\left[A_1(\boldsymbol{k}) - M\omega^2\right]\left[A_2(\boldsymbol{k}) - \mu\omega^2\right] = |B(\boldsymbol{k})|^2. \tag{3.1.9}$$

Thus, each value of the quasi-wave vector in our model of a polyatomic lattice corresponds to two frequencies $\omega = \omega_\alpha(\boldsymbol{k})$, $\alpha = 1, 2$, i.e., the dispersion law has two branches of the dependencies ω on $\boldsymbol{k}$. The typical distinctions between these two branches can be shown on considering their limiting properties.

The most essential differences are observed at small $\boldsymbol{k}$ ($ak \ll 1$). In this case, as we have already seen in sect.1, expansion of the function $A_1(\boldsymbol{k})$ in powers of the quasi-wave vector starts with the quadratic terms

$$A_1(\boldsymbol{k}) = Ms_0^2k^2.$$

Due to the above-mentioned properties of the matrix $\alpha_2(\boldsymbol{n})$ the expansion of the function $A_2(\boldsymbol{k})$ starts with a zero term and does not contain the first power of $\boldsymbol{k}$

$$A_2(\boldsymbol{k}) = A_2(0) - b(\boldsymbol{\kappa})k^2,$$

where, by definition,

$$A_2(0) = \sum \alpha_2(\boldsymbol{n}) = \mu\omega_0^2 \neq 0, \qquad \boldsymbol{k} = k\boldsymbol{\kappa}, \tag{3.1.10}$$

and

$$b(\boldsymbol{\kappa}) = \frac{1}{2}\sum_{\alpha,\beta=1}^{3}(\boldsymbol{\kappa}\boldsymbol{a}_\alpha)(\boldsymbol{\kappa}\boldsymbol{a}_\beta)\sum_{n}\alpha_2(\boldsymbol{n})n_\alpha n_\beta.$$

Finally, to simplify the analysis, we assume that the matrix $\beta(\boldsymbol{n})$ is symmetric $\beta(\boldsymbol{n}) = \beta(-\boldsymbol{n})$. Then, using (3.1.4), we have

$$B(\boldsymbol{k}) = -\frac{1}{2}\left[\sum_{\alpha,\beta=1}^{3}\boldsymbol{\kappa}\boldsymbol{a}_\alpha\boldsymbol{\kappa}\boldsymbol{a}_\beta\sum_{n}\beta(\boldsymbol{n})n_\alpha n_\beta\right]k^2.$$

Hence, $|B(\boldsymbol{k})|^2 \sim (ak)^4$ at $ak \ll 1$ and, in the main approximation with respect to the small parameter ak (3.1.9) is split into two independent ones

$$\omega = \omega_1^2(\boldsymbol{k}) \equiv s_0^2(\boldsymbol{k})k^2; \tag{3.1.11}$$

$$\omega = \omega_2^2 \equiv \omega_0^2 - (b(\boldsymbol{\kappa})/\mu)k^2, \quad \omega_0^2 = A_2(0)/\mu. \tag{3.1.12}$$

The dispersion law (3.1.11) gives the above-discussed dispersion of sound vibrations. Indeed, substitute (3.1.11) into (3.1.7) and perform the limit transition $k \to 0$. Since $A_2(0) \neq 0$ we have $\xi_0 = 0$. Thus, under long-wave vibrations with the dispersion law (3.1.11), the unit cell centers of mass vibrate with the relative position of atoms in a pair remaining unchanged. Therefore, using (3.1.6), we get $u(\boldsymbol{n}) = u_0 e^{i\omega t}$, $\xi(\boldsymbol{n}) = 0$.

A feature of the dispersion law (3.1.12) is that the corresponding vibrations with an infinitely large wavelength have the finite frequency ω_0. It follows from (3.1.7) that at $\boldsymbol{k} = 0$ this vibration is

$$u(\boldsymbol{n}) = 0, \quad \xi(\boldsymbol{n}) = \xi_0 e^{-i\omega t}. \tag{3.1.13}$$

Under such crystal vibrations the centers of mass of the unit cells are at rest and the motion in the lattice is reduced to relative vibrations inside the unit cells. The presence of vibrations such as (3.1.13) distinguishes a diatomic crystal lattice from a monatomic one.

For arbitrary $\boldsymbol{k}$, the form of the dispersion law is strongly dependent on the dynamical matrix properties. In simple models it is generally observed that $\omega_1(\boldsymbol{k}) < \omega_2(\boldsymbol{k})$. The dependence $\omega = \omega_\alpha(\boldsymbol{k})$ along a certain "good" direction in the reciprocal lattice has the form of a plot in Fig. 3.1, where b is the period of a reciprocal lattice in the chosen direction. The low-frequency branch of the dispersion law ($\omega < \omega_m$) describes the *acoustic* vibrations, and the high-frequency one ($\omega_1 < \omega < \omega_2$) the *optical* vibrations of a crystal. Thus, the polyatomic crystal lattice, apart from acoustic vibrations (A) also has optical vibrations (O).

The generally accepted name for high-frequency branches of the vibrations is explained by the fact that in many crystals they are optically observed. In the NaCl ion crystal, the unit cell contains two different ions whose relative displacement changes the dipole moment of the unit cell. Consequently, the vibrations connected with relative ion displacements interact intensively with an electromagnetic field and may, thus, be studied by optical methods.

When $\omega_1(\boldsymbol{k}) < \omega_2(\boldsymbol{k})$ for a fixed direction of $\boldsymbol{k}$, the spectrum of optical vibration frequencies is separated by a finite gap from the spectrum of acoustic vibration frequencies. However, it is possible that for some $\boldsymbol{k}$ the condition $\omega_1(\boldsymbol{k}) > \omega_2(\boldsymbol{k})$ is fulfilled. Then, in $\boldsymbol{k}$-space there are points of degeneracy where the acoustic and optical vibration frequencies coincide and the plots of two branches are tangential or intersect. The simplest plot of the dispersion law for a diatomic lattice is given in Fig. 3.1 ($\frac{1}{2}b$ indicates the Brillouin zone boundary).

Regardless of the complicated form of the dispersion law for the optical branch, the corresponding frequencies always lie in a band of finite width and its ends are generally the extremum points for the function $\omega = \omega_2(\boldsymbol{k})$. The latter means that the dispersion law near the ends of the optical band is quadratic.

The characteristic optical vibration frequencies have the same order of magnitude as the limiting frequency of acoustic vibrations, i.e., $\omega_0 \sim \omega_1 \sim \omega_m \sim 10^{13}\ \mathrm{s}^{-1}$.

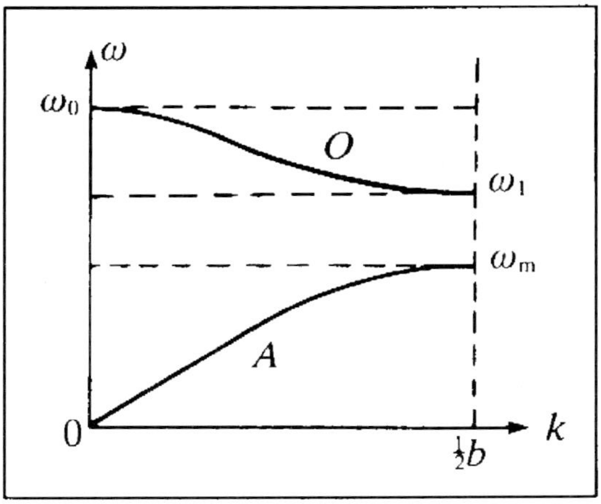

Fig. 3.1 Acoustic and optical branches of the dispersion law.

But in the crystal due to some specific physical reasons the frequencies of some optical vibrations may increase markedly. In particular, this may occur in so-called *molecular crystals*.

Let us consider a crystal lattice of two-atom molecules positioned at its sites. The optical branch of vibrations of such a crystal describes intramolecular motions weakly dependent on intermolecular interactions. These motions are characterized by frequencies close to the eigenfrequencies of a free molecule. Then, $\omega_0 \sim \omega_1 \gg \omega_m$, $\omega_0 - \omega_1 \ll \omega_0$ and the optical branch of vibrations has a narrow high-lying frequency range.

Let $\xi(\boldsymbol{n})$ be a coordinate of intramolecular vibrations at the point $\boldsymbol{n}$. Then the assumption of a strong atomic interaction inside a molecule is reduced to the conditions

$$\alpha_2(0) \gg |\alpha_2(\boldsymbol{n})|, \quad \alpha_2(0) \gg \frac{\mu}{M}|\alpha_1(\boldsymbol{n})|; \quad \alpha_2(0) \gg |\beta_2(\boldsymbol{n})|, \boldsymbol{n} \neq 0.$$

Under these conditions the equations for intramolecular vibrations with frequencies $\omega \sim \omega_0$ have the form

$$\mu \frac{d^2\xi(\boldsymbol{n})}{dt^2} = -\sum_{n'} \alpha_2(\boldsymbol{n} - \boldsymbol{n}')\xi(\boldsymbol{n}'), \tag{3.1.14}$$

and reduce to the equations for one scalar quantity $\xi(\boldsymbol{n})$.

Thus, a scalar model may also be used for qualitative analysis of the optical lattice vibrations. For this, it suffices to replace the requirement (3.1.4) for the matrix $\alpha_1(\boldsymbol{n})$ with the condition

$$\frac{1}{\mu}\sum \alpha_2(\boldsymbol{n}) = \omega_0^2 \neq 0, \tag{3.1.15}$$

determining the long-wavelength limit for the frequency of optical vibrations.

3.2 General Analysis of Vibrations of Polyatomic Lattice

We now consider the complete equations of three-dimensional vibrations of a crystal whose unit cell contains q atoms. We number different atoms in one unit cell by the index s and denote by $\boldsymbol{u}_s(\boldsymbol{n})$ the displacement vector of the s-th atom in a unit cell with number-vector $\boldsymbol{n}$, e. g., the ($\boldsymbol{n}, s$ atom). The Lagrangian function of small crystal vibrations is

$$L = \frac{1}{2}\sum m_s \left(\frac{du_s(\boldsymbol{n})}{dt}\right)^2 - \frac{1}{2}\sum \alpha_{ss'}^{ik}(\boldsymbol{n}-\boldsymbol{n}')u_s^i(\boldsymbol{n})u_{s'}^k(\boldsymbol{n}'), \tag{3.2.1}$$

where the matrix of atomic force constants has the obvious properties

$$\alpha_{ss'}^{ik}(\boldsymbol{n}) = \alpha_{s's}^{ki}(-\boldsymbol{n}), \quad i,k = 1,2,3, \tag{3.2.2}$$

and a summation over all indices is assumed. The equations of motion are obtained from (3.2.1) by the usual method

$$m_s \frac{d^2 u_s^i(\boldsymbol{n})}{dt^2} = -\sum_{n',s'} \alpha_{ss'}^{ik}(\boldsymbol{n}-\boldsymbol{n}')u_{s'}^k(\boldsymbol{n}'), \tag{3.2.3}$$

The invariance of (3.2.1) relative to the motion of a crystal as a whole ($u_s^k(\boldsymbol{n}) = u_0^k$) leads to the following constraints imposed on the force matrix elements

$$\sum_{n,s'} \alpha_{ss'}^{ik}(\boldsymbol{n}) = \sum_{n,s} \alpha_{ss'}^{ik}(\boldsymbol{n}) = 0. \tag{3.2.4}$$

The relations (3.2.4) follow directly from (3.2.3). For stationary vibrations with the frequency ω, we have

$$\sum_{n',s'} \alpha_{ss'}^{ik}(\boldsymbol{n}-\boldsymbol{n}')u_{s'}^k(\boldsymbol{n}') - \omega^2 m_s u_s^i(\boldsymbol{n}) = 0. \tag{3.2.5}$$

We choose the solution to (3.2.5) as above in the form

$$\boldsymbol{u}_s(\boldsymbol{n}) = \boldsymbol{u}_s e^{i\boldsymbol{k}\boldsymbol{r}(\boldsymbol{n})}, \quad s = 1,2,3,\ldots,q,$$

and get for $\boldsymbol{u}_s$ the following system of homogeneous algebraic equations

$$\sum_{s'} A_{ss'}^{ij}(\boldsymbol{k})u_{s'}^j - \omega^2 m_s u_s^i = 0, \tag{3.2.6}$$

in which

$$A_{ss'}^{ij}(\boldsymbol{k}) = \sum_n \alpha_{ss'}^{ij}(\boldsymbol{n})e^{-i\boldsymbol{k}\boldsymbol{r}(\boldsymbol{n})}.$$

Transform (3.2.5), (3.2.6) to a canonical form of the equations of some eigenvalue problem. For this we introduce instead of the displacement vectors, the new variables

$$v_s(\boldsymbol{n}) = \sqrt{m_s}u_s(\boldsymbol{n}),$$

and denote

$$\tilde{A}_{ss'} = \frac{A_{ss'}}{\sqrt{m_s m_{s'}}}. \tag{3.2.7}$$

The symbolic matrix form of vibration equations both in the site, (3.2.5), and the $\boldsymbol{k}$-representation (3.2.6) reduces to the standard one

$$\omega^2 v - \tilde{\boldsymbol{A}} v = 0, \tag{3.2.8}$$

where v stands for a column of $3q$ elements; $\tilde{\boldsymbol{A}}$ is the quadratic $(3q \times 3q)$ matrix.

The compatibility condition for the system (3.2.8) determines the relationship between the vibration frequency ω and the quasi-wave vector $\boldsymbol{k}$

$$\boldsymbol{Det}\left\|\omega^2 \boldsymbol{I} - \tilde{A}(\boldsymbol{k})\right\| = 0. \tag{3.2.9}$$

Since (3.2.8) is an equation of degree $3q$ with respect to ω^2, its roots determine $3q$ branches of the dispersion law for crystal vibrations.

To denote the displacements corresponding to different branches of vibrations, we write the solution to (3.2.8) in the form

$$v_s(\boldsymbol{n}, t) = \boldsymbol{e}_s(\boldsymbol{k}, \alpha) e^{i(\boldsymbol{kr}(\boldsymbol{n}) - \omega t)}, \tag{3.2.10}$$

where α is the number of a branch of vibrations; $\boldsymbol{e}_s(\boldsymbol{k}, \alpha)$ is the unit vector of polarization of this branch. Equation (3.2.9) means that atoms with different numbers s, i. e., atoms belonging to various sublattices (various Bravais lattices) can vibrate in different ways.

We impose on the real polarization vectors the requirement $\boldsymbol{e}_s(\boldsymbol{k}, \alpha) = \boldsymbol{e}_s(-\boldsymbol{k}, \alpha)$ and choose them so as to satisfy the following normalization conditions

$$\sum_{s=1}^{q} e_s(\boldsymbol{k}, \alpha) e_s(\boldsymbol{k}, \alpha') = \delta_{\alpha,\alpha'}. \tag{3.2.11}$$

The orthogonality (3.2.11) of polarization vectors belonging to different branches of vibrations generalizes the properties (1.3.2) of polarization vectors of a monatomic lattice and realizes the linear independence of the proper solutions (3.2.10). We separate the time multiplier $e^{i\omega t}$ and determine the normalized solutions (3.2.10) by analogy with (1.3.3)

$$\boldsymbol{\varphi}_{k\alpha}(\boldsymbol{n}, s) = \frac{\boldsymbol{e}_s(\boldsymbol{k}, \alpha)}{\sqrt{N}} e^{i\boldsymbol{kr}(\boldsymbol{n})},$$

implying standard normalization conditions such as (1.3.4).

Coming back to the eigenvibrations of the crystal it is necessary to remember (3.2.7). Therefore, the eigenvibrations of a complex lattice have the form

$$\boldsymbol{\psi}_{k\alpha}(\boldsymbol{n}, s) = \frac{1}{\sqrt{N m_s}} \boldsymbol{e}_s(\boldsymbol{k}, \alpha) e^{i\boldsymbol{kr}(\boldsymbol{n})}. \tag{3.2.12}$$

The orthogonality and normalization properties of the functions (3.2.12), based on the conditions (3.2.11), are such that

$$\sum_{ns} m_s \boldsymbol{\psi}^*_{\boldsymbol{k}\alpha}(\boldsymbol{n}, s)\boldsymbol{\psi}_{k'\alpha'}(\boldsymbol{n}, s) = \delta_{\boldsymbol{k}\boldsymbol{k}'}\delta_{\alpha\alpha'}.$$

The eigenvalues, i. e., the squared eigenfrequencies corresponding to the functions (3.2.12), are determined by the dispersion law (3.2.9). Unfortunately, a consistent analysis of the dispersion laws for a polyatomic crystal lattice that are determined as solutions to equations (3.2.9) is difficult. But we can easily perform a qualitative study where the guidelines will be the properties of vibrations in a two-component crystal model.

We proceed from the limiting case $\boldsymbol{k} = 0$. Introduce the displacement of the center of mass of a unit cell $\boldsymbol{u}(\boldsymbol{n})$:

$$M\boldsymbol{u} = \sum_{s=1}^{q} m_s \boldsymbol{u}_s, \quad M = \sum_{s=1}^{q} m_s, \tag{3.2.13}$$

and the sum (3.2.6) over all s

$$\omega^2 M u^i = \sum_{ss'} A^{ij}_{ss'}(\boldsymbol{k}) u^j_{s'}. \tag{3.2.14}$$

We now put $\boldsymbol{k} = 0$ on the r.h.s. of (3.2.14) and see that it vanishes due to (3.2.4)

$$\omega^2(0) M u^i = \sum_{ss'} A^{ij}_{ss'}(0) u^j_{s'} = \sum_{s'} \left(\sum_{ns} \alpha^{ij}_{ss'}(\boldsymbol{n}) \right) u^j_{s'} = 0.$$

Thus, if $\boldsymbol{u} \neq 0$ equations (3.2.6) have solutions whose frequency vanishes together with the value of the quasi-wave vector. Since there are three independent components of the unit cell center of mass there exist three branches of vibrations where, for $\boldsymbol{k} = 0$ ($\lambda = \infty$), the unit cell of a polyatomic lattice moves as a single whole with $\omega = 0$.

One may show that for $ak \ll 1$ the linear dispersion law holds for these branches

$$\omega(\boldsymbol{k}) = s_\alpha(\boldsymbol{\kappa})k, \quad \boldsymbol{\kappa} = \frac{\boldsymbol{k}}{k}, \quad \alpha = 1, 2, 3,$$

where $s_\alpha(\boldsymbol{\kappa})$ are the three sound velocities in the corresponding anisotropic medium.

Thus, in a polyatomic crystal lattice there are always three acoustic branches of vibrations. The long-wave vibrations for these branches coincide with ordinary sound vibrations of a crystal.

Equations (3.2.6) also have other solutions corresponding to $\omega \neq 0$ at $\boldsymbol{k} = 0$. In order to obtain such vibrations, we exclude from our discussion the unit cell vibrations discussed above. For this purpose, we put directly in (3.2.6) $\boldsymbol{k} = 0$

$$\sum_{s'} A^{ij}_{ss'}(0) u^j_{s'} = \omega^2 m_s u^i_s, \tag{3.2.15}$$

and impose on the atom displacement the requirement

$$\sum_s m_s \boldsymbol{u}_s = 0. \tag{3.2.16}$$

It is easily seen that (3.2.16) is the compatibility condition of (3.2.15) for $\omega \neq 0$.

Using (3.2.16), (3.2.15) can be reduced to a set of $3q - 3$ homogeneous linear algebraic equations to determine the same number of independent displacements. Equating the determinant of this system to zero, we obtain an equation of degree $3q - 3$ relative to ω^2 that yields $3q - 3$ nonzero and generally different frequency values at $\boldsymbol{k} = 0$

$$\omega_\alpha(0) = \omega_\alpha^0 \neq 0, \quad \alpha = 4, 5, 6, \ldots, 3q.$$

The frequency of these vibrations with $\boldsymbol{k} = 0$ is nonzero because there occurs a finite relative displacement of atoms in one unit cell, requiring finite energy.

The presence of nonzero frequency for the maximum long-wave vibrations is typical for the optical branches of a crystal. Therefore, we can conclude that in a polyatomic crystal lattice there are $3q - 3$ optical branches of vibrations. Since, at $\boldsymbol{k} = 0$, the condition (3.2.16) is valid for all these vibrations the unit cell center of mass remains fixed under relevant optical vibrations. Thus, the limiting long-wave optical vibrations of a polyatomic lattice are vibrations of various monatomic sublattices (various Bravais lattices) relative to another.

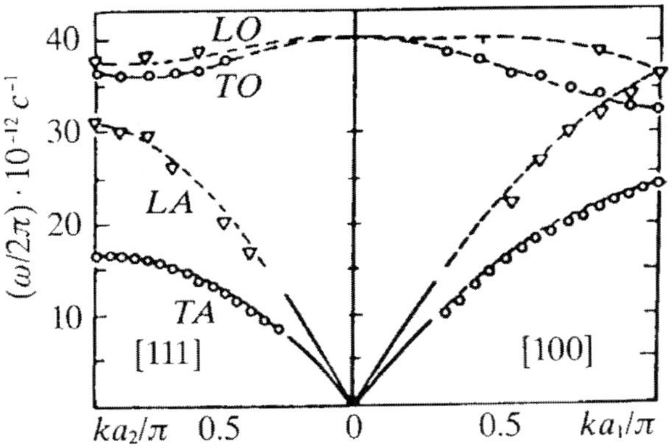

Fig. 3.2 Experimental dispersion curves of acoustic (A) and optical (O) vibrations for diamond in the directions [100] and [111] (Warren et al. 1965).

The real spectrum of vibrations of a crystal with two atoms in the unit cell is shown in Fig. 3.2, where the dispersion curves for diamond with two wave-vector directions are given. The plots show the acoustic branches (LA is the longitudinal acoustic and TA is the transverse acoustic branch) and optical branches (LO is the longitudinal optical, and TO is the transverse optical branch). Since both the chosen directions of the vector $\boldsymbol{k}$ are symmetric directions in the reciprocal lattice, all transverse modes prove to be doubly degenerate.

In conclusion, we note that the presence of optical branches of vibrations is easily taken into account by introducing the normal coordinates of a crystal. Indeed, the transition to a polyatomic lattice is formally the adding of an extra index s (the atom number in the unit cell). The expansion of arbitrary displacements $\boldsymbol{u}_s(\boldsymbol{n})$ in the functions (3.2.12) remains the same as above:

$$\boldsymbol{u}_s(\boldsymbol{n}) = \sum_{k\alpha} Q(\boldsymbol{k},\alpha)\boldsymbol{\psi}_{k\alpha}(\boldsymbol{n},s), \tag{3.2.17}$$

where $Q(\boldsymbol{k},\alpha)$ are the complex normal coordinates of a polyatomic lattice. The Hamiltonian function of small crystal vibrations (2.6.8) as well as the Hamiltonian function expressed through the real normal coordinates and momenta (2.11.14) also retain their previous form. However, the summation over α is now from 1 to $3q$.

3.3 Molecular Crystals

A crystal with a polyatomic lattice whose unit cell has a group of atoms interacting one with another stronger than with the atoms of neighboring groups is said to be a *molecular crystal*. The atoms from the chosen group are assumed to form an individual molecule, with the surrounding lattice producing an insignificant effect on its internal motion. Generally, such a crystal consists of molecules of the substance whose structure differs insignificantly from their structure in a gaseous phase. The space lattice of a molecular crystal is, as a rule, polyatomic and its unit cell often contains several molecules. Since the molecules of certain complex chemical compounds (e. g., organic ones) include a great number of atoms the linear dimension of the unit cell (identity periods) of molecular crystals may be hundreds of Angstroms.

The optical branch of molecular crystal vibrations that is responsible for intramolecular motions of strongly coupled atoms and, thus, having very high frequencies can be described as shown in Section 3.1. Such vibrations also involve the covalent atomic bonds in a molecule, and are studied as a rule independently of low-frequency types of vibrations. They represent a separate form of crystal motions and are conventionally called the *internal modes* of vibrations.

It is clear that the internal modes of vibrations do not exhaust all forms of motions of molecular crystals. There exist molecular motions that do not practically deform the covalent intramolecular bonds. These are the rotations of a molecule as a single whole relative to the unit cell, more exactly, around a certain crystallographic axis. These motions ("swings" of molecules) are often called *librations*, implying the classifications of mechanical motions of a top.

Thus, in a molecular crystal, apart from internal modes, other physically different types of motions are possible. Therefore, special terms for the corresponding vibrations are introduced. The displacements of molecular centers of mass determine the *translational* vibrations and the molecular librations manifest themselves in the *orientational* vibrations. The interaction between molecular librations is comparable in intensity with the interaction of displacements of their centers of mass. Therefore, the frequencies of the corresponding vibrations have the same order of magnitude. In other words, the orientational vibrations are not specific in terms of frequencies, but they are optically observed and, thus, specific.

In describing the vibrations of molecular crystals N_2, O_2, CO_2, etc., it is assumed that rigid linear molecules are located at the lattice sites. The assumption of the rigidity of molecules means neglecting high-frequency intramolecular vibrations and allows one to represent a linear molecule as dumb-bells having two angular degrees of freedom only. There are five degrees of freedom per unit cell in such a model: the three degrees of freedom of translational vibrations described by the vector $\boldsymbol{u}(\boldsymbol{n})$ and the two degrees of freedom of librational motions that are described by two angles $\theta(\boldsymbol{n})$ and $\varphi(\boldsymbol{n})$ that give the space orientation of linear molecules.

Restricting ourselves to a two-component model, we shall characterize a molecular state by the displacement of its center of mass $u(\boldsymbol{n})$ and by the rotation angle around a certain axis $\varphi(\boldsymbol{n})$. If the displacements of molecules as well as their librations are small, the linear equations for $u(\boldsymbol{n})$ and $\varphi(\boldsymbol{n})$ will be found by directly rewriting (3.1.3) with the replacement of $\xi(\boldsymbol{n})$ by $\varphi(\boldsymbol{n})$, and of the reduced mass μ by the molecular inertia moment I:

$$\begin{aligned} m\frac{d^2u(\boldsymbol{n})}{dt^2} &= -\sum_{n'}\left[\alpha_1(\boldsymbol{n}-\boldsymbol{n}')u(\boldsymbol{n}') + \beta(\boldsymbol{n}-\boldsymbol{n}')\varphi(\boldsymbol{n}')\right], \\ I\frac{d^2\varphi(\boldsymbol{n})}{dt^2} &= -\sum_{n'}\left[\alpha_2(\boldsymbol{n}-\boldsymbol{n}')\varphi(\boldsymbol{n}') + \beta(\boldsymbol{n}'-\boldsymbol{n})u(\boldsymbol{n}')\right]. \end{aligned} \tag{3.3.1}$$

All the results follow from (3.1.3) and discussed in Section 3.1 are automatically extended to the conclusions from (3.3.1). In particular, the dispersion relations are found as the solutions to the algebraic equation (3.1.9), where μ must be replaced by I.

Let us analyze in more detail the peculiarities of the dispersion law of a molecular crystal. Suppose the molecules are placed at the sites of a primitive cubic lattice. Now we take into account the interaction of the nearest molecules only. The nonzero elements of $\alpha(\boldsymbol{n})$ and $\beta(\boldsymbol{n})$ matrices will then be $\alpha(0), \alpha(\boldsymbol{n}_0)$ and $\beta(0)$, $\beta(\boldsymbol{n}_0)$, where $\boldsymbol{n}_0$ is the radius number of any one of the six nearest neighbors. We assume the matrix $\beta(\boldsymbol{n})$ to be symmetric, and from (3.1.4), (3.1.10) (because the lattice is highly symmetric) we get

$$\alpha_1(0) + 6\alpha_1(\boldsymbol{n}_0) = 0; \qquad \beta(0) + 6\beta(\boldsymbol{n}_0) = 0;$$

$$\alpha_2(0) + 6\alpha_2(\boldsymbol{n}_0) = I\omega_2^0. \tag{3.3.2}$$

We choose the coordinate axes along the four-fold symmetry axes and substitute (3.3.2) into (3.1.8)

$$\begin{aligned} A_1(\boldsymbol{k}) &= \frac{1}{3}\alpha_1(0)(3 - \cos ak_x - \cos ak_y - \cos ak_z); \\ B(\boldsymbol{k}) &= \frac{1}{3}\beta(0)(3 - \cos ak_x - \cos ak_y - \cos ak_z); \\ A_2(\boldsymbol{k}) &= \alpha_2(0) - \frac{1}{3}(\alpha_2(0) - I\omega_0^2)(\cos ak_x - \cos ak_y - \cos ak_z). \end{aligned} \tag{3.3.3}$$

Using the explicit expressions (3.3.3) and solving (3.1.9), we find two dispersion relations

$$\begin{aligned} 2mI\omega_{\pm}^2(\boldsymbol{k}) &= IA_1(\boldsymbol{k}) + mA_2(\boldsymbol{k}) \\ &\pm \{[IA_1(\boldsymbol{k}) - mA_2(\boldsymbol{k})]^2 + 4mIB^2(\boldsymbol{k})\}^{1/2}. \end{aligned} \tag{3.3.4}$$

We already know that for small k ($ak \ll 1$) the function $B^2(\boldsymbol{k}) \sim (ak)^4$, and in (3.3.4) it can be omitted. The long-wave dispersion laws will then coincide with (3.1.11), (3.1.12), but we write them in a more general form

$$2mI\omega_{\pm}^2(\boldsymbol{k}) = IA_1(\boldsymbol{k}) + mA_2(\boldsymbol{k}) \pm |mA_2(\boldsymbol{k}) - IA_1(\boldsymbol{k})| \,. \tag{3.3.5}$$

Assume now that the interaction of translational and orientational vibrations is small for all $\boldsymbol{k}$. Since the function $B(\boldsymbol{k})$ is responsible for this interaction we omit the last term in (3.3.4), i. e., we assume that the dispersion laws are determined completely by (3.3.5). If $IA_1(\boldsymbol{k}) < mA_2(\boldsymbol{k})$ for all $\boldsymbol{k}$, this conclusion allows one to interpret the results. The resulting dispersion laws of acoustic and optical vibrations

$$\omega_A^2(\boldsymbol{k}) = \frac{1}{m}A_1(\boldsymbol{k}) = \frac{\alpha_1(0)}{3m}(3 - \cos ak_x - \cos ak_y - \cos ak_z)\,, \tag{3.3.6}$$

$$\begin{aligned} \omega_O^2(\boldsymbol{k}) = \frac{1}{I}A_2(\boldsymbol{k}) = \quad & \frac{\alpha_2(0)}{I} = \frac{1}{3}\left(\omega_0^2 - \frac{\alpha_2(0)}{I}\right) \\ \times \quad & (\cos ak_x + \cos ak_y + \cos ak_z)\,, \end{aligned} \tag{3.3.7}$$

are described by the plots in Fig. 3.1, if the vector $\boldsymbol{k}$ is along the direction [111]. In this case the point k_B corresponds to the values $ak_x = ak_y = ak_z = \pi$ and the limiting short-wave frequencies are equal to

$$\omega_m = \sqrt{2\alpha_1(0)/m}\,, \quad \omega_1 = \sqrt{(2\alpha_2(0)/I) - \omega_0^2}. \tag{3.3.8}$$

By assumption, $\omega_m < \omega_1$ (because the molecule has a small moment of inertia). The positive B^2 in (3.3.4) may only decrease ω_m and increase ω_1. Thus, the condition

$\omega_m < \omega_1$ is not violated by taking into account the interaction of translational and orientational vibrations.

We have already noted that the characteristic frequencies of translational and librational waves have the same order of magnitude: $\alpha_1(0) \sim m\alpha_2(0)$. Hence, the crystals for which $I\alpha_1(0) > m\alpha_2(0)$ and $\omega_1 < \omega_m$ are quite possible. The plots of the dispersion laws (3.3.6), (3.3.7) along the direction [111] will then intersect at a certain point k_* (Fig. 3.3a), so that a *crossover situation* arises. Formally, following (3.3.4), the dispersion laws of acoustic (low-frequency) and optical (high-frequency) vibrations are

$$\omega_A^2(\boldsymbol{k}) = \begin{cases} \dfrac{1}{m} A_1(\boldsymbol{k}), & 0 < k < k_*\,, \\[2ex] \dfrac{1}{I} A_2(\boldsymbol{k}), & k_* < k < k_{\mathrm{B}}\,, \end{cases}$$

$$\omega_O^2(\boldsymbol{k}) = \begin{cases} \dfrac{1}{I} A_2(\boldsymbol{k}), & 0 < k < k_*\,, \\[2ex] \dfrac{1}{m} A_1(\boldsymbol{k}), & k_* < k < k_{\mathrm{B}}\,, \end{cases} \tag{3.3.9}$$

As a result of the interaction at the point $k = k_*$, a discontinuous change in the polarizations of the vibrations of two branches occurs. Actually, it follows from (3.1.7) at $B(\boldsymbol{k}) = 0$ that the dispersion law $m\omega^2 = A_1(\boldsymbol{k})$ refers to purely translational vibrations ($\phi = 0$), and the dispersion law $I\omega^2 = A_2(\boldsymbol{k})$ to the orientational vibrations ($u = 0$). The jump-like change in the vibration polarizations and the breaks appearing in the plots of the dispersion laws of acoustic and optical vibrations are the result of disregarding the interaction of translational and librational vibrations at $I\alpha_1(0) > m\alpha_2(0)$. The inclusion of even a small value of B^2 in (3.3.4) eliminates both misunderstandings, as intersection point vanishes, the plots in the vicinity of $k = k_*$ move apart, the dispersion laws become regular (Fig. 3.3b) and the polarizations transform continuously.

The above-described peculiarities of the vibration spectrum are typical for some molecular crystals mentioned above.

3.4 Two-Dimensional Dipole Lattice

A two-dimensional lattice can be made up of particles adsorbed onto an atomically smooth face of some crystals. If charged particles (ions) are adsorbed onto a metal surface each of them manifests itself as an electric dipole perpendicular to the surface. This is connected with the fact that the electrostatic charge field near the conductor (metal) plane surface is equivalent to a Coulomb charge field and its mirror reflec-

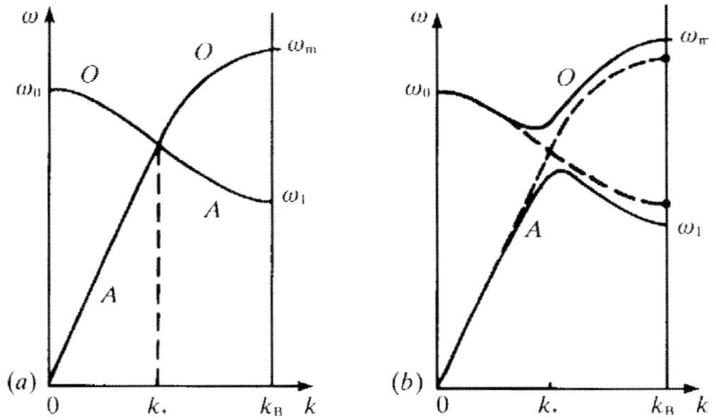

Fig. 3.3 The "cross situation": (*a*) intersection of dispersion branches; (*b*) removal of degeneracy.

tion in the plane surface (the opposite sign charge). Therefore, the adsorbed charged particles interact as parallel dipoles. When adsorbed particles are dense enough they are ordered and form a two-dimensional crystal that is the simplest realization of a 2D *dipole lattice*. The adsorbed particles, however, interact strongly enough with a substrate, such that this crystal can be regarded as two-dimensional only in terms of the geometry of its lattice.

Examples of systems whose dynamics under certain approximations are equivalent to 2D vibrations seem to be of greater interest. Under certain conditions a 2D crystal forms from electrons on a liquid helium surface at low temperatures. Electrons that have been (forced) pressed by an electric field to the liquid helium surface behave generally as a gas (gas of interacting particles), but at low enough temperatures *Wigner crystallization* may occur in the electron gas concerned and a 2D electron crystal appears. Electrons over the thick layer surface of dielectric helium interact almost like point charges. Thus, although the electrons on the liquid helium surface create a crystal lattice, the latter may not be a dipole. Only when a thin helium film is formed on a metal substrate is the electron interaction at large distances similar to the dipole interaction.

Finally, a 2D crystal may be formed by a system of magnetic bubbles. They can be obtained in a thin ferromagnetic film whose magnetic anisotropy axis is perpendicular to the plane of the film. In a strong enough magnetic field $\boldsymbol{H}$ directed along the normal to the film, a ferromagnetic film is in a single domain state. The magnetization $\boldsymbol{M}$ coincides in its direction with $\boldsymbol{H}$. However, if the external field is not very strong, cylindrical "islands" where the magnetization $\boldsymbol{M}$ is directed opposite to the external magnetic field appear in the film (Fig. 3.4). These are just cylindrical magnetic domains or bubbles. They are generally observed in films of thickness $10^{-4} - 10^{-3}$ cm and have a diameter of the same order or even less. But in any case the bubbles prove to be macroscopic formations whose dimension is much larger than the thickness of the domain boundary dividing the regions with different orientation of magnetization. A considerable "surface" energy concentrated at the domain boundary provides a circular form of the bubble cross section.

The peculiarity of the bubbles is their great mobility and the ability to move easily along the film. But moving bubbles create around themselves a dynamical magnetization field with certain inertia. As a rule, the inertia of the inhomogeneous magnetization field is attributed, in such cases, to its source, the bubble. As a result it appears that the bubble may be regarded as some particle in a 2D crystal with the definite effective mass m^*. It is clear that the bubble is an isolated magnetic dipole in the background of a uniformly magnetized film (its magnetic moment equals $\mu = 2MhS$, where h is the film thickness; S is the bubble cross-sectional area). Therefore, it is affected by the action of both inhomogeneous external magnetic field and the forces of magnetic dipole interaction with the other bubbles. The dipole interaction results in a repulsion of the same magnetic dipoles. Therefore, in a film of finite area the bubbles may form a periodic lattice. If the magnetic properties of a film are isotropic in its plane, a hexagonal (or trigonal) lattice is stable.

Let the lattice constant a be much larger that the bubble diameter ($a^2 \gg S$) and the plate thickness ($a \gg h$). The interaction energy of two bubbles at the points $\mathbf{R}$ and $\mathbf{R}'$ may then be represented as the dipole–dipole interaction energy $V(\mathbf{R} - \mathbf{R}')$, where

$$V(\mathbf{R}) = \frac{\mu^2}{R^3}. \qquad (3.4.1)$$

It is assumed in (3.4.1) that the bubble magnetic moments μ are strictly perpendicular to the plane of the film and precession deviations are absent.

Let us consider small translational vibrations[1] of a bubble lattice. We introduce $u_i(\mathbf{n})$, $i = 1, 2$ as the displacement vector of a bubble located at the point with 2D number $\mathbf{n}(n_1, n_2)$ that numbers unit cells. The equilibrium distance between the bubble centers in a static lattice is denoted by a.

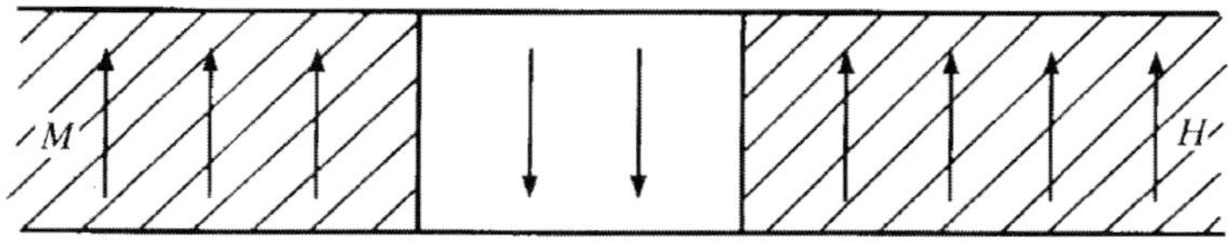

Fig. 3.4 Distribution of magnetization in a film with bubbles.

Let the bubbles experience small displacements from the equilibrium lattice points: $\mathbf{R}_n = \mathbf{r}(\mathbf{n}) + \mathbf{u}(\mathbf{n})$. Then, within the approximation quadratic in $\mathbf{u}(\mathbf{n})$ the total inter-

1) Analyzing only the vibrations of the bubble gravity centers we assume that at the frequencies we are interested in the values and the direction of a magnetic moment μ remain unchanged, the domain pulsation and precession motion of its magnetic moment do not exist, with these motions taking place at higher frequencies.

action energy of the bubbles is given by

$$U = \frac{1}{2}\sum_{\mathbf{R}\neq\mathbf{R}'} V(\mathbf{R}-\mathbf{R}') = U_0 + \frac{1}{4}\sum\left(\frac{\partial^2 V}{\partial X_i \partial X_k}\right)_0 \times [u_i(\mathbf{n}) - u_i(\mathbf{n}')]\,[u_k(\mathbf{n}) - u_k(\mathbf{n}')]\,, \tag{3.4.2}$$

where U_0 is the bubble equilibrium lattice energy.

The simplest version of the pair interaction energy (3.4.1) allows one to describe consistently the low-frequency vibrations of a dipole lattice. The expression (3.4.2) is easily reduced to a standard form (2.1.11) by denoting

$$\alpha^{ik}(\mathbf{n}) = -\left(\frac{\partial^2 V}{\partial X_i \partial X_k}\right)_0 = \frac{3\mu^2}{r^7(\mathbf{n})}\left[r^2(\mathbf{n})\delta_{ik} - 5x_i(\mathbf{n})x_k(\mathbf{n})\right], \quad \mathbf{n} > 0; \qquad \alpha^{ik}(0) = \sum_{\mathbf{n}\neq 0}\alpha^{ik}(\mathbf{n}). \tag{3.4.3}$$

By calculating the elements of the matrix of atomic force constants (3.4.3), the problem of harmonic vibrations of the bubble plane lattice is actually solved.

The long-wave dipole lattice vibrations should be described by two-dimensional equations of elasticity theory of the type (2.8.6), (2.8.7), where elasticity moduli are calculated using (2.8.5), (2.8.11). With the presence of the six-fold symmetry axis (hexagonal dipole lattice), the 2D symmetrical fourth-rank tensor c^{iklm} can be presented as $(i, k = 1, 2)$

$$c^{iklm} = A\delta_{ik}\delta_{lm} + B(\delta_{il}\delta_{km} + \delta_{im}\delta_{kl}). \tag{3.4.4}$$

To calculate A and B we will use (1.8.5). We will convolute over the pair of indices i and k in (2.8.5) and (3.4.4) and then over k and l. Thus, we get

$$A = \frac{3\mu^2}{16}Q, \qquad B = \frac{15\mu^2}{16}Q, \qquad Q = \sum_{\mathbf{n}\neq 0}\frac{1}{r^3(\mathbf{n})}. \tag{3.4.5}$$

In a two-dimensional lattice, the sum involved in (3.4.5) converges. Therefore, it is easily estimated by replacing the sum with a two-dimensional integral:

$$Q = \sum_{\mathbf{n}\neq 0}\frac{1}{r^3(\mathbf{n})} \sim \frac{1}{S_0}\int\frac{dx\,dy}{r^3} \sim \frac{1}{a^2}\int_a^\infty\frac{2\pi dr}{r^2} = \frac{2\pi}{a^3}.$$

Analogous to (3.4.4), the expression for a 2D tensor of elastic moduli λ_{iklm} in an hexagonal lattice can be written as $(i, k = 1, 2)$:

$$\lambda^{iklm} = \lambda\delta_{ik}\delta_{lm} + G(\delta_{il}\delta_{km} + \delta_{im}\delta_{kl}), \tag{3.4.6}$$

where G is the shear modulus, λ is the *Lamé* coefficient of a two-dimensional elastic medium. We now make use of (2.8.11) replacing V_0 with S_0, the 2D lattice unit cell area. We then obtain $\lambda S_0 = 2B - A = 9A$, $GS_0 = A$. Thus, the elastic moduli of a 2D dipole lattice are $\lambda = 9A/S_0 \sim 2\pi\mu^2/a^5$.

Since the two-dimensional dipole lattice has both a total compression modulus and shift modulus, longitudinal and transverse elastic waves can propagate in it. The squares of longitudinal, c_l^2 and transverse c_t^2, wave velocities are determined by the formulae known from elasticity theory

$$c_l^2 = \frac{\lambda + 2G}{m^*} S_0; \qquad c_t^2 = \frac{G}{m^*} S_0. \tag{3.4.7}$$

Irrespective of specific elastic wave velocity values, the ratio of their squares in the model of a bubble lattice is given by

$$\left(\frac{c_l}{c_t}\right)^2 = 11. \tag{3.4.8}$$

The number obtained is due to the choice of the pair interaction energy in the form of (3.4.1). Therefore, the relation (3.4.8) should be preserved for any plane dipole lattice of a similar type.

3.5 Optical Vibrations of a 2D Lattice of Bubbles

Using the simplest model we considered in the previous section the low-frequency (i. e., acoustic) vibrations of a 2D dipole lattice. Generalizing the model one can also study the optical vibrations of a 2D dipole lattice.

We consider two independent generalizations of the above model. First, for the bubble lattice, we take into account the magnetic domain pulsations connected with symmetric extensions and compressions of the region of "inverted" orientation of the magnetization vector $\boldsymbol{M}$.

With existing pulsations there appears a new dynamic variable – the bubble radius – denoted as $\boldsymbol{R}$. Setting up the equations for the new dynamic variable, we assume as before that the bubble radius is much larger than the thickness of a domain boundary dividing the regions with opposite magnetization directions. The bubble inertia will then be concentrated almost in its cylinder surface and the surface kinetic energy can be written as $(1/2)\eta v_n^2$, where η is the surface effective mass density of the domain boundary; v_n is the boundary motion velocity normal to its surface ($v_n = dR/dt$). If the bubbles move translationally with velocity V, then $v_n = V\cos\varphi (0 < \varphi < 2\pi)$ and the kinetic energy of its translational motion is determined by

$$E_{\text{kin}}^{tr} = \pi R h \eta \langle v_n^2 \rangle = \frac{\pi}{2} R h V^2 = \frac{1}{2} m^* V^2. \tag{3.5.1}$$

If there are small pulsations under which the boundary moves uniformly at all points of the bubble surface with velocity $v_n = V$, then the kinetic energy is given by

$$E_{\text{kin}}^{\text{pul}} = \pi \eta h R v_n^2 = \pi \eta h R V^2. \tag{3.5.2}$$

Comparing (3.5.1), (3.5.2) we conclude that the effective mass of symmetric pulsations is twice as large as the effective mass of translation motion, m^*. Therefore, the kinetic energy of pulsations can be expressed through dR/dt in the form

$$E_{\text{kin}}^{\text{pul}} = m^* \left(\frac{dR}{dt}\right)^2 .$$

Since the bubble pulsations change the bubble radius, they are associated with a certain increase in the potential energy of the system. Under small vibrations the potential energy of an individual bubble depends quadratically on $R - R_0$ (R_0 is the equilibrium bubble radius)

$$U^{\text{pul}} = m^* \omega_0^2 (R - R_0)^2, \tag{3.5.3}$$

where ω_0 is the pulsation eigenfrequency. In a lattice, the equilibrium radius R_0 is different from the equilibrium radius of an isolated bubble, since the magnetodipole repulsion in a static lattice simulates a decrease in the equilibrium radius of an individual bubble.

We introduce $\xi = R - R_0$ and calculate, in the approximation quadratic in $\xi(\boldsymbol{n})$, the change in the magnetic-dipole interaction energy in the bubble lattice. We represent the pulsating magnetic moment of a bubble at the $\boldsymbol{n}$-th site as

$$\mu(\boldsymbol{n}) = \mu \left[1 + 2\frac{\xi}{R_0} + \left(\frac{\xi}{R_0}\right)^2\right],$$

where μ is an equilibrium value of the magnetic moment. Then, using (3.4.1), we write

$$\begin{aligned} U_{\text{md}} = & \frac{1}{2}\sum_{\boldsymbol{n}\neq\boldsymbol{n}'} \frac{\mu(\boldsymbol{n})\mu(\boldsymbol{n}')}{r^3(\boldsymbol{n}-\boldsymbol{n}')} = \frac{1}{2}N\mu^2 Q + \frac{2\mu^2}{R_0} Q \sum \xi(\boldsymbol{n}) \\ + & \left[\frac{\mu}{R_0}\right]^2 Q \sum \xi^2(\boldsymbol{n}) + \frac{1}{2}\left[\frac{2\mu}{R_0}\right]^2 \sum_{\boldsymbol{n}\neq\boldsymbol{n}'} \frac{\xi(\boldsymbol{n})\xi(\boldsymbol{n}')}{r^3(\boldsymbol{n}-\boldsymbol{n}')}, \end{aligned} \tag{3.5.4}$$

where Q is determined by (3.4.5); N is the number of sites in the lattice.

The first term in (3.5.4) is included in the ground-state energy of the lattice and does not contribute to the equation of the motion for $\xi(\boldsymbol{n})$. The second term is responsible for the renormalization of the equilibrium bubble radius. We do not give here the calculations of this renormalization (see Problem 1), keeping in mind that the equilibrium bubble radius in a lattice is different from that of an isolated bubble. The third term in (3.5.4) contributes to the renormalization of the homogeneous pulsation frequency and, finally, the fourth term determines the pulsation wave dispersion.

Using (3.5.3), (3.5.4) we introduce in a standard way the equation of motion for pulsations

$$2m^* \frac{d^2\xi(\boldsymbol{n})}{dt^2} = -\left[2m^*\omega_0^2 + 2Q\left(\frac{\mu}{R_0}\right)^2\right]\xi(\boldsymbol{n}) - \sum_{\boldsymbol{n}\neq\boldsymbol{n}'} \beta(\boldsymbol{n}-\boldsymbol{n}')\xi(\boldsymbol{n}'), \tag{3.5.5}$$

where

$$\beta(\boldsymbol{n}) = \left(\frac{2\mu}{R_0}\right)^2 \frac{1}{r^3(\boldsymbol{n})}. \tag{3.5.6}$$

Transforming (3.5.5), taking into account the definition Q, gives

$$m^* \frac{d^2\xi(\boldsymbol{n})}{dt^2} = -m^*\omega_r^2\xi(\boldsymbol{n}) + \frac{1}{2}\sum_{\boldsymbol{n}'} \beta(\boldsymbol{n}-\boldsymbol{n}')[\xi(\boldsymbol{n}) - \xi(\boldsymbol{n}')], \tag{3.5.7}$$

where ω_r is the renormalized pulsation frequency,

$$\omega_r^2 = \omega_d^2 + \omega_0^2, \ \ \omega_d^2 = \frac{3Q}{m^*}\left(\frac{\mu}{R_0}\right)^2. \tag{3.5.8}$$

Finally, the relation (3.5.7) is written for small pulsation vibrations of the bubble lattice. For this equation the relation between the pulsations and the translational vibration of the lattice concerned was neglected. The equation for small translational vibrations was obtained, without taking into account the bubble pulsations. The relation between translational and pulsation vibrations is easily allowed for (in the expansion (3.5.4) it suffices to include the terms proportional to the bubble displacements), so the reader is invited to do this (see Problem 2).

The dispersion law for pulsation vibrations is

$$\omega^2(\boldsymbol{k}) = \omega_r^2 - \frac{1}{2m^*}\sum \beta(\boldsymbol{n})[1 - e^{-i\boldsymbol{kr}(\boldsymbol{n})}]. \tag{3.5.9}$$

It is clear that $\omega_0 = \omega_r \neq 0$. Thus, the pulsation vibrations are the optical vibrations. Furthermore, it follows from the definition (3.5.6) that $\beta(\boldsymbol{n}) > 0$; hence, the second term in (3.5.9) leads to a frequency decrease with increasing k. Thus, $\omega = \omega_r$ is the largest frequency in the spectrum of pulsation vibrations.

Let us analyze the dispersion law (3.5.9) in the long-wavelength limit ($ak \ll 1$). We denote

$$B(\boldsymbol{k}) = \sum \beta(\boldsymbol{n})[e^{-i\boldsymbol{kr}(\boldsymbol{n})} - 1], \tag{3.5.10}$$

and replace the sum over the lattice with the integral in the lattice plane using the polar coordinates (r, φ) associated with the $\boldsymbol{k}$ vector direction

$$\begin{aligned} B(\boldsymbol{k}) &= \left(\frac{2\mu}{R_0}\right)^2 \frac{1}{S_0} \int_0^{L/2} \frac{dr}{r} \int_0^{2\pi} (e^{-ikr\cos\varphi} - 1)\, d\varphi \\ &= \left(\frac{2\mu}{R_0}\right)^2 \frac{2\pi}{S_0} \int_0^{L/2} \frac{dr}{r^2}[J_0(kr) - 1], \end{aligned}$$

where S_0 is the unit cell area of a bubble lattice; L is the dimension (diameter) of the plane lattice; $J_0(x)$ is the first-order Bessel function (with zero index).

Replacing the integration variable results in

$$B(\mathbf{k}) = \left(\frac{2\mu}{R_0}\right)^2 \frac{2\pi k}{S_0} \Phi\left(\frac{kL}{2}\right), \tag{3.5.11}$$

where

$$\Phi(z) = \frac{1}{z}[1 - J_0(z)] + J_1(z) - \int_0^z J_0(x)\, dx.$$

Taking into account the possible values of the quasi-wave vector components (1.5.3), we conclude that the product kL either equals zero ($B(0) = 0$) or $kL > 2\pi$. But in the second case $\Phi(kL/2)$ does not differ in order of magnitude from its limiting value $\Phi(\infty) = -1$ and rapidly approaches it with increasing k.[2] Therefore, as we are interested in the finite interval of small values k ($ak \ll 1$), we can write at $k \neq 0$

$$B(\mathbf{k}) = \left(\frac{2\mu}{R_0}\right)^2 \frac{2\pi k}{S_0} \Phi(\infty) = -\frac{2\pi}{S_0}\left(\frac{2\mu}{R_0}\right)^2 k.$$

Thus, the long-wave dispersion law looks like

$$\omega^2 = \omega_r^2 - \frac{\pi}{m^* S_0}\left(\frac{2\mu}{R_0}\right)^2 k. \tag{3.5.12}$$

The frequency in (3.5.12) is independent of the direction of the 2D vector $\mathbf{k}$, this being a result of the symmetry of the hexagonal lattice.

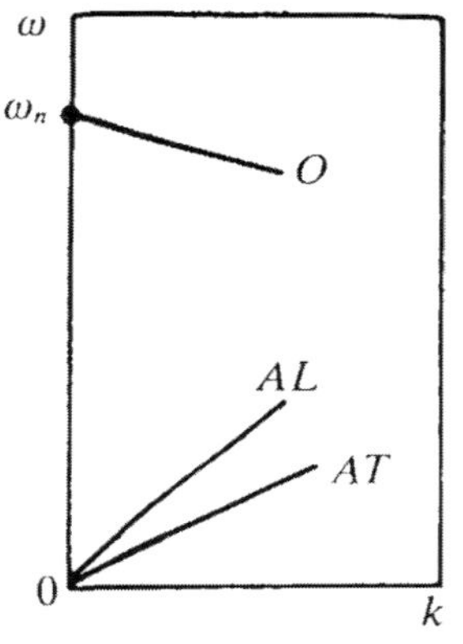

Fig. 3.5 Dispersion laws of the long-wave translational (acoustic) and pulsation (optical) vibrations of the 2D bubble lattice.

It follows from (3.5.12) that the pulsation waves for $ak \ll 1$ have nonzero velocity

$$v = \lim_{k \to 0} \frac{\partial \omega}{\partial k} = -\frac{1}{4m^* \omega_r}\frac{\partial B}{\partial k} = -\frac{\pi}{2m^* S_0 \omega_r}\left(\frac{2\mu}{R_0}\right)^2,$$

2) The value of $\Phi(2\pi)$ differs from $\Phi(\infty) = -1$ by 15 per cent.

directed opposite to the vector $\boldsymbol{k}$. The plots of long-wave dispersion laws of translational and pulsation waves in a bubble lattice are shown schematically in Fig. 3.5.

The dispersion law peculiarities that manifest themselves in a nonanalytic dependence of the frequency on the wave-vector components at $\boldsymbol{k} \to 0$ are due to the magnetostatic character of the dipole-interaction forces making up a lattice of bubbles.

3.6 Long-Wave Librational Vibrations of a 2D Dipole Lattice

We now proceed to the second possible generalization of a two-dimensional dipole lattice model. Assume that the centers of gravity of hard electric dipoles, oriented in the equilibrium state, perpendicular to the lattice plane (parallel to the z-axis) are fixed in the sites of a symmetric (triangular or quadratic) lattice. Each dipole can make librational vibrations and the libration frequency of an isolated dipole is equal to ω_0.

If the librational vibration angles are small, the change in the z-projection of a dipole moment is of the order of its square projection onto the lattice plane. Thus, let us denote by $\boldsymbol{d}_0$ a dipole moment and let $e\xi$ be its component in the lattice plane. It then follows from the condition $d^2 = d_0^2 =$ constant that

$$d_z - d_0 = -e^2 \left(\xi_x^2 + \xi_y^2\right) / 2d_0 \ll e\xi.$$

We calculate the dipole energy of the lattice in the approximation quadratic in ξ. The interaction energy of two dipoles $\boldsymbol{d}$ and $\boldsymbol{d}'$ at the points $\boldsymbol{n}$ and $\boldsymbol{n}'$ is equal to

$$V(\boldsymbol{n}-\boldsymbol{n}') = \frac{1}{R^5}[R^2\boldsymbol{d}\boldsymbol{d}' - 3(\boldsymbol{R}\boldsymbol{d})(\boldsymbol{R}\boldsymbol{d}')], \qquad \boldsymbol{R} = r(\boldsymbol{n}-\boldsymbol{n}'). \tag{3.6.1}$$

A scalar product $\boldsymbol{d}\boldsymbol{d}'$ in (3.6.1) for the approximation quadratic in ξ is

$$\boldsymbol{d}(\boldsymbol{n})\boldsymbol{d}(\boldsymbol{n}') = d_0^2 + e^2\boldsymbol{\xi}(\boldsymbol{n})\boldsymbol{\xi}(\boldsymbol{n}') - \frac{e^2}{2d_0}\left[\xi^2(\boldsymbol{n}) + \xi^2(\boldsymbol{n}')\right]. \tag{3.6.2}$$

Using (3.6.1), (3.6.2) and keeping in mind that the vector $\boldsymbol{R}$ lies in the lattice plane, we get, similar to (3.5.4), the following expression for the dipole energy

$$\begin{aligned} U = \; & \frac{1}{2}\sum V(\boldsymbol{n}-\boldsymbol{n}') \\ = \; & \frac{1}{2}d_0^2 NQ - \frac{e^2}{2}\sum \xi^2(\boldsymbol{n}) + \frac{e^2}{2}\sum D_{ik}(\boldsymbol{n}-\boldsymbol{n}')\xi_i(\boldsymbol{n})\xi_k(\boldsymbol{n}'), \end{aligned} \tag{3.6.3}$$

where Q is determined, as before, by (3.4.5), and with

$$D_{ik}(\boldsymbol{n}) = \frac{1}{R^5(\boldsymbol{n})}\left[R^2(\boldsymbol{n})\delta_{ik} - 3X_i(\boldsymbol{n})X_k(\boldsymbol{n})\right]. \tag{3.6.4}$$

The third term on the r.h.s. of (3.6.3) is traditional and the sign of the second term is unusual. It describes the energy decrease when dipoles deviate from the z-axis. This is due to the fact that at a fixed distance between the centers of two parallel dipoles, the mutual orientation along the line connecting them is the most advantageous energetically, rather than when they are perpendicular to it. But introducing the frequency ω_0, we assume the presence of forces keeping the lattice in an equilibrium state with the dipoles parallel to the z-axis. The dipole interaction acts against these forces.

We now write the kinetic energy of a vibrating dipole. If we consider a dipole as a symmetric top with moment of inertia I the kinetic energy of its small librational vibrations is given by

$$E_{\text{kin}} = \frac{1}{2} m^* \left[\left(\frac{d\xi_x}{dt} \right)^2 + \left(\frac{d\xi_y}{dt} \right)^2 \right], \tag{3.6.5}$$

where $m^* = e^2 I / d_0^2 \alpha$ is the effective mass.

The simple form of the kinetic (3.6.5) and potential (3.6.3) energies of the lattice allows one to write down the equations of collective librational motion that take into account the dipole eigenvibrations

$$m^* \frac{d^2 \xi_i(\boldsymbol{n})}{dt^2} = -[m^* \omega_0^2 - e^2 Q] \xi_i(\boldsymbol{n}) - \sum_{\boldsymbol{n} \neq \boldsymbol{n}'} \beta_{ik}(\boldsymbol{n} - \boldsymbol{n}') \xi_k(\boldsymbol{n}'), \tag{3.6.6}$$

where

$$\beta_{ik}(\boldsymbol{n}) = e^2 D_{ik}(\boldsymbol{n}), \qquad i, k = 1, 2. \tag{3.6.7}$$

We now represent (3.6.6) in a form analogous to (3.5.7)

$$\frac{d^2 \xi_i(\boldsymbol{n})}{dt^2} = -\omega_1^2 \xi_i(\boldsymbol{n}) + \frac{1}{m^*} \sum_{\boldsymbol{n}'} \beta_{ik}(\boldsymbol{n} - \boldsymbol{n}') \left[\xi_k(\boldsymbol{n}) - \xi_k(\boldsymbol{n}') \right], \tag{3.6.8}$$

where

$$\omega_1^2 = \omega_0^2 - \frac{3}{2m^*} e^2 Q. \tag{3.6.9}$$

In (3.6.8), (3.6.9) we use the relation, obvious for a symmetric lattice,

$$\sum D_{ik}(\boldsymbol{n}) = Q \delta_{ik} - 3 \sum \frac{X_i(\boldsymbol{n}) X_k(\boldsymbol{n})}{R^5(\boldsymbol{n})} = \left(Q - \frac{3}{2} Q \right) \delta_{ik} = -\frac{1}{2} Q \delta_{ik}.$$

It follows from (3.6.9) that the dipole–dipole interaction lowers the frequency of homogeneous librational vibrations of a two-dimensional dipole lattice. Since the lattice must be stable with respect to these vibrations it is necessary that $\omega_1^2 > 0$. In other words, the stability condition for a dipole lattice of the type concerned is given by the inequality

$$\frac{3}{2m^*} e^2 Q < \omega_0^2, \tag{3.6.10}$$

which imposes a restriction on the dipole density (i. e., on the lattice period a) since $Q \sim 2\pi / a^3$.

To analyze the dispersion law of librational vibrations we introduce the tensor functions

$$B_{ij}(\boldsymbol{k}) = \sum \beta_{ij}(\boldsymbol{n}) \left[e^{-i\boldsymbol{k}\boldsymbol{r}(\boldsymbol{n})} - 1\right], \tag{3.6.11}$$

which in the long-wave approximation can be written in terms of two-dimensional space integrals. In a symmetric lattice these integrals may be calculated in a specific form of Cartesian coordinates with the x-axis directed along the vector $\boldsymbol{k}$

$$B_{xx}(\boldsymbol{k}) = \frac{e^2}{S_0} \int \frac{dr}{r^2} \oint (1 - \cos^2\varphi)[e^{-ikr\cos\varphi} - 1]\, d\varphi, \tag{3.6.12}$$

$$B_{yy}(\boldsymbol{k}) = \frac{e^2}{S_0} \int \frac{dr}{r^2} \oint (1 - 3\sin^2\varphi)[e^{-ikr\cos\varphi} - 1]\, d\varphi, \tag{3.6.13}$$

$$B_{xy}(\boldsymbol{k}) = -\frac{e^2}{S_0} \int \frac{dr}{r^2} \oint \cos\varphi \sin\varphi)[e^{-ikr\cos\varphi} - 1]\, d\varphi.$$

It is clear that $B_{xy}(\boldsymbol{k}) = 0$. Thus, longitudinal vibrations whose dispersion law is determined by the function $B_l = B_{xx}(\boldsymbol{k})$, and the transverse vibrations whose dispersion law is given by $B_t = B_{yy}(\boldsymbol{k})$ are independent. Their dispersion laws are as follows

$$\omega_l^2 = \omega_1^2 + \frac{1}{m^*} B_l(\boldsymbol{k}); \qquad \omega_t^2 = \omega_1^2 + \frac{1}{m^*} B_t(\boldsymbol{k}). \tag{3.6.14}$$

We note that the limiting frequencies (at $\boldsymbol{k} = 0$) of longitudinal and transverse optical 2D lattice vibrations are the same. The behavior of the dispersion laws near the limiting frequency ω_1 is determined by the form of the functions $B_l(\boldsymbol{k})$ and $B_t(\boldsymbol{k})$.

Repeating the arguments used for the treatment of the limiting behavior of the functions (3.6.12), (3.6.13) at $ak \ll 1$ we can write them as

$$\begin{aligned} B_l(\boldsymbol{k}) &= \frac{e^2 k}{S_0} \int_0^\infty \frac{dx}{x^2} \oint (1 - 3\cos^2\varphi)(e^{-ix\cos\varphi} - 1)\, d\varphi, \\ B_t(\boldsymbol{k}) &= \frac{e^2 k}{S_0} \int_0^\infty \frac{dx}{x^2} \oint (1 - 3\sin^2\varphi)(e^{-ix\cos\varphi} - 1)\, d\varphi. \end{aligned} \tag{3.6.15}$$

It follows from the definition of the Bessel functions $J_0(x)$ and $J_1(x)$ that

$$\begin{aligned} &\oint \cos^2\varphi(e^{-ix\cos\varphi} - 1)\, d\varphi = -\left(\pi + 2\pi\frac{d^2 J_0(x)}{dx^2}\right) \\ &= 2\pi\frac{dJ_1(x)}{dx} - \pi = 2\pi\left(J_0(x) - \frac{1}{2}J_1(x) - \frac{1}{2}\right). \end{aligned} \tag{3.6.16}$$

Using this expression one can show that

$$\int_0^\infty \frac{dx}{x^2} \oint \cos^2\varphi(e^{-ix\cos\varphi} - 1)\, d\varphi = -\frac{4\pi}{3}. \tag{3.6.17}$$

Substituting (3.6.17) into (3.6.15) we get

$$B_l(\boldsymbol{k}) = 2\pi \frac{e^2 k}{S_0}, \qquad B_t(\boldsymbol{k}) = 0.$$

Thus, if we restrict ourselves only to the linear terms of the expansion in powers of k, the dispersion laws (3.6.14) will take the form

$$\omega_l^2 = \omega_1^2 + \frac{2\pi e^2}{m^* S_0} k, \qquad \omega_t^2 = \omega_1^2. \tag{3.6.18}$$

It is meaningless to write down the next terms of the expansion in powers of k without taking into account the librational and translational motions of a dipole lattice.

As in the case of the dispersion law for pulsation vibrations of a bubble lattice (3.5.12), the nonanalyticity of the dispersion law (3.6.18) considered as a function of $\boldsymbol{k}$ is connected, for $\boldsymbol{k} \to 0$, with a slow decay of the coefficients $\beta_{ik}(\boldsymbol{n})$ in an infinite sum (3.6.11). Even the first derivative of $B_{ij}(\boldsymbol{k})$ with respect to $\boldsymbol{k}$ is determined by the sum having no absolute convergence. This explains the singularity of this function as $\boldsymbol{k} \to 0$.

Although the nonanalyticity of the dispersion law (3.6.18) seems to be insignificant its appearance is important. While discussing the general properties of the dispersion law it was noted that similar nonanalyticity is observed only in the points of $\boldsymbol{k}$-space where there is degeneracy and where, going over from one branch of the spectrum to another, it is possible to preserve the continuity of the group velocity vector $v = \partial\omega/\partial\boldsymbol{k}$. In the given case such a possibility is absent and one might think that the dispersion law for small $\boldsymbol{k}$ has been derived incorrectly, and this is really so. We have neglected the retardation of the electromagnetic interaction and used the static expression for the energy of the dipole interaction (3.6.1), although we have taken into account the interaction of very distant pairs of moving dipoles. Taking into account the finite velocity of electromagnetic wave propagation affects the form of the dipole pair interaction energy and results in a restricted dispersion law in the region of small k. This situation is discussed in detail in the next section.

3.7
Longitudinal Vibrations of 2D Electron Crystal

Let us analyze the simplest model for long-wavelength vibrations of a two-dimensional electron crystal formed due to Wigner crystallization on a liquid helium surface or any other realization of a 2D electron crystal. We consider a system of electrons and ions with mass m and M, respectively, with opposite, but equal in absolute value, charges e. The entire system is neutral, if the number of ions in the volume unit equals the number of electrons. We disregard the fact that the ions form a lattice, i. e., we shall treat them as a liquid. This simple model is called the *jelly* model. As it is more convenient to study a purely electron crystal, taking into account $m/M \ll 1$, we further simplify the jelly model by assuming $M = \infty$. In this model the continuous distribution of a positive charge is time independent and coincides with the equilibrium one that provides stability of an electron crystal.

We suppose that electrons may be displaced only in the crystal plane and denote by $\xi_i(\boldsymbol{n})$ the two-dimensional vector ($i = 1, 2$) of electron displacement at the site $\boldsymbol{n}$. Let $\xi \ll a$; now, using the symmetry of an electron lattice, we expand the Coulomb energy of the interaction between electrons and the positive charge of an ion liquid in powers of ξ:

$$U = U_0 - \frac{e^2}{2} \sum D_{ik}(\boldsymbol{n} - \boldsymbol{n}') \left[\xi_i(\boldsymbol{n}) - \xi_i(\boldsymbol{n}')\right] \left[\xi_k(\boldsymbol{n}) - \xi_k(\boldsymbol{n}')\right], \qquad (3.7.1)$$

where U_0 is the energy of an equilibrium crystal. The matrix D_{ik} is given by (3.6.4).

Using (3.7.1), it is easy to write down the equation of motion of an electron crystal

$$\frac{d^2\xi_i(\boldsymbol{n})}{dt^2} = \frac{1}{m} \sum_{n} \beta_{ik}(\boldsymbol{n} - \boldsymbol{n}') \left[\xi_k(\boldsymbol{n}) - \xi_k(\boldsymbol{n}')\right], \qquad (3.7.2)$$

with the matrix β_{ik} coincident with (3.6.7).

Equation (3.7.2) is different from (3.6.8) only in the fact that $\omega_0 = 0$. Consequently, in the approximation ($ak \ll 1$) linear in k, the dispersion law for longitudinal vibrations of an electron crystal can be obtained using (3.6.18)

$$\omega_l = (2\pi n_s e^2/m)^{1/2} \sqrt{k}, \quad n_s = 1/S_0, \qquad (3.7.3)$$

where n_s is the electron density (the number of electrons per unit crystal area).

It turns out that, essentially, the optical longitudinal vibrations of an electron crystal have zero limiting frequency ($\omega(0) = 0$), i. e., they have a gapless frequency spectrum. This is a direct result of the fact that the electron system is two-dimensional. The long-wave dispersion law (3.7.3) coincides with the frequencies of plasma vibrations of a 2D electron plasma: $\omega_l^2 = \omega_{\mathrm{pl}}^2(\boldsymbol{k}) = 2\pi n_s e^2 k/m$.

However, if we make an attempt to use the results (3.6.18) to find the frequencies of long-wave transverse vibrations of an electron crystal, we see that the approximation, linear in $\boldsymbol{k}$, that resulted in (3.6.18) is insufficient. When $\omega_1 = 0$, the function $B_t(\boldsymbol{k})$ should be calculated with more accuracy, i. e., taking into account the terms quadratic in k. To make these calculations, it is necessary to replace the sums (3.6.11) by the integrals (3.6.12), (3.6.13), i. e., to go over from a discrete to a continuum description of an electron crystal.

We consider a triangular lattice that has a six-fold symmetry axis and is, thus, elastically isotropic. We put a chosen electron in the center of a circle of radius a and unit cell area ($S_0 = \pi a^2$). The sums that determine the functions $B(\boldsymbol{k})$ in the long-wave approximation ($ak \ll 1$) may then be represented, with a good accuracy in the form

$$B(\boldsymbol{k}) = e^2 \sum_{n \neq 0} \frac{f(\boldsymbol{r}(\boldsymbol{n}), \boldsymbol{k})}{r^3(\boldsymbol{n})} = \frac{e^2}{S_0} \int_a^\infty \frac{dr}{r^2} \oint f(\boldsymbol{r}, \boldsymbol{k})\, d\varphi$$

$$= \frac{e^2}{S_0} \left(\int_0^\infty \frac{dr}{r^2} - \int_0^a \frac{dr}{r^2} \right) \oint f(\boldsymbol{r}, \boldsymbol{k})\, d\varphi.$$

We make use of this and take the definition (3.6.15) into account, and also the property of the function $B_t(\boldsymbol{k})$ for small k:

$$B_t(\boldsymbol{k}) = -\frac{e^2 k}{S_0} \int_0^{ak} \frac{dx}{x} \oint (1 - 3\sin^2(\varphi))(e^{ix\cos\varphi} - 1)\, d\varphi. \qquad (3.7.4)$$

We substitute (3.6.16) into (3.7.4) retaining in the integral the leading term of the expansion in powers of x:

$$B_t(\boldsymbol{k}) = \pi e^2 a k^2 / (8 S_0).$$

We now take the second relation (3.6.14) for $\omega_1 = 0$ and write the dispersion law for transverse vibrations of an electron crystal

$$\omega_t = s_t k, \;\; s_t^2 = 0.125 \frac{e^2 \sqrt{\pi n_3}}{m}. \qquad (3.7.5)$$

So the transverse vibrations of an electron crystal have the character of sound waves with a large velocity determined by the Coulomb electron interaction[3] ($m s_t^2 \sim e^2/a$). For the realizations of a 2D electron crystal on a helium surface, the limiting period of the lattice is $a \sim 10^{-5} - 10^{-4}$ cm, so that the velocity of a transverse sound may attain the values $s_t \sim 10^5 - 10^6$ cm/s.

The presence of transverse sound vibrations in an electron crystal distinguishes it from an electron liquid (plasma). Therefore, observation of such waves is a direct proof of the crystallization of a 2D electron system.

We come back, however, to an unusual form of the dispersion relation for longitudinal waves in an electron crystal. The group velocity of longitudinal vibrations with the dispersion law (3.7.3) tends to infinity as $k \to 0$. This nonphysical result is due to neglecting the electromagnetic wave retardation. In describing the electron interaction, we considered only the electrostatic potential energy (3.7.1). When this energy is calculated, the dominant contribution comes from the terms corresponding to large distances $R(\boldsymbol{n} - \boldsymbol{n}')$ when the electromagnetic wave retardation is significant.

3) The exact calculation of 2D lattice sums gives in s_t^2 the numerical multiplier 0.138 that differs from the approximate result (3.7.5) by 10 per cent.

We assume that the dispersion law (3.6.18) is valid only for those k at which the group wave velocity is small compared with that of light c, i. e., under the condition

$$ak \gg \frac{e^2}{amc^2}. \tag{3.7.6}$$

The condition (3.7.6) allows, in principle, the existence of a wide interval of wavelengths simultaneously satisfying the conditions that the long-wave approximation be applicable and the dispersion law (3.7.3) be valid.

Although the region of the dispersion law applicability (3.7.3) is rather wide, the resulting nonphysical singularity for $k \to 0$ stimulates us to clarify how this singularity will vanish in a more consistent calculation. We consider a plane 2D electron crystal in an unbounded 3D medium with dielectric constant $\varepsilon = 1$ and consider the problem of vibrations of a 2D electron system from a different point of view. It is known that the accelerating electric charges radiate electromagnetic waves. In the quasi-static approximation used above, the radiation is absent. But it is necessary to make sure that there is also no radiation in the volume with dynamic effects taken into account, i. e., it is necessary to prove that the electromagnetic wave is localized in space near a 2D electron crystal. This is possible if the dispersion law of electromagnetic vibrations connected with electron crystal vibrations is incompatible with the dispersion law $\omega = ck$, where c is the light velocity in vacuum (in the medium with $\varepsilon = 1$).

Not taking into account the specific dielectric properties of a surrounding medium, we assume that it provides the electrons move only in the crystal plane (the plane xOy). We shall describe the electromagnetic field by means of scalar (φ) and vector ($\boldsymbol{A}$) potentials. The medium with a 2D electron crystal has a specific plane (the plane $z = 0$) with trapped but movable electric charges that have surface charge density $\rho_s = -en_s\partial\xi_k/\partial x_k$ ($k = 1, 2$) and the surface current density $\boldsymbol{j}_s = en_s\partial\boldsymbol{\xi}/\partial t$. Therefore, the equations for the potentials φ and $\boldsymbol{A}$, in the long-wave approximation are

$$\begin{aligned} \Delta\varphi - \frac{1}{c^2}\frac{\partial^2\varphi}{\partial t^2} &= 4\pi e n_s \frac{\partial \xi_k}{\partial x_k}\delta(z), \quad k = 1, 2; \\ \Delta\boldsymbol{A} - \frac{1}{c^2}\frac{\partial \boldsymbol{A}}{\partial t^2} &= -4\pi e n_s \frac{\partial \boldsymbol{\xi}}{\partial t}\delta(z), \end{aligned} \tag{3.7.7}$$

where $\delta(z)$ is the delta-function, and the dependence of $\boldsymbol{\xi}$ on time and coordinates (x and y) is found from an obvious equation for the electron motion in the continuum approximation

$$m\frac{\partial^2\boldsymbol{\xi}}{\partial t^2} = e\boldsymbol{E}|_{z=0} \equiv -e\left(\text{grad}\,\varphi - \frac{\partial \boldsymbol{A}}{\partial t}\right)\Big|_{z=0}, \tag{3.7.8}$$

where $\boldsymbol{E}$ in the mean electric field. Using the linear approximation the mean magnetic field acting on the electron may not be taken into account in the Lorentz force.

It is evident that (3.7.7), (3.7.8) admit the solutions

$$\begin{aligned} \varphi &= \varphi_0 e^{-q|z|} e^{ik\zeta - i\omega t}, \quad \mathbf{A} = \mathbf{A}_0 e^{-q|z|} e^{ik\zeta - i\omega t}, \; \xi = \xi_0 e^{ik\zeta - i\omega t} \\ q^2 &= k^2 - \omega^2/c^2, \qquad \zeta = (x, y). \end{aligned} \tag{3.7.9}$$

The presence of the delta-like right-hand sides in (3.7.3) is equivalent to jumps of the space derivative of the potentials with respect to the coordinate z at $z = 0$. Therefore,

$$\varphi_0 = -2\pi i \frac{e n_s}{q} \mathbf{k}\xi_0, \quad \mathbf{A}_0 = -2\pi i \frac{e n_s \omega}{q c} \xi. \tag{3.7.10}$$

We substitute (3.7.10) into (3.7.8) to obtain

$$m\omega^2 \xi_0 = \frac{2\pi e^2 n_s}{q} \left[\mathbf{k}(\mathbf{k}\xi_0) - \frac{\omega^2}{c^2} \xi_0 \right]. \tag{3.7.11}$$

Multiplying both parts of (3.7.11) by the vector $\mathbf{k}$, we write implicitly the dispersion law for longitudinal (plasma) vibrations

$$\omega^2 = \frac{2\pi e^2 n_s}{m} q \equiv \frac{2\pi e^2 n_s}{m} \sqrt{k^2 - \frac{\omega^2}{c^2}}. \tag{3.7.12}$$

Solving (3.7.12) for ω^2, we renormalize the dispersion law for longitudinal vibrations to include the electromagnetic wave retardation

$$\omega^2 = 2(ck_*)^2 \left[\sqrt{1 + \left(\frac{k}{k_*}\right)^2} - 1 \right], \tag{3.7.13}$$

where $k_* = \pi e^2 n_s / mc^2$ is the characteristic wave vector whose value is shared by two regions of a different behavior of the dispersion law (3.7.13). A plot of the dispersion law for longitudinal vibrations of a 2D electron crystal is given in Fig. 3.6.

For $k \ll k_*$, we get the dispersion law of electromagnetic waves in a medium with $\varepsilon = 1$: $\omega = ck$. Thus, the maximum long-wave longitudinal vibrations of an electron crystal ($k \ll k_*$) prove to be consistent with field vibrations: for $\omega \to ck$ the parameter $q \to 0$ and the electromagnetic wave penetrates deep inside the medium. For $k \gg k_*$ we come back to the dispersion law (3.7.3). Thus, shorter-wave vibrations (but under the condition $ak \ll 1$) have the dispersion law that differs greatly from that for electromagnetic waves in the medium. Therefore, the corresponding field vibrations cannot exist in the bulk far from a 2D electron crystal and they are localized near it.

However, it should be noted that the region where the dispersion law (3.7.13) is essentially renormalized is not important in experiments with a 2D electron crystal on the helium surface. Indeed, with existing electron densities $n_s \sim 10^8 - 10^{10}$ cm^{-2}, we have $k_* \sim 10^{-5} - 10^{-3}$ cm^{-1}, so that the characteristic wavelength $\lambda_* = 2\pi/k_*$ is too large for laboratory conditions.

Finally, we dwell briefly on the renormalization of the dispersion law for transverse sound vibrations. Equation (3.7.11) admits no transverse vibrations (for $q > 0$) localized near a 2D electron crystal. This is due to the fact that (3.7.11) is obtained in the continuum approximation, completely ignoring the periodic structure of a 2D electron crystal. Both the derivation of the dispersion law (3.7.5) and its renormalization can be obtained only on the basis of a certain discrete model.

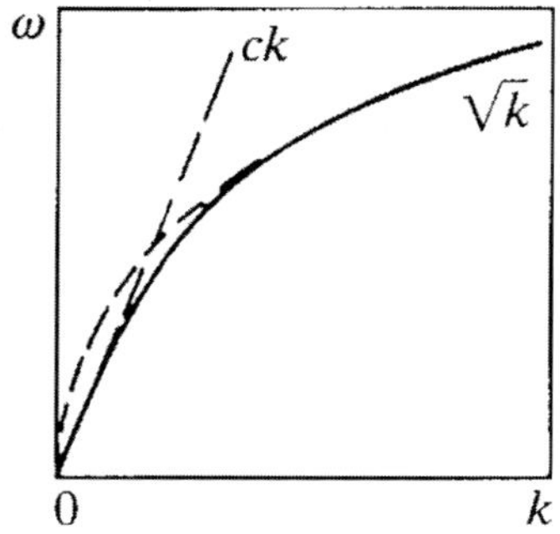

Fig. 3.6 Transformation of limiting long-wave dispersion law for 2D plasma vibrations.

3.8 Long-Wave Vibrations of an Ion Crystal

Let us discuss the equations for the vibrations of a 3D lattice composed of ions. The equations of motion of a crystal (2.8.6) in the long-wave approximation and the elasticity theory equations (2.8.7) were obtained using the assumption that the interaction of vibrating atoms decreases rapidly with increasing distance between them. Thus, expanding the displacement fields in a series, we can use only its leading terms and reduce the equation of crystal motion to a system of second-order differential equations. However, our experience of studying the optical vibrations of two-dimensional dipole lattices shows that this procedure is not always possible. If a polyatomic lattice does not consist of neutral atoms, but of electrically or magnetically active dipoles or ions (electrons), the interaction potential between them decreases weakly with distance and the standard method of expansion in a series over the displacement gradients is inapplicable.

Apart from that, under lattice vibrations the moving ions excite dynamical electromagnetic fields in a crystal whose retardation effect is important. But, for long-wave 3D crystal vibrations the electromagnetic interaction is easily taken into account by means of the Maxwell equations for a continuous medium.

A macroscopic approach allows one to divide the ionic interactions (for simplicity we assume these to be point charges) into two types with an essentially different physical origin. The first type is represented by the interactions (determined unambiguously by local conditions) that are described by the matrix of atomic force constants $\alpha(\boldsymbol{n})$. These interactions as well as the atomic ones are responsible for the elastic properties of nonionic crystals. The second type is generated by the action of the mean electric field in a crystal on the ion charge (the effect of the mean magnetic field on moving ions can be neglected[4]). This type, also involving the Coulomb ion interaction, cannot be determined by local conditions only and is described by introducing macroscopic fields.

We now realize the program proposed. In the ionic crystal, equations of the type (1.8.4) should be written for each ion sublattice separately. If $\boldsymbol{u}_s(\boldsymbol{n})$ is the displacement of the $\boldsymbol{n}$-th site of the sublattice with number s, we have:

$$m_s \frac{d^2 u_s^i}{dt^2} = \sum_{s'} \{ C_{ss'}^{ik} u_{s'}^k + C_{ss'}^{ikl} \nabla_l u_{s'}^k + C_{ss'}^{iklm} \nabla_l \nabla_m u_{s'}^k \} + e_s E^i, \tag{3.8.1}$$

where e_s is the s-th ion charge in the unit cell (by virtue of electric neutrality, $\sum_{s'} e_s = 0$), $\boldsymbol{E}$ is the averaged electric field strength in a crystal. The tensors of second, third and fourth rank $C_{ss'}$, are obviously connected with the elements of the matrix of atomic force constants in a polyatomic lattice. It follows from their definition and the equality (3.2.4) that

$$\sum_{s} C_{ss'}^{ik} = \sum_{s'} C_{ss'}^{ik} = \sum_{ns} \alpha_{ss'}^{ik}(\boldsymbol{n}) = 0. \tag{3.8.2}$$

Furthermore, in the ionic crystal where each ion is an inversion center, $C_{ik}^{ikl} \equiv 0$. For crystals such as NaCl, we shall further use the last identity. We introduce by (3.2.13) the displacement of the center of mass of a unit cell $\boldsymbol{u}$, as well as the relative displacement of ions $\boldsymbol{\xi}_s = \boldsymbol{u}_s - \boldsymbol{u}$. We keep on the r.h.s. of (3.8.1), the lowest-order space derivatives of the vector $\boldsymbol{u}$ and independent displacements $\boldsymbol{\xi}_s$

$$m_s \frac{d^2 u_s^i}{dt^2} = \sum_{s'} \{ C_{ss'}^{ik} \xi_{s'}^k + C_{ss'}^{iklm} \nabla_l \nabla_m u^k \} + e_s E^i. \tag{3.8.3}$$

In this approximation, the dynamic equations for the vectors $\boldsymbol{u}$ and $\boldsymbol{\xi}_s$ are separated

$$\begin{aligned} M_s \frac{d^2 u_s^i}{dt^2} &= c^{iklm} \nabla_k \nabla_l u^k + e_s E^i, \quad c^{iklm} = \sum_{ss'} C_{ss'}^{iklm}; \\ m_s \frac{d^2 \xi_s^i}{dt^2} &= \sum_{s'} C_{ss'}^{ik} \xi_{s'}^k + e_s E^i. \end{aligned} \tag{3.8.4}$$

Equations (3.8.3) should be solved together with the system of macroscopic Maxwell equations whose form is well known. We note some points that refer to

4) Even in the region of limiting vibration frequencies ($\omega \sim 10^{13} c^{-1}$) the ion motion velocities are less than the sound velocities in a crystal ($v < s \sim 10^5$ cm/s) and the Lorentz force generated by the magnetic field is vanishingly small compared to the electric force.

writing and solving the Maxwell equations. First, ionic crystals are dielectric (insulator) crystals and have no marked magnetic properties. Thus, the only quantity that refers to the characteristics of the medium properties and is involved in Maxwell equations is the crystal polarization vector $\boldsymbol{P}$. We remind ourselves that the vector $\boldsymbol{P}$ is included in the definition of electric induction of a dielectric $\boldsymbol{D}$: $\boldsymbol{D} = \boldsymbol{E} + 4\pi\boldsymbol{P}$. Then, since we consider the ions as point charges, we should relate the vector $\boldsymbol{P}$ only with the displacement of the ions from their equilibrium positions

$$\boldsymbol{P} = \frac{1}{V_0}\sum_s e_s \boldsymbol{u}_s \equiv \frac{1}{V_0}\sum e_s \boldsymbol{\xi}_s. \tag{3.8.5}$$

We analyze the equations for crystal ion vibrations by considering NaCl-type crystals that possess cubic symmetry and contain only two ions in the unit cell. Each of the ions is a center of symmetry (Fig. 0.2). In this case (3.8.4) will be much simplified

$$\begin{aligned} m_1\ddot{\xi}_1^i &= C_{11}^{ik}\xi_1^k + C_{12}^{ik}\xi_2^k + e_1 E^i; \\ m_2\ddot{\xi}_2^i &= C_{21}^{ik}\xi_{21}^k + C_{22}^{ik}\xi_2^k + e_2 E^i. \end{aligned} \tag{3.8.6}$$

By virtue of the cubic symmetry of a crystal and the properties (3.8.2) all the elements of the matrix $C_{ss'}^{ik}$, are expressed through one scalar quantity:

$$C_{ss'}^{ik} = C_{ss'}\delta_{ik}, \;\; C_{11} = C_{22} = -C_{12} = -C_{21} = -\alpha_0. \tag{3.8.7}$$

To be specific, let the subscript 1 refer to a positive ion ($e_1 = e$, $e_2 = -e$). Then (3.8.6), taking into account (3.8.7), will be

$$\begin{aligned} m_1\frac{d^2\boldsymbol{\xi}_1}{dt^2} &= -\alpha_0(\boldsymbol{\xi}_1 - \boldsymbol{\xi}_2)e\boldsymbol{E}; \\ m_2\frac{d^2\boldsymbol{\xi}_2}{dt^2} &= -\alpha_0(\boldsymbol{\xi}_2 - \boldsymbol{\xi}_1)e\boldsymbol{E}. \end{aligned} \tag{3.8.8}$$

Equations (3.8.8) can be replaced by a single equation for the relative displacement $\boldsymbol{\xi} = \boldsymbol{\xi}_1 - \boldsymbol{\xi}_2$:

$$\mu\frac{d^2\boldsymbol{\xi}}{dt^2} = -\alpha_0\boldsymbol{\xi} + e\boldsymbol{E}, \tag{3.8.9}$$

where μ is the reduced mass of a unit cell ($\mu = m_l m_2/M$). The crystal polarization is expressed through the displacement ξ

$$\boldsymbol{P} = \frac{e}{V_0}\boldsymbol{\xi}. \tag{3.8.10}$$

Equations (3.8.9), (3.8.10) and also the Maxwell equations describe the combined (coupled) optical ion crystal vibrations and the electromagnetic field vibrations. Thus, the above system of equations allows one to take into account the interaction of a free electromagnetic field and independent optical vibrations of a crystal. The interaction between electromagnetic waves and the optical eigenvibrations of a crystal is especially large when their frequencies and the wave vectors almost coincide (under the resonance conditions). As a result of such an interaction, collective excitations of a new type known as *polariton vibrations* appear. This is why the long-wave ion crystal vibrations are analyzed within a macroscopic theory of polaron excitations.

It is known from electrodynamics that the displacement field ξ is excluded from Maxwell equations by incorporating the dielectric permeability of the medium[5], ε, relating the Fourier time components of the vectors $\boldsymbol{D}$ and $\boldsymbol{E}$:

$$\boldsymbol{D}(\omega) = \varepsilon(\omega)\boldsymbol{E}(\omega). \tag{3.8.11}$$

To obtain the dielectric permeability, we consider the harmonic vibrations of all fields. For stationary vibrations with frequency ω, (3.8.9) is transformed to the algebraic one

$$(\omega_0^2 - \omega^2)\boldsymbol{\xi} = \frac{e}{\mu}\boldsymbol{E}, \tag{3.8.12}$$

where $\omega_0^2 = \alpha_0/\mu$ is the square of the characteristic frequency that is the limiting optical frequency for a crystal where there is no interaction with an electromagnetic field.

If $\omega \neq \omega_0$, we obtain trivially from (3.8.10)–(3.8.12) an expression for the dielectric permeability

$$\varepsilon(\omega) = 1 + \frac{\omega_{\text{pl}}^2}{\omega_0^2 - \omega^2} \equiv \frac{\omega_0^2 + \omega_{\text{pl}}^2 - \omega^2}{\omega_0^2 - \omega^2}, \tag{3.8.13}$$

where $\omega_{\text{pl}}^2 = 4\pi e^2/\mu V_0$ is the square of the so-called plasma frequency (ω_{pl} is the frequency of the eigenvibrations of an ionic 3D plasma whose ions interact only through a macroscopic Coulomb field), μ is the reduced mass of a pair of ions with opposite signs, V_0 is the volume of this pair. If we take $V_0 \sim 10^{-23}$ cm^3 and assume $\mu \sim 5 \times 10^{-23}$ g, it is easily seen that $\omega_{\text{pl}} \sim 10^{13} c^{-1}$. Therefore, we may assume that $\omega_{\text{pl}} \sim \omega_0$.

With formula (3.8.13) for dielectric permeability, we can directly write the dispersion law of transverse vibrations for a crystal–electromagnetic field system (the possible trivial decomposition of electromagnetic vibrations into transverse and longitudinal ones results from the symmetry of a cubic crystal). In fact, the transverse electromagnetic vibrations (electromagnetic waves) in a medium with dielectric permeability $\varepsilon(\omega)$ have the dispersion law

$$\omega = \frac{ck}{\sqrt{\varepsilon(\omega)}}, \tag{3.8.14}$$

5) As the NaCl-type crystal has cubic symmetry, the dielectric permeability tensor of such a medium reduces to a single constant.

where c is the light velocity in vacuum. Substituting (3.8.13) into (3.8.14), we write the dispersion law for transverse vibrations as

$$c^2k^2 = \omega^2 \frac{\omega^2 - \omega_0^2 - \omega_{\mathrm{pl}}^2}{\omega^2 - \omega_0^2}. \qquad (3.8.15)$$

The longitudinal electromagnetic vibrations with nonzero frequency are possible in the medium only when[6] $\boldsymbol{D} = 0$, when $\boldsymbol{E} = -4\pi \boldsymbol{P}$. But for $\boldsymbol{E} \neq 0$, it follows from (3.8.11) that the frequencies of the corresponding vibrations are zeros of the functions $\varepsilon(\omega)$: $\varepsilon(\omega) = 0$. Thus, longitudinal vibrations are possible only at frequencies $\omega = \omega_1$, $\omega_l^2 = \omega_0^2 + \omega_{\mathrm{pl}}^2$. Therefore, their dispersion law reduces to a trivial dependence; the frequency coincident with the value ω_l is constant.

The dispersion laws for noninteracting electromagnetic waves and optical crystal vibrations can be written in a simpler form by formally excluding the interaction in the system concerned, i. e., by setting $e = 0$. The Maxwell equations will then reduce to the field equations in vacuum and the dispersion law for electromagnetic waves will be of a simple form

$$\omega = ck. \qquad (3.8.16)$$

The dispersion law for optical vibrations follows from (3.8.12) with $e = 0$

$$\omega = \omega_l. \qquad (3.8.17)$$

The limiting simplicity of this dispersion law follows from the condition $ak \ll 1$. We note that the plots of the dispersion laws (3.8.16), (3.8.17) (Fig. 3.7a, straight lines 1 and 2, respectively) intersect. At the intersection point of the plots the frequencies and wavelengths of vibrations of different nature coincide and the above-mentioned cross situation arises. Therefore, when even a small interaction between vibrations occurs, resonance considerably affecting the electromagnetic and mechanical processes of the system concerned is observed.

Indeed, we come back to the dispersion law of transverse vibrations (3.8.15) where we have taken into account the interaction required. If $\omega \ll \omega_0$, (3.8.15) gives the dispersion law for electromagnetic waves in the medium with a certain static dielectric permeability

$$\omega = \frac{ck}{\sqrt{\varepsilon(0)}}, \qquad \varepsilon(0) = 1 + \left(\frac{\omega_{\mathrm{pl}}}{\omega_0}\right).$$

Under such low-frequency vibrations, the lattice manages to adapt to the electromagnetic field, causing a decrease in the electromagnetic wave velocity only.

If $\omega \gg \omega_l$ the dispersion law for electromagnetic waves in vacuo (3.8.16) follows from (3.8.15). The electromagnetic waves with such frequencies that arise in a crystal

6) The condition $\boldsymbol{D} = 0$ for $\omega \neq 0$ is a result of Maxwell's equation for the dielectric

$$\operatorname{curl} \boldsymbol{H} = \frac{1}{c}\frac{\partial \boldsymbol{D}}{\partial t} = -i\frac{\omega}{c}\boldsymbol{D},$$

and the longitudinal character of vibrations ($\operatorname{curl} \boldsymbol{H} = 0$).

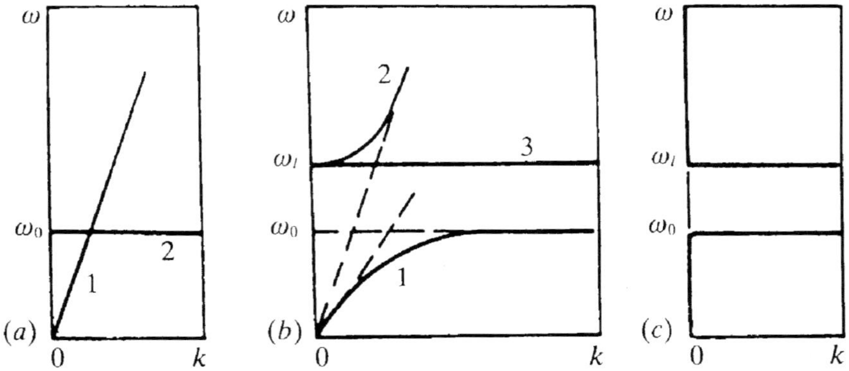

Fig. 3.7 Dispersion law of polariton vibrations: (*a*) dispersion curves of independent field and crystal vibrations when there is no interaction; (*b*) removal of degeneracy; (*c*) vibration frequencies when the retardation ($c = \infty$) is neglected.

do not force the lattice to move because of its inertia, so that the crystal does not react to the wave transmission.

In the frequency range $\omega \sim \omega_0$, the dispersion law ceases to be linear (Fig. 3.7b). A radical rearrangement of the dispersion law at frequencies $\omega \sim \omega_0$ testifies to a resonance character of the interaction between the electromagnetic field and the optical vibrations at $\omega = \omega_0$.

It follows from (3.8.13) that at frequencies that do not correspond to the crystal vibrations concerned, the dielectric permeability becomes negative (total wave reflection from a crystal).

Finally, we consider an extreme form of the dispersion law for transverse vibrations in a special case when the retardation of the electromagnetic waves are entirely neglected ($c \to \infty$). In this case for $\mathbf{k} \neq 0$, (3.8.15) yields the relation $\omega = \omega_0$ that is the same as the dispersion law of optical vibrations (3.8.17) for the lattice noninteracting with the field. For $\omega \neq \omega_0$, (3.8.15) is consistent only for $\mathbf{k} = 0$.

The longitudinal vibrations are retardation independent (they are quasi-static), so that the plots of the dispersion laws at $c = \infty$ have the form shown in Fig. 3.7c.

The difference in the plots in Fig. 3.7b is manifest at wavelengths for which $ak \lessapprox a(\omega_0/c) \sim s/c$, where s is the sound velocity in a crystal. As the ratio $s/c \sim 10^{-5}$, the retardation effects the crystal vibration spectrum in a small part of the allowed interval of $\mathbf{k}$. Thus, in all problems where the vibrations with wavelengths satisfying the condition $s/c \ll ak \ll 1$ are dominant, the retardation is insignificant and the optical vibrations in an ionic crystal should be associated with the frequency ω_0 for transverse mechanical vibrations, and with the frequency ω_l for longitudinal vibrations, caused by a purely static electric field.

The ratio between transverse and longitudinal vibration frequencies is related to macroscopic characteristics of the ionic crystal, namely, the limiting values of its dielectric permeability. In fact, it directly follows from (3.8.13) that

$$\left(\frac{\omega_l}{\omega_0}\right)^2 = \frac{\varepsilon(0)}{\varepsilon(\infty)}. \qquad (3.8.18)$$

Although, using (3.8.13) it is possible to conclude that $\varepsilon(\infty) = 1$. But this property of the model concerned is not used in (3.8.18). If we reject the point charges model and take into account the electric structure of ions that allows one to consider them being polarizable in an external field, we come to the conclusion that $\varepsilon(\infty) \neq 1$, but in this case too (3.8.18) remains valid[7].

3.8.1
Problems

1. Find the dispersion relation for longitudinal vibrations of a 1D chain with equidistant oscillators. Discuss the possible existence of similar 1D systems with an inhomogeneous ground state.

Hint. Assume each atom possesses the energy $U_n = (1/2)m\omega_0^2 u_n^2$ and take into account the interaction of nearest neighbors only.

Solution. The dispersion law is

$$\omega^2 = \omega_0^2 + \frac{2\alpha(0)}{m}\sin^2\frac{ak}{2},$$

if $\alpha(0) < 0$ and $2|\alpha(0)| > m\omega_0^2$ then for a certain $k = k_0$ in the range $0 < k < \pi/a$ the vibration frequency ω vanishes. Thus, such a system has an equilibrium superlattice with period $b = \pi/k_0$.

2. Find the long-wave dispersion relation for librational vibrations of a 2D dipole lattice, taking into account the electromagnetic wave retardation.

3. Find the long-wave dispersion relation for a 2D electron crystal in a magnetic field $\boldsymbol{H}$ perpendicular to the crystal plane.

Solution.

$$\omega^2 = \left(\frac{eH}{mc}\right)^2 + \frac{Ak}{m^*(k)}.$$

7) By the frequency $\omega = \infty$ we mean, in the last case, the frequency satisfying the condition $\omega \gg \omega_l$, but still remaining small compared to the characteristic frequencies of the interatomic electron motion.

4
Frequency Spectrum and Its Connection with the Green Function

4.1
Constant-Frequency Surface

A set of the polarization vectors $\boldsymbol{e}(\boldsymbol{k},\alpha)$ of crystal vibrations and the dispersion law

$$\omega = \omega_\alpha(\boldsymbol{k}) \tag{4.1.1}$$

provide full information on the character of vibrational states of the lattice. However, to understand some phenomena generated by lattice vibrations, the geometric (more exactly, topographic) representation of the dispersion law seems to be more instrumental than its analytical notation (4.1.1). This can be done by the introduction and analysis of so-called isofrequency surfaces.

Isofrequency surfaces are constructed independently for each branch of vibrations, so that in writing the dispersion law (4.1.1), we omit the index α, implying one of the branches.

An *isofrequency surface* or a *constant-frequency* surface is the surface in $\boldsymbol{k}$-space described by

$$\omega(\boldsymbol{k}) = \omega, \qquad \omega = \text{constant}. \tag{4.1.2}$$

Let us elucidate how the form of isofrequency surfaces changes when the frequency ranges from $\omega = 0$ up to the maximum possible one. For small frequencies the dispersion law of a crystal coincides with the dispersion law of sound waves, so that the isofrequency surface is determined by

$$k = \frac{\omega}{s(\boldsymbol{\kappa})}, \qquad \boldsymbol{\kappa} = \frac{\boldsymbol{k}}{k}, \tag{4.1.3}$$

where $s(\boldsymbol{\kappa})$ is the sound velocity. Expression (4.1.3) is a reason to introduce another name of the isofrequency surface for low-frequency sound vibrations, namely, a *slowness surface*.

As the sound velocity in a crystal is a finite quantity for all directions of $\boldsymbol{\kappa}$, it follows from (4.1.3) that the corresponding slowness surfaces are *closed* (and similar to each

The Crystal Lattice: Phonons, Solitons, Dislocations, Superlattices, Second Edition. Arnold M. Kosevich

ISBN: 3-527-40508-9

other). If we look at the figure where these surfaces are crossed by the plane that goes through a point $\boldsymbol{k} = 0$ (Fig. 4.1a), we see that the frequencies ω_1 and ω_2 satisfy the condition $\omega_1 < \omega_2$ (an isofrequency surface for a smaller frequency is inside that for larger frequency). The slowness surfaces for $\omega \to 0$ are not necessarily convex, they can be "pillow"-shaped (Fig. 4.1b).

The fact that the constant-frequency surface even of long-wave crystal vibrations may be other than convex leads to some interesting physical results. The parts where the cross section of an isofrequency surface is convex are isolated from those where it is concave by the points with zero cross-sectional curvature (Fig. 4.1b). In a 3D $\boldsymbol{k}$-space, the convex parts of an isofrequency surface are isolated from the concave ones by the lines along which the Gaussian (total) surface curvature vanishes. As an example, we consider a constant-frequency surface for one of the transverse vibration modes in a germanium crystal, where the larger part of the mode is convex. Since the Ge crystal has cubic symmetry, only one octant is shown in Fig. 4.2. The thick lines illustrate the geometrical position of the points with zero Gaussian curvature. The characteristic directions, for which the curvature is concave, are also shown.

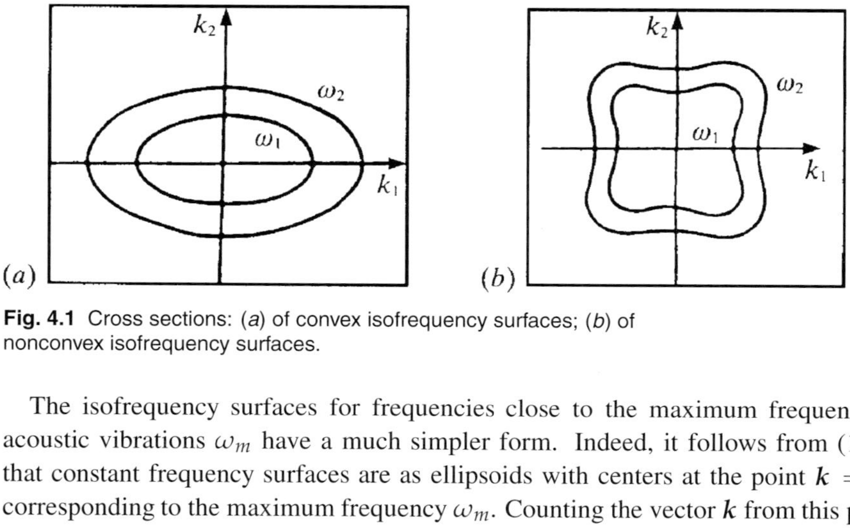

Fig. 4.1 Cross sections: (*a*) of convex isofrequency surfaces; (*b*) of nonconvex isofrequency surfaces.

The isofrequency surfaces for frequencies close to the maximum frequency of acoustic vibrations ω_m have a much simpler form. Indeed, it follows from (1.4.4) that constant frequency surfaces are as ellipsoids with centers at the point $\boldsymbol{k} = \boldsymbol{k}_m$ corresponding to the maximum frequency ω_m. Counting the vector $\boldsymbol{k}$ from this point, we get the following equation for an ellipsoid

$$\frac{1}{2}\gamma_{ij}k_ik_j = \omega_m - \omega, \qquad \omega_m - \omega \ll \omega_m. \tag{4.1.4}$$

A set of ellipsoids (4.1.4) crossed by the plane that goes through their common center may be compared with those in Fig. 4.1a, but now $\omega_1 > \omega_2$ (an isofrequency surface for larger frequency is placed inside that for smaller frequency).

On clearing up the topology of isofrequency surfaces at the boundaries of the eigenfrequency band of acoustic vibrations, we note that the points where $\omega = 0$ and $\omega = \omega_m$ repeat periodically in a reciprocal space due to the periodicity properties dispersion law. To image the corresponding geometrical picture, we consider a symmetrical enough crystal whose vibration frequencies take maximum values only at the unit cell vertices of a reciprocal lattice.

Let the k_1Ok_2 plane in reciprocal space correspond to a certain "good" crystallographic direction. We denote by b_1 and b_2 the reciprocal lattice vectors along the axes k_1, k_2 and assume that $\omega = 0$ at the point $k_1 = k_2 = 0$, and $\omega = \omega_m$ at $k_1 = (1/2)b_1$, $k_2 = (1/2)b_2$. The points $(0,0)$, $(0,b_2)$, $(b_1,0)$, (b_1,b_2) are then surrounded by the closed isofrequency surfaces "expanding" with increasing ω ($\omega \ll \omega_m$, and the closed surfaces "compressing" with the frequency increasing ($\omega - \omega_m \ll \omega_m$) will surround the points $((1/2)b_1, (1/2)b_2)$ (Fig. 4.3). The properties of closed surfaces expanding and compressing with rising ω cannot transform into each other without changing their topology. In order to pass from the surface of one set to that of another set, it is necessary to "unscrew" the surface. But the closed surface cannot be unscrewed. Thus, apart from closed isofrequency surfaces there exist isofrequency surfaces of another geometrical type, e. g., the isofrequency surfaces ranging within $\omega_1 < \omega < \omega_2$ (Fig. 4.3). These surfaces passing continuously from one unit cell of a reciprocal lattice to another one are called *open*.

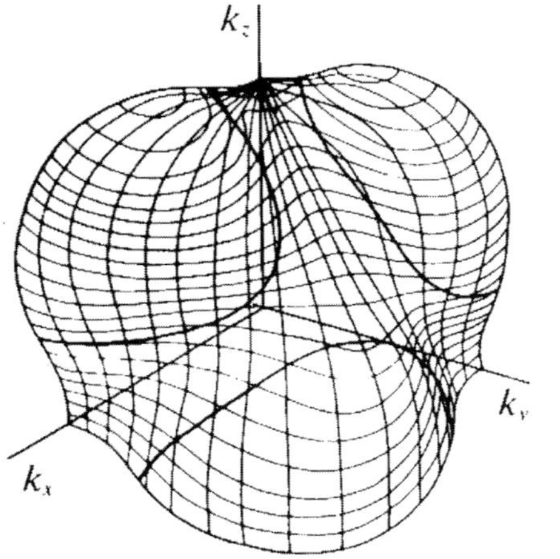

Fig. 4.2 Isofrequency surface for one of the branches of Ge crystal vibrations.

Generally, points of the type $(0, (1/2)b_2)$ or $(b_1, (1/2)b_2)$ through which the isofrequency surfaces are transformed into open ones are the *conical* points in a 3D k-space. We can imagine the form of constant-frequency surfaces of real crystal vibrations studying the results of calculations of isofrequency surfaces for a face-centered cubic lattice of Al (Walker, 1956). Figure 4.4 shows the cross sections of the isofrequency surfaces of a longitudinal branch vibration in Al inside the same Brillouin zone. The fractional numbers near the cross section lines show the value ω/ω_m for the branch concerned. The conical points are not singled out, but they are positioned

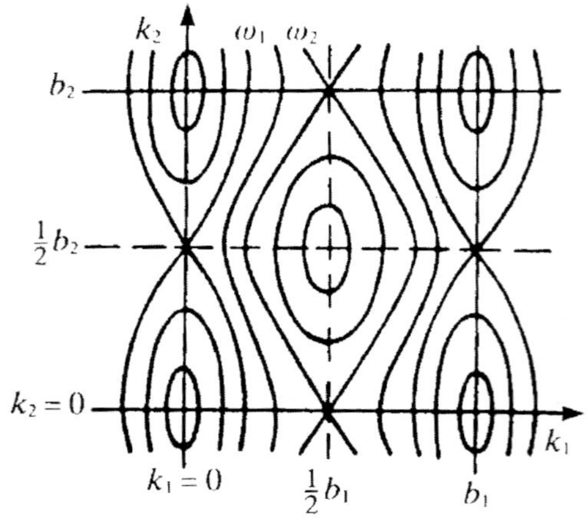

Fig. 4.3 Cross section of isofrequency surfaces of Ge crystal vibrations through the plane $k_z = 0$.

in the frequency range $0.8 < \omega/\omega_m < 0.825$. Near such points the dispersion law is written as

$$\omega = \omega_k + \gamma_1 k_1^2 + \gamma_2 k_2^2 - \gamma_3 k_3^2, \tag{4.1.5}$$

where the constant coefficients $\gamma_1, \gamma_2, \gamma_3$ have the same signs, the vector $\boldsymbol{k}$ is measured from the conical points, the axis k_3 is directed along the axis of the corresponding cone. The value ω_k at the conic point is determined by a specific form of the dynamical matrix $\boldsymbol{A}(\boldsymbol{k})$ of the crystal.

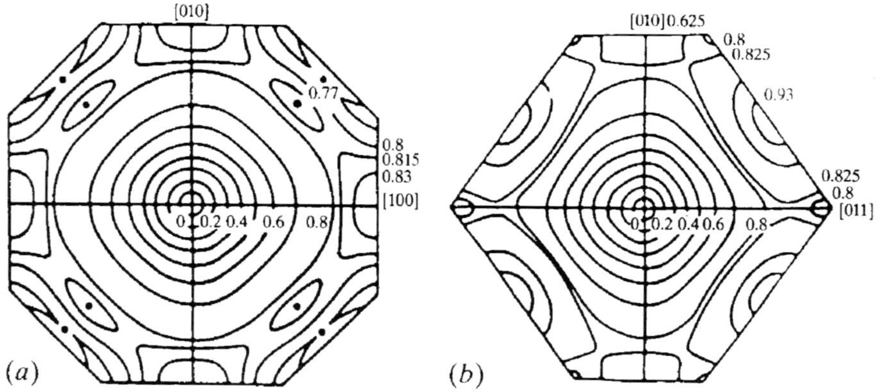

Fig. 4.4 Cross section of calculated isofrequency surfaces of a longitudinal branch of vibrations of Al (Walker 1956): (*a*) the plane (100); (*b*) the plane (110).

The equation for isofrequency surfaces resulting from (4.1.5) determines a set of hyperboloids. If all the coefficients γ_α are positive when $\omega < \omega_c$, we obtain double-banded hyperboloids, and when $\omega > \omega_c$ single-banded ones (Fig. 4.5,

$\omega_1 < \omega_c < \omega_2$). Similarly, we can trace the specific forms of isofrequency surfaces of optical branches of crystal vibrations. No matter how complicated the dispersion law of an optical branch may be, the corresponding frequencies are always placed in a band of finite width and the edges of this band are the extremum points for the function $\omega = \omega_\alpha(\boldsymbol{k}), \quad \alpha = 4, 5 \ldots$. This means that the dispersion law near the optical band edges is quadratic, and the isofrequency surfaces are ellipsoids of the type (4.1.4) where the matrix γ_{ij} is positively definite in the vicinity of the maximum frequency, and it is negatively definite in the vicinity of the minimum frequency.

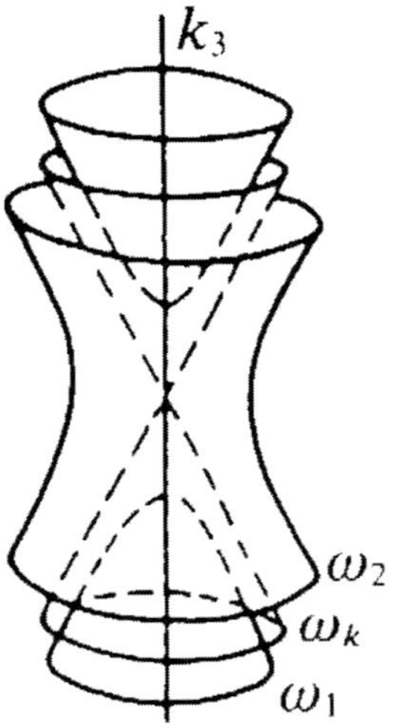

Fig. 4.5 The form of isofrequency surfaces near the conical point.

Concluding the geometrical analysis of constant-frequency surfaces, we note that the direction of the wave group velocity is totally dependent on the form of an isofrequency surface that passes through the point $\boldsymbol{k}$. Indeed, the group velocity is determined by (1.4.10), i. e., by the gradient in $\boldsymbol{k}$-space of the function $\omega(\boldsymbol{k})$. The gradient is directed along the normal to the surface of constant level of the function for which it is calculated, so that the velocity $\boldsymbol{v}(\boldsymbol{k})$ is directed along the normal to an isofrequency surface going through the point $\boldsymbol{k}$.

4.2 Frequency Spectrum of Vibrations

To explain some properties of crystals, one does not need to have exhaustive information on vibrations such as provided by the dispersion law. It suffices to know only the frequency distribution of the vibrations. In view of this, the vibration density concept or the frequency distribution function is introduced.

Let us use the notation $\varepsilon = \omega^2$. The fraction of vibrations of the α-th branch dn_α whose frequencies are in the interval $(\varepsilon,\ \varepsilon + d\varepsilon)$ can be written as

$$dn_\alpha = N g_\alpha(\varepsilon) d\varepsilon, \tag{4.2.1}$$

where N is the number of unit cells in the crystal[1]. The function $g_\alpha(\varepsilon)$ is called the *density of states*. Since for every branch of vibrations, this function is calculated independently, the index α will be omitted in future. Obviously, the following formula holds

$$g(\varepsilon) = \frac{1}{N} \sum \delta\left(\varepsilon - \omega^2(\boldsymbol{k})\right), \tag{4.2.2}$$

where the dependence $\omega^2(\boldsymbol{k})$ is determined by the crystal dispersion law, and the summation is over all physically nonequivalent values of this vector (in one cell of the reciprocal lattice or in the first Brillouin zone).

We transform (4.2.2) transforming the summation to the integration by the rule (1.5.4) and, on changing the integration variables, we have

$$g(\varepsilon) = \frac{V_0}{(2\pi)^3} \int \delta\left(\varepsilon - \omega^2(\boldsymbol{k})\right) d^3\boldsymbol{k} = \frac{V_0}{(2\pi)^3} \int \delta\left(\varepsilon - \omega^2\right) \frac{d\omega^2\, dS_k}{\left|\dfrac{\partial \omega^2(\boldsymbol{k})}{\partial \boldsymbol{k}}\right|}, \tag{4.2.3}$$

where V_0 is the unit cell volume of the crystal, dS_k is an element of the surface area $\omega^2(\boldsymbol{k}) = \omega^2 = \text{const}$.

On performing the integration over ω in (4.2.3), we obtain the ultimate formula for the density of vibrational states

$$g(\varepsilon) = \frac{V_0}{(2\pi)^3} \oint\limits_{\omega^2(\boldsymbol{k})=\varepsilon} \frac{dS_k}{\left|\dfrac{\partial \omega^2}{\partial \boldsymbol{k}}\right|}, \tag{4.2.4}$$

where the integration is carried out over the isofrequency surface $\omega^2(\boldsymbol{k}) = \varepsilon =$ constant.

In many cases it is convenient to use the distribution function in frequencies ω, rather than in squared frequencies ε. If we write the fraction of vibrations whose frequencies lie in the interval $(\omega,\ \omega + d\omega)$ in the form of $dn = \nu(\omega)\, d\omega$, then $\nu(\omega)$ will determine the *frequency spectrum*.

The normalization of the function $\nu(\omega)$ is different from that of the function $g(\varepsilon)$, namely, $\int \nu(\omega)\, d\omega = N$.

A simple relation exists between the functions $\nu(\omega)$ and $g(\varepsilon)$:

$$\nu(\omega) = 2Ng\omega g(\omega^2), \tag{4.2.5}$$

1) In a monatomic lattice, the number of unit cells coincides with the number of atoms in the crystal. However, in a polyatomic lattice, this is not so, and to preserve in (4.2.1) the normalization of distribution function of any branch $\int g_\alpha(\varepsilon) d\varepsilon = 1$ we will always use N as the number of unit cells in the crystal.

and for the functions $\nu(\omega)$ one can write a formula similar to (4.2.4):

$$\nu(\omega) = \frac{V_0}{(2\pi)^3} \oint\limits_{\omega(\boldsymbol{k})=\omega} \frac{dS_k}{v}, \tag{4.2.6}$$

where the velocity definition (1.4.10) is taken into account.

If we are not interested in the frequency spectrum, but in the total number of vibrations whose frequencies are less that the fixed frequency ω, this number can be established from

$$n(\omega) = \int\limits_0^{\omega} \nu(\omega)\, d\omega = N \int\limits_0^{\omega^2} g(\varepsilon) d\varepsilon = \frac{V\Omega(\omega)}{(2\pi)^3}, \tag{4.2.7}$$

where $\Omega(\omega)$ is the volume in $\boldsymbol{k}$-space inside an isofrequency surface $\omega(\boldsymbol{k}) = \omega$.

Apart from (4.2.4) or (4.2.5) that are useful in analytically calculating the frequency spectrum for a given dispersion law, there are other ways of writing it, which are instrumental in solving some questions that refer to the vibration spectrum. Thus, coming back to the initial definition (4.2.2), we consider it from another point of view. We use one of the Dirac δ-function definitions $\delta(x)$ and write the following chain of equations:

$$\delta(x) = \frac{1}{\pi} \lim_{\gamma \to +0} \frac{\gamma}{x^2 + \gamma^2} = \frac{1}{\pi} \lim_{\gamma \to +0} \operatorname{Im} \frac{1}{x - i\gamma}, \tag{4.2.8}$$

where Im is the symbol for an imaginary part of a corresponding complex number.

Then, we write the relation resulting from (4.2.8) as

$$\delta(x) = \frac{1}{\pi} \lim_{\gamma \to +0} \operatorname{Im} \frac{d}{dx} \ln(x - i\gamma), \tag{4.2.9}$$

and use it to transform (4.2.2):

$$Ng(\varepsilon) = \frac{1}{\pi} \lim_{\gamma \to +0} \operatorname{Im} \frac{d}{d\varepsilon} \sum_{k} \ln\left[\varepsilon - \omega^2(\boldsymbol{k}) - i\gamma\right]. \tag{4.2.10}$$

We rearrange the order of operations in (4.2.10) and write

$$Ng(\varepsilon) = \frac{1}{\pi} \lim_{\gamma \to +0} \frac{d}{d\varepsilon} \operatorname{Im} \ln D(\varepsilon - i\gamma), \tag{4.2.11}$$

where

$$D(\varepsilon) = \prod_{k} \left[\varepsilon - \omega^2(\boldsymbol{k})\right]. \tag{4.2.12}$$

It is easily seen that the definition (4.2.12) coincides with that of (1.2.8) in a scalar model. In the general case,

$$D(\varepsilon) = \prod_{k,\alpha} \left[\varepsilon - \omega_\alpha^2(\boldsymbol{k})\right], \tag{4.2.13}$$

where the function $\omega_\alpha^2(\boldsymbol{k})$ describes the dispersion law of the α-th branch of vibrations, and the formula (4.2.11) then gives the total density of vibrational states of the crystal.

We rewrite the definition of (4.2.9) taking all vibration branches in account:

$$g(\varepsilon) = \frac{1}{\pi} \lim_{\gamma \to +0} \operatorname{Im} \frac{1}{N} \sum_{\boldsymbol{k},\alpha} \frac{1}{\varepsilon - i\gamma - \omega_\alpha^2(\boldsymbol{k})}. \qquad (4.2.14)$$

The last formula plays a major role in establishing the relation between the density of vibrational states and the Green function.

4.3
Analysis of Vibrational Frequency Distribution

A specific form of the function $g(\epsilon)$ or $\nu(w)$ as well as the dispersion law $\omega(\boldsymbol{k})$ is different for different crystals. We can thus discuss general concepts concerned only with the behavior of the vibration density near some special points. Such points are primarily the eigenfrequency band edges.

We begin by analyzing the distribution functions for the acoustic branch of vibrations in the extremely low-frequency region $\omega \ll \omega_m$, where the dispersion law is the simplest:

$$\omega = s(\boldsymbol{\kappa})k. \qquad (4.3.1)$$

It follows from (4.3.1) that the group velocity of the vibration wave $\boldsymbol{v} = \partial\omega/\partial\boldsymbol{k}$ is dependent only on the direction of the vector $\boldsymbol{k}$ (and is independent of its value). In studying the general properties of the function $\nu(\omega)$, one can put $s = s_0 = \text{constant}$; then $dS_k = k^2 dO = (\omega^2/s_0^2)dO$, where dO is an element of a solid angle in $\boldsymbol{k}$-space. As a result, for the frequency spectrum we have:

$$\nu(\omega) = \frac{V\omega^2}{2\pi^2 s_0^3}. \qquad (4.3.2)$$

In an anisotropic case ($s = s(\boldsymbol{\kappa})$) the parameter s_0^3 is obtained by certain averaging of the function $s(\boldsymbol{\kappa})$ in all directions of the vector $\boldsymbol{k}$.

Thus we have shown that in the low-frequency region ($\omega \ll \omega_m$) the function $\nu(\omega)$ is proportional to the frequency squared. The latter is a direct consequence of the linear character of the dependence $\omega = \omega(\boldsymbol{k})$ in the low-frequency region.

We write (4.3.2) in a somewhat different form admitting an elementary estimate of the order of value of the function $\nu(\omega)$ at low frequencies, namely

$$\nu(\omega) = \frac{N\omega^2}{2\pi^2(s_0/a)^3} \sim N\frac{\omega^2}{\omega_m^3}.$$

In this region using (4.2.5) for the function $g(\varepsilon)$, we find

$$g(\varepsilon) = \frac{V_0}{2\pi^2 s_0^3}\sqrt{\varepsilon}, \qquad \varepsilon \ll \omega_m^2. \qquad (4.3.3)$$

The appearance of a so-called root singularity is connected with the fact that the dispersion law written as $\varepsilon = \varepsilon(\boldsymbol{k}) = \omega^2(\boldsymbol{k})$ is quadratic for small $\boldsymbol{k}$. From (4.3.3) follows an estimate of the value of the vibration density: $g(\varepsilon) \sim \sqrt{\varepsilon}/\omega_m^3$.

Near the high-frequency boundary of the continuous spectrum ($\omega_m - \omega \ll \omega_m$), determining the reference origin for the vector $\boldsymbol{k}$ as for (4.1.4), the dispersion law can be written as

$$\omega = \omega_m - \frac{1}{2}\gamma_{ij}k_i k_j, \tag{4.3.4}$$

and the equation for isofrequency surfaces is

$$k^2(\boldsymbol{\kappa}) = \frac{2(\omega_m - \omega)}{\gamma_{ik}\kappa_i\kappa_k}.$$

In the simplest (isotropic) version $\omega = \omega_m - (1/2)\gamma k^2$ and then $v = -\gamma k = -\sqrt{2\gamma(\omega_m - \omega)}$ and also $dS_k = k^2 dO = 2(\omega_m - \omega)dO/\gamma$. As a result, for the frequency spectrum we obtain

$$\nu(\omega) = \frac{2V}{\pi^2(2\gamma)^{2/3}}\sqrt{\omega_m - \omega}. \tag{4.3.5}$$

In the general case, the coefficient before the root in (4.3.5) is obtained by averaging in directions of the vector $\boldsymbol{k}$. According to the above, the root singularity of the frequency spectrum near the upper spectrum boundary, described by (4.3.5), is a result of the quadratic dispersion law (4.3.4).

The density of states $g(\varepsilon)$ near the high-frequency band edge of eigenfrequencies has the form

$$g(\varepsilon) = V_0\frac{\sqrt{\omega_m^2 - \varepsilon}}{(2\pi)^2(\gamma\omega_m)^{3/2}}, \qquad \omega_m^2 - \varepsilon \ll \omega_m^2. \tag{4.3.6}$$

Comparing (4.3.3), (4.3.6), we conclude that the density of states $g(\varepsilon) = \text{constant} \times \sqrt{|\varepsilon - \varepsilon^*|}$, where ε^* determines the position of any one of the boundaries of the continuous spectrum of squared frequencies.

Apart from the continuous spectrum boundaries, the vicinities of frequencies dividing isofrequency surfaces of different topology can also be analyzed. We restrict ourselves to the case when the "boundary" isofrequency surface $\omega = \omega_c$ has a conical point near which the dispersion law is given by (2.1.5). We assume that outside a small neighborhood of the conical point, all isofrequency surfaces of a thin layer near $\omega = \omega_c$ are regular and the velocity v does not vanish on them. The specific properties of the density of vibrations that we expect at $\omega = \omega_c$ may only be associated with the contribution of vibrations corresponding to the small conical point neighborhood. Thus, we draw a pair of planes $k_3 = \pm K_3$ at such a distance from the conical point where the isofrequency surfaces still have the form of hyperboloids (Fig. 4.5), and calculate the fraction of vibrational states $\delta n(\omega)$ in the volume $\delta\Omega(\omega)$ limited by these planes and the hyperboloid $\omega(\boldsymbol{k}) = \omega$. To simplify calculations we set $\gamma_1 = \gamma_2 = \gamma$, i. e., we assume the isofrequency surfaces near the conical point to be rotation hyperboloids. Furthermore, we put $\gamma > 0$ and $\gamma_3 > 0$.

If $\omega < \omega_c$, the volume $\delta\Omega(\omega)$ is determined by

$$\delta\Omega(\omega) = 2\pi \int_{k_3^0}^{K_3} (k_1^2 + k_2^2)\, dk_3 = 2\pi \frac{\gamma_3}{\gamma} \int_{k_3^0}^{K_3} \left[k_3^2 - (k_3^0)^2\right] dk_3,$$

where the factor 2 takes into account the presence of two hyperboloid cavities and k_3^0 determines the hyperboloid vertex: $k_3^0 = \sqrt{\omega_c - \omega}/\sqrt{\gamma_3}$.

Thus,

$$\delta\Omega(\omega) = 2\pi \frac{\gamma_3}{\gamma} \left\{ \frac{1}{3} K_3^3 + K_3 \frac{\omega - \omega_c}{\gamma_3} + \frac{2}{3} \left(\frac{\omega_c - \omega}{\gamma_3} \right)^{3/2} \right\}. \tag{4.3.7}$$

Using (4.3.7) we now take into account (4.2.7) and the definition of the frequency spectrum function $\nu(\omega)$ to calculate the contribution to $\nu(\omega)$ of the states that correspond to the conical point vicinity:

$$\delta\nu(\omega) = \frac{V}{(2\pi)^2 \gamma \sqrt{\gamma_3}} \left\{ \sqrt{\gamma_3} K_3 - \sqrt{\omega_c - \omega} \right\}, \qquad \omega < \omega_k. \tag{4.3.8}$$

We calculate the same function for $\omega > \omega_c$. Since in this case the isofrequency surface has the form of a hyperboloid, for one sheet we have

$$\delta\Omega(\omega) = \int_{-K_3}^{K_3} \pi (k_1^2 + k_2^2)\, dk_3 = 2\pi \frac{\gamma_3}{\gamma} \left\{ \frac{1}{3} K_3^3 + K_3 \frac{\omega - \omega_c}{\gamma_3} \right\}.$$

Consequently, the contribution to the function $\nu(\omega)$ of a small neighborhood of the conical point is independent of frequency:

$$\delta\nu(\omega) = \frac{V}{(2\pi)^2 \gamma} K_3. \tag{4.3.9}$$

The contribution to the density of vibrations near $\omega = \omega_c$ from the vibrational states that correspond to the points in $\boldsymbol{k}$-space, lying outside the conical point vicinity are described by a regular frequency function. Thus, comparing (4.3.9) and (4.3.8), we see that:

1. the function $\nu(\omega)$ is a continuous frequency function at the point $\omega = \omega_c$;
2. its plot has a break at this point;
3. its derivative has an infinite jump.

It is characteristic that on approaching the point $\omega = \omega_c$ from the side corresponding to closed isofrequency surfaces, the derivative $d\nu/d\omega$ becomes infinite and, approaching the same point from the other side, the derivative remains finite.

The same singularity at the point concerned is typical of the function $g(\varepsilon)$:

$$g(\varepsilon) = \begin{cases} g(\varepsilon_c) - A\sqrt{\varepsilon_c - \varepsilon} + O(\varepsilon_c - \varepsilon), & \varepsilon < \varepsilon_c \\ g(\varepsilon_c) + O(\varepsilon_c - \varepsilon), & \varepsilon > \varepsilon_c, \end{cases} \tag{4.3.10}$$

where $\varepsilon_c = \omega_c^2$ is the squared frequency at the critical point, A is a positive constant value, $O(x)$ is a small quantity of the order of x (for small x).

The spectral functions $\nu(\omega)$ and $g(\varepsilon)$ may also have singularities at other points, but the latter are generally associated with frequencies at which an isofrequency surface changes its topology. In the frequency range $(0, \omega_m)$ at least two such frequencies exist. These frequencies separate the "layer" of open isofrequency surfaces from closed surfaces and, as a rule, they determine the surfaces with conical points. Hence, at least two critical frequencies can be expected at which the spectral functions possess singularities as described (Fig. 4.6).

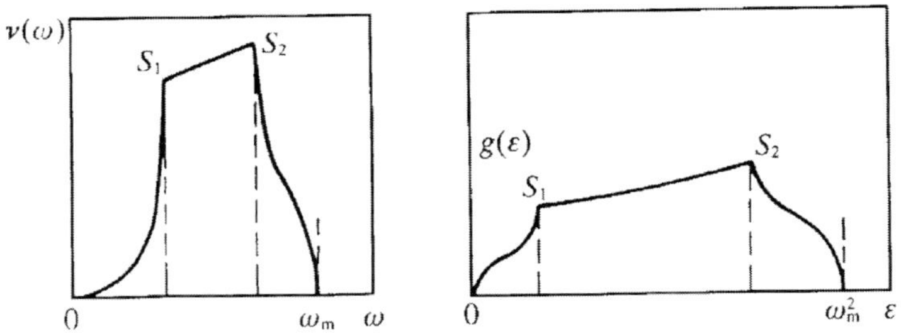

Fig. 4.6 Typical frequency spectrum [the functions $\nu(\omega)$ and $g(\epsilon)$] for the acoustic branch of the dispersion law.

In the vicinity of critical frequencies associated with the conical points, the function $\omega(\mathbf{k})$ is always expressed as (4.1.5), where all the coefficients γ_α ($\alpha = 1, 2, 3$) have the same sign. According to the terminology adopted, these frequencies are called *analytical critical points of type S*. We distinguish between the crystal points of type S_1 when the coefficients are positive and those of type S_2 when γ_α are negative. The standard form of the singularities of the functions $\nu(\omega)$ or $g(\varepsilon)$ at the S-type points, as well as their number inside the interval $(0, \omega_m)$ is regulated by the *van Hove theorem*: the frequency spectrum of each crystal branch should involve at least one crystal point of both types S_1 and S_2. In the near vicinity of the S_1-type point the frequency spectrum has the form

$$\nu(\omega) = \begin{cases} \nu(\omega_c) - B_1\sqrt{\omega_c - \omega} + O(\omega_c - \omega), & \omega < \omega_c, \\ \nu(\omega_c) + O(\omega_c - \omega), & \omega > \omega_c, \end{cases}$$

where ω_c is the critical frequency, and in the vicinity of the S_2-type point the following expression is valid

$$\nu(\omega) = \begin{cases} \nu(\omega_c) + O(\omega_c - \omega), & \omega < \omega_c, \\ \nu(\omega_c) - B_2\sqrt{\omega - \omega_c} + O(\omega - \omega_c), & \omega > \omega_c, \end{cases}$$

where B_1 and B_2 are positive constants. The function $g(\omega)$ for each branch of vibrations has at least four singular points: two eigenfrequency band edges and two S-type critical points. At all these points, the singularity $g(\varepsilon)$ implies that on the one hand the function of each of these points is regular (in particular, it may vanish identically), and on the other hand it has the form

$$g(\varepsilon) = g(\varepsilon^*) \pm \text{const}\sqrt{|\varepsilon - \varepsilon^*|}\ ,$$

where ε^* denotes the square of the frequency at the corresponding singular point.

Finally, let us discuss the singularities of the optical vibration spectrum. At the optical vibration frequency band edges the dispersion law is quadratic, leading to the "root singularity" for either the frequency spectrum $\nu(\omega)$ or the density of states $g(\varepsilon)$ at both edges of the band of the allowed frequencies. The singularities inside the optical frequency interval are determined by the van Hove theorems, because due to these theorems the optical frequency band does not differ from the acoustic one.

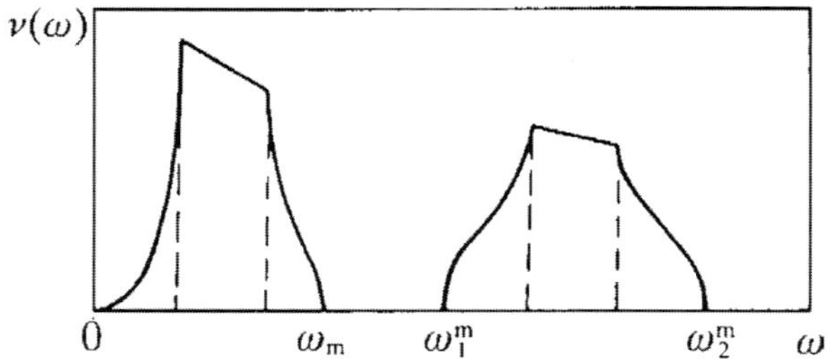

Fig. 4.7 Frequency spectrum functions of the acoustic and optical branches of the polyatomic lattice.

The total distribution of crystal lattice vibrations involves the densities for separate branches. The plot of the total frequency spectrum can be obtained by imposing (summing) plots such as shown in Fig. 4.7 with different ω, ω_{m1}, ω_{m2}, and different positions of the singular points S_1 and S_2.

4.4
Dependence of Frequency Distribution on Crystal Dimensionality

The singularities of the frequency distribution function are transformed fundamentally when going from a 3D lattice to a 2D or 1D lattice.

The Distribution Functions of 2D Crystal Vibrations. The distribution functions of 2D crystal vibrations have singularities similar to 3D crystal singularities. They are at the edges of the eigenfrequency spectrum and coincide with the frequencies that divide the regions of closed and open isofrequency curves (Fig. 4.3). Full analysis of the singularities in a 2D case is little different from that presented in Section 4.3. Therefore we restrict ourselves only to a qualitative characterization of the singularities, emphasizing their difference from those in a 3D crystal. A low-frequency (long-wave) limit of the dispersion law for each branch in the isotropic approximation is

$$\varepsilon = s^2k^2 = s^2(k_x^2 + k_y^2). \tag{4.4.1}$$

The behavior of the density of vibrational states $g(\varepsilon)$ for $\varepsilon \to 0$ is estimated by (4.4.1) using the following chain of equations

$$g(\varepsilon)d\varepsilon = \frac{S_0}{(2\pi)^2}d^2k = \frac{S_0}{(2\pi)^2}\,dk_x\,dk_y \sim a^2k\,dk \sim \left(\frac{a}{s}\right)^2 d\varepsilon\,,$$

where S_0 is the unit cell area of a 2D lattice. Thus, in the two-dimensional case

$$g(\varepsilon) \to g(0) = \text{const} \sim \left(\frac{a}{s}\right)^2 \quad at\ \varepsilon \to 0. \tag{4.4.2}$$

Taking into account the relation between the two frequency distribution functions (4.2.5), we see that

$$\nu(\omega) \to 2N\omega g(0) \sim N\left(\frac{a}{s}\right)^2\omega \qquad \text{as } \omega \to 0. \tag{4.4.3}$$

The behavior of the density of states (4.4.2) and the frequency spectrum (4.4.3) in the extremely low-frequency region differs from (4.3.2) and (4.3.3) for a 3D crystal.

As follows from (4.3.4) near the edge of the high-frequency spectrum we can write $\varepsilon = \varepsilon_m - \gamma k^2$, to obtain $g(\varepsilon)d\varepsilon \sim a^2k\,dk \sim \left(\frac{a^2}{\gamma}\right)d\varepsilon$. Consequently, the limiting behavior of the vibration density is analogous to (4.4.2): $g(\varepsilon) \to g(\varepsilon_m) = \text{constant} \sim \left(\frac{a^2}{\gamma}\right)$ at $\varepsilon \to \varepsilon_m$, i.e., the function $g(\varepsilon)$ has a break at the point $\varepsilon = \varepsilon_m$, since $g(\varepsilon) \equiv 0$ at $\varepsilon > \varepsilon_m$. This singularity of the vibration density also differs from the dependence (4.3.6) that is typical for a 3D crystal.

We examine the vicinity of the frequency ω_c corresponding to the analog of a conical point in a 3D crystal, a point near which the isofrequency curves differ little from hyperboli (Fig. 4.8a). The dispersion law near the frequency ω_c can be written, similar to (4.1.5), as

$$\omega = \omega_c + \gamma_1k_1^2 - \gamma_2k_2^2, \quad \gamma_1\gamma_2 > 0. \tag{4.4.4}$$

We repeat the reasoning used in calculating the functions $g(\varepsilon)$ and $\nu(\omega)$ near the frequency ω_c in a 3D crystal. Let us draw two straight lines $k_2 = \pm Q$ that cut off part of the space near the point we are interested in and calculate a 2D volume (the area Σ

in a 2D $\boldsymbol{k}$-space) limited by these straight lines and the curve $\omega(\boldsymbol{k}) = \omega = \text{constant}$. We set for definiteness $\gamma_1 > 0$ and $\gamma_2 > 0$. For $\omega < \omega_c$ (curve 1 in Fig. 4.8a), we then get

$$\Sigma_1(\omega) = \int dk_x\, dk_y = 4\sqrt{\frac{\gamma_2}{\gamma_1}} \int_{k_0}^{Q} \sqrt{k_2^2 - k_0^2}\, dk_2$$
$$= \frac{4k_0^2\sqrt{\gamma_2}}{\sqrt{\gamma_1}} \int_1^{Q/k_0} \sqrt{y^2 - 1}\, dy\,, \tag{4.4.5}$$

where $k_0^2 = (\omega_c - \omega)/\gamma_2$.

Near the singular point, when $k_0 \ll Q$ we have from (4.4.5):

$$\Sigma_1(\omega) = 2\sqrt{\frac{\gamma_2}{\gamma_1}} \left\{ Q^2 - k_0^2 \ln \frac{Q}{k_0} \right\}. \tag{4.4.6}$$

For $\omega > \omega_c$ (curve 2 in Fig. 4.8a)

$$\Sigma_2(\omega) = 4\sqrt{\frac{\gamma_2}{\gamma_1}} \int_0^{Q} \sqrt{q_0^2 + k^2}\, dk_2 = \frac{4q_0^2\sqrt{\gamma_2}}{\sqrt{\gamma_1}} \int_0^{Q/q_0} \sqrt{y^2 + 1}\, dy\,,$$

where $q_0^2 = (\omega - \omega_c)/\gamma_2$ and, thus, near the singular point ($q_0 \to 0$), we have

$$\Sigma_2(\omega) = 2\sqrt{\frac{\gamma_2}{\gamma_1}} \left\{ Q^2 + q_0^2 \ln \frac{Q}{q_0} \right\}. \tag{4.4.7}$$

The terms in (4.4.6), (4.4.7) that lead to a singular behavior at $|\omega - \omega_c| \to 0$ are

$$\Sigma_1(\omega) = 2\sqrt{\frac{\gamma_2}{\gamma_1}} \left\{ Q^2 - \frac{\omega_c - \omega}{\gamma_2} \log\sqrt{\omega_c - \omega} \right\}, \quad \omega < \omega_k;$$
$$\Sigma_2(\omega) = 2\sqrt{\frac{\gamma_2}{\gamma_1}} \left\{ Q^2 + \frac{\omega - \omega_k}{\gamma_2} \log\sqrt{\omega - \omega_k} \right\}, \quad \omega > \omega_k. \tag{4.4.8}$$

Since the frequency spectrum is

$$\nu(\omega) = \frac{S}{(2\pi)^2} \frac{d\Sigma}{d\omega}\,,$$

where S is the crystal area, from (4.4.8) we directly derive the logarithmic singularity at $\omega \to \omega_c$

$$\nu(\omega) = \frac{S}{(2\pi)^2\sqrt{\gamma_1\gamma_2}} \log\left(\frac{\omega_c}{|\omega - \omega_c|}\right). \tag{4.4.9}$$

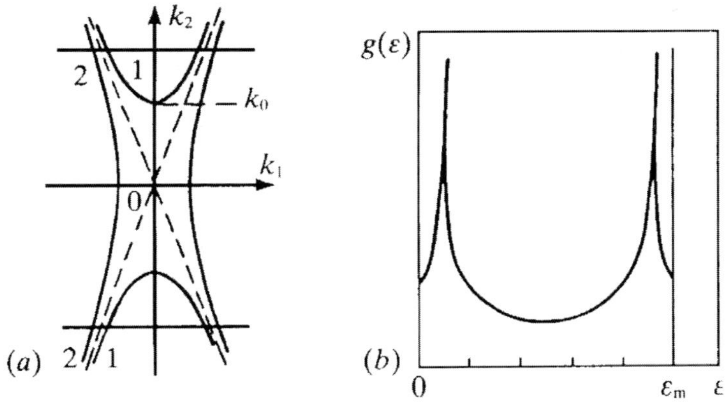

Fig. 4.8 Singularities of the frequency distribution in a 2D crystal: (*a*) isofrequency lines near the critical point of the 2D dispersion law; (*b*) density of states in a 2D lattice with nearest-neighbor interaction.

Using (4.2.5), we see that the vibration density of a 2D crystal $g(\varepsilon)$ also has a logarithmic singularity at $\varepsilon = \varepsilon_c = \omega_c^2$ and its singular part is symmetric in the vicinity of this point:

$$g(\varepsilon) = \frac{S_0}{(2\pi)^2\sqrt{\gamma_1\gamma_2\varepsilon_c}} \log\left(\frac{\varepsilon_c}{|\varepsilon - \varepsilon_c|}\right)^{1/2}. \tag{4.4.10}$$

As indicated in Chapter 3, there should exist no less than two such singular points. If $\gamma_1 = \gamma_2$ two singular points merge to form one singularity.

For the simplest models the vibration density of a 2D crystal can be calculated explicitly, e. g., in a scalar model for a crystal lattice with interaction of nearest neighbors only. The dispersion law of vibrations is simple in this case and, for a rectangular lattice with period a along the x-axis and period b along the y-axis, can be written as

$$\varepsilon = \omega_1^2 \sin^2\frac{ak_x}{2} + \omega_2^2 \sin^2\frac{bk_y}{2}. \tag{4.4.11}$$

We write the definition of the density of vibration states

$$g(\varepsilon) = \frac{ab}{(2\pi)^2}\frac{d}{d\varepsilon}\int dk_x\, dk_y = \frac{ab}{(2\pi)^2}\frac{d}{d\varepsilon}\int k_y(\varepsilon, k)\, dk, \tag{4.4.12}$$

where $k_y(\varepsilon, k_x)$ is the solution to (4.4.11) considered as an equation relative to k_y. The 2D integral in (4.4.12) is calculated in the area limited by the isofrequency curve ε = constant and situated in one unit cell of the $\mathbf{k}$-plane. The integration limits of the last integral in (4.4.12) depend on whether the isofrequency curve intersects the boundaries of the reciprocal lattice unit cell. If such a curve is closed and does not intersect the unit cell boundaries the integration limits are obtained from the condition

$k_y(\varepsilon, k) = 0$. The change of the integration variables in (4.4.12) results in

$$g(\varepsilon) = \frac{1}{(\pi\omega_1)^2} \int_{-x_0(\varepsilon)}^{x_0(\varepsilon)} \frac{dx}{\sqrt{(\xi^2 - \sin^2 x)(\eta^2 - \xi^2 + \sin^2 x)}}, \tag{4.4.13}$$

where $\xi^2 = \varepsilon/\omega_1^2$, $\eta^2 = (\omega_2/\omega_1)^2$ and $\sin x_0 = \xi$.

We know that the isofrequency curve is closed at frequencies that are near edges of the vibration spectrum. Thus, (4.4.13) describes the behavior of the vibration distribution in these parts of the spectrum. The density of states $g(\varepsilon)$ is shown in the form of a total elliptic integral of the first kind $K(k)$:

$$g(\varepsilon) = \frac{2}{\pi^2\omega_1\omega_2} K\left(\frac{\sqrt{\varepsilon(\varepsilon_m - \varepsilon)}}{\omega_1\omega_2}\right), \tag{4.4.14}$$

where $\varepsilon_m = \omega_1^2 + \omega_2^2$. As $K(0) = \pi/2$, (4.4.14) agrees with (4.4.2). Calculating the vibration distribution in the middle part of the eigenfrequency spectrum, we set for definiteness $\eta > 1$. Then for $\omega_1^2 < \varepsilon < \omega_2^2$ the isofrequency curve intersects the reciprocal lattice unit cell boundaries $k_x = \pm\pi/a$, and the density of states is determined by (4.4.13) with $x_0 = \pi/2$. Simple calculation shows that

$$g(\varepsilon) = \frac{2}{\pi^2\sqrt{\varepsilon(\varepsilon_m - \varepsilon)}} K\left(\frac{\omega_1\omega_2}{\sqrt{\varepsilon(\varepsilon_m - \varepsilon)}}\right). \tag{4.4.15}$$

We note that (4.4.5) and (4.4.15) are symmetrical relative to permutation of ω_1 and ω_2. Thus, their applicability conditions are: (4.4.14) is valid for $0 < \varepsilon(\varepsilon_m - \varepsilon) < \omega_1^2\omega_2^2$, and (4.4.15) – for $\varepsilon(\varepsilon_m - \varepsilon) > \omega_1^2\omega_2^2$ (Fig. 4.8b).

The elliptic integral $K(k)$ has a logarithmic singularity at the point $k = 1$. This singularity is typical for the function $g(\varepsilon)$ at $\varepsilon = \omega_1^2$ and $\varepsilon = \omega_2^2$. The behaviors of the functions (4.4.14), (4.4.15) near these points can be considered as a partial case of the general relation (4.4.10).

One-Dimensional Structure. In the case of a linear chain where only the nearest neighbors interact, an analog of (4.4.11) is the dispersion law

$$\varepsilon = \varepsilon_m \sin^2 \frac{ak}{2}, \tag{4.4.16}$$

where ε_m is the upper-band boundary of the possible frequency squares. It is important that the dependence of the frequency ω on k given by (4.4.16), determines a monotonic function in the range of values of the quasi-wave vector $(0, \pi/a)$. This means that in the internal points of the interval $(0, \omega_m)$, the frequency distribution functions $\nu(\omega)$ and $g(\varepsilon)$ of a 1D crystal with the dispersion law (4.4.16) have no singularities. Indeed, from the definition of the vibration density $g(\varepsilon)$ and (4.4.16) it follows that

$$g(\varepsilon) = \frac{1}{\pi\sqrt{\varepsilon(\omega_m^2 - \varepsilon)}}. \tag{4.4.17}$$

Thus, the spectral characteristics of a 1D crystal differ from those of a 3D (and 2D) one in that they have no singularities inside the continuous spectrum band. At the ends of the continuous spectrum interval the singularities are more pronounced than those of 3D and 2D crystals. It is clear from (4.4.17) that for $\varepsilon \to \varepsilon_*$, where ε_* is a continuous spectrum edge,

$$g(\varepsilon) \sim \frac{1}{\sqrt{|\varepsilon - \varepsilon_*|}}. \tag{4.4.18}$$

The frequency spectrum function of a 1D crystal with the nearest-neighbor interactions is

$$\nu(\omega) = \frac{2N}{\pi\sqrt{\omega_m^2 - \omega^2}}, \tag{4.4.19}$$

where N is the number of atoms in the chain.

If the dispersion law of a lD crystal is a nonmonotonic function $\omega(k)$ such as in Fig. 4.9a the spectral characteristics of 1D crystal vibrations at the points $\omega = \omega_1$ and $\omega = \omega_2$, will have distinctive singularities. Indeed, suppose that near $k = k_1$ and $\omega = \omega_1$ the dispersion law is then given by

$$\omega^2 = \omega_1^2 - \gamma_1^2 (k - k_1)^2.$$

The contribution of this part of the dispersion law to the density of states is then easily determined to be

$$\delta g\varepsilon = \frac{a}{\pi}\frac{dk}{d\varepsilon}\begin{cases} \dfrac{a}{2\pi\gamma_1\sqrt{\varepsilon_1 - \varepsilon}}, & \varepsilon < \varepsilon_1; \\ 0, & \varepsilon > \varepsilon_1. \end{cases}$$

Similarly, if we write the dispersion law near the points $k = k_2$ and $\omega = \omega_2$ in the form of $\omega^2 = \omega_2^2 - \gamma_2^2(k - k_2)^2$, the corresponding contribution to the vibration density is

$$\delta g\varepsilon = \begin{cases} 0, & \varepsilon < \varepsilon_2 = \omega_2^2; \\ \dfrac{a}{2\pi\gamma_2\sqrt{\varepsilon - \varepsilon_2}}, & \varepsilon > \varepsilon_2. \end{cases}$$

A plot of the density of states for a 1D crystal in the case when the interval $(0, \omega_m)$ has two singular points $\varepsilon = \varepsilon_1$ (maximum) and $\varepsilon = \varepsilon_2$ (minimum) is given in Fig. 4.9b.

4.5 Green Function for the Vibration Equation

The singularities of the spectrum of crystal vibrations can be analyzed, apart from the geometric method, by the analytical one, which is based on the mathematical structure of the equations of crystal lattice motion. The so-called Green dynamic

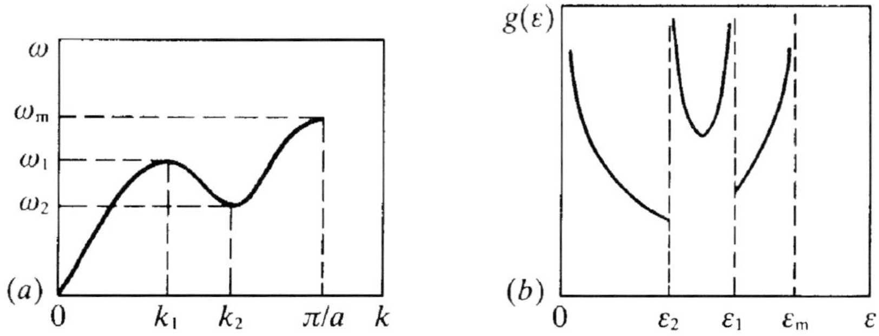

Fig. 4.9 Spectrum of vibrations of a 1D crystal: (*a*) dispersion curve; (*b*) singularity of the density of states.

function is used for this purpose. It is defined as follows. The equations of free vibrations of a monatomic crystal lattice (1.1.12) are represented as

$$m\frac{d^2u(\boldsymbol{n},t)}{dt^2}+\sum_{n'}\boldsymbol{A}(\boldsymbol{n}-\boldsymbol{n}')u(\boldsymbol{n}',t)=0. \qquad (4.5.1)$$

The *Green function* for unbounded crystal vibrations or the Green function of (4.5.1) is called a solution to an inhomogeneous equation vanishing at infinity

$$m\frac{d^2}{dt^2}G_{ik}(\boldsymbol{n},t|\boldsymbol{n}_0)+\sum_{n'}\alpha^{ij}(\boldsymbol{n}-\boldsymbol{n}')G_{jk}(\boldsymbol{n}',t|\boldsymbol{n}_0)=-m\delta_{ik}\delta(t)\delta_{\boldsymbol{n}\boldsymbol{n}_0}, \qquad (4.5.2)$$

where $\delta(t)$ is the Dirac δ-function.

By virtue of space homogeneity of an unbounded crystal, the solution to the equation depends on the difference $\boldsymbol{n}-\boldsymbol{n}_0$: $G(\boldsymbol{n},t|\boldsymbol{n}_0)=G(\boldsymbol{n}-\boldsymbol{n}_0,t)$. The Green function of any linear equation allows one to find a partial solution to an inhomogeneous equation describing the induced motion of the system concerned. If the motion of crystal atoms is determined by the force $mf^i(\boldsymbol{n},t)$:

$$m\frac{d^2}{dt^2}u^i(\boldsymbol{n},t)+\sum_{n'}\alpha^{ik}(\boldsymbol{n}-\boldsymbol{n}')u^k(\boldsymbol{n}',t)=mf^i(\boldsymbol{n},t), \qquad (4.5.3)$$

the partial solution to this equation is

$$u^i(\boldsymbol{n},t)=-\sum_{n'}\int G_{ik}(\boldsymbol{n}-\boldsymbol{n}',t-t')f^k(\boldsymbol{n}',t)\,dt'. \qquad (4.5.4)$$

However, use of the Green function is not limited to (4.5.4) alone. The Green function gives rich information on the properties of a system whose free motion is described by a homogeneous equation (4.5.1).

In a scalar model, the Green function $G(\boldsymbol{n},t)$ is a solution to the equation

$$m\frac{d^2}{dt^2}G(\boldsymbol{n},t)+\frac{1}{m}\sum_{n'}\alpha(\boldsymbol{n}-\boldsymbol{n}')G(\boldsymbol{n}',t)=-\delta(t)\delta_{\boldsymbol{n}0}. \qquad (4.5.5)$$

We represent $G(\boldsymbol{n}, t)$ as the Fourier integral

$$\begin{aligned} G(\boldsymbol{n}, t) &= \frac{1}{2\pi} \int_{-\infty}^{+\infty} G_{\omega^2}(\boldsymbol{n}) e^{-i\omega t}\, d\omega; \\ G_{\omega^2}(\boldsymbol{n}) &= \int_{-\infty}^{+\infty} G(\boldsymbol{n}, t) e^{i\omega t}\, dt, \end{aligned} \tag{4.5.6}$$

and call the function $G_\varepsilon(n)$, where $\varepsilon = \omega^2$, the Green function of stationary crystal vibrations. This function is defined as an appropriate solution to the equation with behavior at infinity

$$\varepsilon G_\varepsilon(\boldsymbol{n}) - \frac{1}{m} \sum_{n'} \alpha(\boldsymbol{n} - \boldsymbol{n}') G_\varepsilon(\boldsymbol{n}') = \delta_{\boldsymbol{n}0}. \tag{4.5.7}$$

The Green function of stationary vibrations is dependent on ε, i. e., it depends only on the frequency squared. Thus, judging from (4.5.6), (4.5.7), the frequency sign in the argument of the time Fourier transformation of the Green function is of no importance. However, in fact this is not so. By analyzing (4.5.1), (4.5.7), it is easy to obtain the Green function for the vibrations of an ideal crystal. We perform the following Fourier transformation of the function $G_\varepsilon(\boldsymbol{n})$

$$\begin{aligned} G_\varepsilon(\boldsymbol{n}) &= \frac{1}{N} \sum_k G(\varepsilon, \boldsymbol{k}) e^{i\boldsymbol{k}\boldsymbol{r}(\boldsymbol{n})}; \\ G(\varepsilon, \boldsymbol{k}) &= \sum_n G_\varepsilon(\boldsymbol{n}) e^{-i\boldsymbol{k}\boldsymbol{r}(\boldsymbol{n})}, \end{aligned} \tag{4.5.8}$$

determining the Green function in $(\varepsilon, \boldsymbol{k})$ representation. We now perform an analogous transformation of (4.5.7)

$$\left[\varepsilon - \omega^2(\boldsymbol{k})\right] G(\varepsilon, \boldsymbol{k}) = 1, \tag{4.5.9}$$

where, as before, the function $\omega(\boldsymbol{k})$ gives the dispersion law of a crystal. It follows from (4.5.9) that the Fourier components of the Green function are

$$G(\varepsilon, \boldsymbol{k}) = \frac{1}{\varepsilon - \omega^2(\boldsymbol{k})}. \tag{4.5.10}$$

The formula (4.5.10) determines the Green function in $(\varepsilon, \boldsymbol{k})$ representation. For some methods of theoretical calculation of ideal crystal properties, this representation is fundamental. As (4.5.10) is important, we write it in the general case when various vibration branches in a crystal exist

$$G_{ij}(\varepsilon, \boldsymbol{k}) = \sum_{\alpha=1}^{3} \frac{e_i(\boldsymbol{k}, \alpha) e_j(\boldsymbol{k}, \alpha)}{\varepsilon - \omega_\alpha^2(\boldsymbol{k})}. \tag{4.5.11}$$

The form of the functions (4.5.10), (4.5.11) and, in particular, the appearance of characteristic denominators is not specific only for the crystal vibrations but reflects the general features of the Green function for systems with collective excitations. We thus analyze the function $G(\varepsilon, \boldsymbol{k})$.

The function $G(\varepsilon, \boldsymbol{k})$ regarded as a function of the variable ε has a pole at the point $\varepsilon = \omega^2(\boldsymbol{k})$, i.e., at a point where ω coincides with the frequency of one, the eigenvibrations. We, therefore, arrive at the following important property of the Green function. The poles of the Fourier components of the Green function are determined by the spectrum of crystal eigenfrequencies or, in other words, by its dispersion law.

Using (4.5.8), we obtain the Green function for stationary vibrations (in a scalar model) in the form

$$G_\varepsilon(\boldsymbol{n}) = \frac{1}{N} \sum_k \frac{e^{i\boldsymbol{k}\boldsymbol{r}(\boldsymbol{n})}}{\varepsilon - \omega^2(\boldsymbol{k})}. \qquad (4.5.12)$$

Assuming quasi-continuity of the spectrum of the $\boldsymbol{k}$-vector values we may rewrite (4.5.12) in the form of an integral:

$$G_\varepsilon(\boldsymbol{n}) = \frac{V_0}{(2\pi^3)} \int \frac{e^{i\boldsymbol{k}\boldsymbol{r}(\boldsymbol{n})} d^3k}{\varepsilon - \omega^2(\boldsymbol{k})}, \qquad (4.5.13)$$

where V_0 is the unit cell volume.

The generalization of (4.5.12) is obvious in view of (4.5.11):

$$G_\varepsilon^{ij}(\boldsymbol{n}) = \frac{1}{N} \sum_{k,\alpha} \frac{e_i(\boldsymbol{k},\alpha) e_j(\boldsymbol{k},\alpha)}{\varepsilon - \omega_\alpha^2(\boldsymbol{k})} e^{i\boldsymbol{k}\boldsymbol{r}(\boldsymbol{n})}. \qquad (4.5.14)$$

Changing (4.5.14) to an integral form is accomplished analogously to the transformation from (4.5.12) to (4.5.13).

We now return to a scalar model. If the value of the parameter ε does not get into the band of crystal eigenfrequency squares (in our case $\varepsilon > \omega_m^2$) the formula (4.5.13) unambiguously determines a certain function $\boldsymbol{n}$ dependent on the parameter ε.

However, the case $0 < \varepsilon < \omega_m^2$, i.e., when the frequency ω is in the continuous spectrum interval, is more interesting. As the Fourier components of the Green function have a pole, the integral (4.5.13) is meaningless (it diverges). More exactly, it is senseless in its literal interpretation when the parameter ε is considered to be real. However, this singularity of the vibrational system behavior is typical for any resonance system when damping (dissipation) of the eigenvibrations is neglected, and it results in infinitely large amplitudes of vibrations as soon as the frequency of the excitation force coincides with one of the eigenfrequencies of the system. It is known how to overcome this difficulty. It is necessary to take into account at least small damping of the eigenvibration existing in the system. For unbounded systems with distributed parameters, whose eigenvibrations have the form of waves with a continuous frequency spectrum, equations such as (4.5.13) can be regularized by choosing special conditions at infinity. Formally, this reduces to the fact that the parameter ε

should be regarded as complex rather than real, but with a vanishingly small imaginary part. Adding even a small imaginary part to ε removes the divergence of the integral (4.5.13), but generates some new problems[2].

In discussing the general properties of the function $G(\varepsilon, \boldsymbol{k})$, we focus on one purely mathematical point. Let us represent the equation for stationary vibrations (1.1.17) in the matrix (operator) form $\varepsilon u - \frac{1}{m}\boldsymbol{A}u = 0$. Equation (4.5.7) for the Green function of stationary vibrations can be written in the same style: $\varepsilon \boldsymbol{G} - \frac{1}{m}\boldsymbol{A}\boldsymbol{G} = \boldsymbol{I}$, where $\boldsymbol{I}$ is the unit operator.

The solution to this equation can be easily found

$$\boldsymbol{G}(\varepsilon) = \left\{\varepsilon \boldsymbol{I} - \frac{1}{m}\boldsymbol{A}\right\}^{-1}, \tag{4.5.15}$$

if we introduce an inverse operator notation (the operator $\boldsymbol{M}^{-1}$ is inverse to $\boldsymbol{M}$).

The Green function in the matrix (or operator) form (4.5.15) is sometimes called a *resolvent*.

The formula (4.5.6) does not specify the representation where the Green function is written, but it implies that the matrix $\boldsymbol{G}(\varepsilon)$ is diagonal in a representation in which the operator $\boldsymbol{A}$ is diagonal. For a vibrating crystal this is a $\boldsymbol{k}$-representation where (4.5.10), (4.5.11) are derived.

4.6 Retarding and Advancing Green Functions

We return to (4.5.6), substitute (4.5.13) in it, and write the Green function as a function of time

$$G(\boldsymbol{n}, t) = \frac{V_0}{(2\pi)^4} \iint \frac{e^{i\boldsymbol{k}\boldsymbol{r}(\boldsymbol{n}) - i\omega t}}{\omega^2 - \omega^2(\boldsymbol{k}) - i\gamma} d\omega d^3k. \tag{4.6.1}$$

To regularize the integral (4.6.1) we have just explicitly separated a small imaginary component $i\gamma$.

Let us consider only the frequency transformation in (4.6.1), resulting in the integral

$$G_k(t) = \int_{-\infty}^{\infty} \frac{e^{-i\omega t}\, d\omega}{\omega^2 - \omega^2(\boldsymbol{k}) - i\gamma}. \tag{4.6.2}$$

The coordinate representation of the Green function is determined by an obvious transformation involved in (4.6.1)

$$G(\boldsymbol{n}, t) = \frac{V_0}{(2\pi)^3} \int G_k(t) e^{i\boldsymbol{k}\boldsymbol{r}(\boldsymbol{n})} d^3k. \tag{4.6.3}$$

2) In particular, the Fourier component of the Green function becomes dependent on the sign of ω.

The integral (4.6.2) is easily calculated by the residues method. The integration should be performed over a closed path consisting of the real axis of the variable ω and an arc of the infinitely remote semicircle. As the convergence of the integral and its vanishing for an infinite semicircle is provided by the exponential factor, for $t < 0$ we should close the path above the real axis, and for $t > 0$ below it. As a result, for $t < 0$ the integral (4.6.2) is determined by residues in the upper half-plane of a complex variable ω, and for $t > 0$ this integral equals the sum of residues in the lower half-plane.

Since the imaginary shift in the complex plane ω is determined by parameter γ, the integration result in (4.6.2) is determined completely by the sign of the parameter γ. The relation between the signs of γ and ω is then very important. In (4.6.1) and (4.6.2), no assumptions were made with regard to the sign of γ, so that the ambiguity in choosing the dependence of the sign of γ on the sign of ω indicates the ambiguity in determination of the Green function by (4.6.1).

Let us assume that the sign of γ changes together with the sign of the frequency, i. e., we admit $\gamma = \gamma_0 \operatorname{sign}(\omega)$. The Fourier component of the Green function for which a new notation $G(\omega, \boldsymbol{k})$ was used, to avoid confusion with (4.5.10), then takes the form

$$G(\omega, \boldsymbol{k}) = \frac{1}{\omega^2 - \omega^2(\boldsymbol{k}) - i\gamma \operatorname{sign}(\omega)}. \tag{4.6.4}$$

It is readily seen that for $\gamma_0 < 0$ (4.6.4) determines a function of the complex variable ω analytic in the upper semi-plane and at $\gamma_0 > 0$, in the low semiplane. Integrating in (4.6.2), we find the time dependence of the Green functions generated by the Fourier components (4.6.4). For $\gamma_0 < 0$, we have

$$G_{\boldsymbol{k}}^{R}(t) = \begin{cases} 0, & t < 0; \\ -\dfrac{\sin[\omega(\boldsymbol{k})t]}{\omega(\boldsymbol{k})}, & t > 0. \end{cases} \tag{4.6.5}$$

By using the transformation (4.6.3), we get a *retarding Green function* $G_R(\boldsymbol{n}, t)$. This function describes the perturbation arising at any crystal point, with an instantaneous force applied at the coordinate origin. At infinity, it corresponds to waves diverging from the origin.

When $\gamma_0 > 0$, we have to determine

$$G_{\boldsymbol{k}}^{A}(t) = \begin{cases} -\dfrac{\sin[\omega(\boldsymbol{k})t]}{\omega(\boldsymbol{k})}, & t < 0; \\ 0, & t > 0 \end{cases} \tag{4.6.6}$$

to derive by means of (4.6.3) an advancing Green function $G_A(\boldsymbol{n}, t)$. The advancing Green function describes the crystal perturbations for $t < 0$ that generate a force concentrated at the origin at time $t = 0$. At infinity, it corresponds to waves converging to the origin. The choice between waves diverging and converging at infinity is the complementary physical factor that can be formulated as additional conditions at infinity and makes the sign of the parameter γ_0 meaningful.

Finally, we consider the Green function for stationary vibrations for which there is no energy flow at infinity (diverging and converging waves have the same amplitude, thus, generating stagnant waves). However, for the stationary vibrations understood literally, the time Fourier component of the Green function is determined by the frequency square and does not depend on the sign of ω. Thus, the imaginary addition $i\gamma$ in (4.6.1) for the corresponding Green function should also be independent of the sign ω, i. e., γ may be regarded as independent of ω.

We set $\gamma = |\gamma| > 0$ for all ω. Then using (4.5.10), we introduce the parameter γ just into the argument of the Green function (4.5.10) and, instead of (4.6.4), we obtain

$$G(\omega, \mathbf{k}) = G(\varepsilon - i\,|\gamma|\,, \mathbf{k}). \tag{4.6.7}$$

We substitute (4.6.7) into (4.6.2) and integrate over frequency. As a result we obtain

$$G_{\mathbf{k}}(t) = \frac{i}{2\omega(\mathbf{k})} \begin{cases} e^{-i\omega(\mathbf{k})t}, & t < 0; \\ e^{+i\omega(\mathbf{k})t}, & t > 0. \end{cases} \tag{4.6.8}$$

The Green function (4.6.8), of course, has a singularity at $t = 0$, but is nonzero for any other t.

4.7 Relation Between Density of States and Green Function

In the definition (4.2.14) of the function $g(\varepsilon)$, we compare the sum on the r.h.s. of (4.2.14) with the expression (4.5.14) for the Green function. It is clear that the density of states can be directly determined by the imaginary part of the Green function for stationary vibrations with the coinciding arguments ($\mathbf{n} = 0$)

$$g(\varepsilon) = \frac{1}{\pi} \lim_{\gamma \to +0} \operatorname{Im} G^{ll}_{\varepsilon - i\gamma}(0). \tag{4.7.1}$$

In determining (4.7.1), we used the property of the polarization vectors (1.3.2).

Thus the crystal vibrational density is determined by the imaginary part of the Green function for stationary vibrations taken at zero. As all Green functions have the same singularities, the vibrational density can be associated with the imaginary part of any of them, e. g., be determined through a retarding Green function. We again consider only one branch of vibrations (more exactly a scalar model) and analyze the $(\omega, \mathbf{k})$-representation (4.6.4) for the function $G^R_\omega(\mathbf{n})$. We take into account the relation

$$\lim_{\gamma \to 0} \operatorname{Im} \frac{1}{\omega^2 - \omega^2(\mathbf{k}) + i\,|\gamma|\operatorname{sign}\omega} = \lim_{\gamma \to +0} \frac{\gamma}{(\omega + i\gamma)^2 - \omega^2(\mathbf{k})},$$

and, by analogy with the discussion above, we obtain

$$g(\omega^2) = \frac{1}{\pi} \lim_{\gamma \to +0} G^R_{\omega + i\gamma}(0). \tag{4.7.2}$$

To elucidate the meaning of the operations indicated in (4.7.1) and (4.7.2), we move over from a summation to an integration in (4.2.14) for a scalar model

$$g(\varepsilon) = \frac{1}{\pi} \frac{V_0}{(2\pi)^3} \lim_{\gamma \to +0} \operatorname{Im} \int \frac{d^3k}{\varepsilon - i\gamma - \omega^2(k)}, \tag{4.7.3}$$

where the integration is over one unit cell in $\boldsymbol{k}$-space.

If we now change the integration variable $z = \omega^2(\boldsymbol{k})$ in (4.7.3), the formula will transform into the obvious equality

$$g(\varepsilon) = \frac{1}{\pi} \lim_{\gamma \to +0} \operatorname{Im} \int_0^{\omega_m^2} \frac{g(z)\,dz}{\varepsilon - i\gamma - z}.$$

Indeed, for the real function $g(\varepsilon)$ and the real variables ε and z, the following identity always holds

$$\lim_{\gamma \to +0} \int \frac{g(z)\,dz}{\varepsilon \pm i\gamma - z} = \text{P.V.} \int \frac{g(z)\,dz}{\varepsilon - z} \mp i\pi g(\varepsilon), \tag{4.7.4}$$

where P.V. means the principal value of the integral.

Relations (4.7.1), (4.7.2), important in real systems, are valid only for a homogeneous crystal. In order to analyze the crystal vibration spectrum with defects breaking the space homogeneity of the system, we generalize these formula to an inhomogeneous case and clarify some formal mathematical points resulting in a relationship such as (4.7.1).

We consider a certain problem on the eigenvalues for a linear Hermitian operator $\hat{\mathcal{L}}$ in an unbounded crystal

$$\hat{\mathcal{L}}\varphi - \lambda\varphi = 0. \tag{4.7.5}$$

The Green function of (4.7.5) is said to be the function

$$G_\lambda(\boldsymbol{n}, \boldsymbol{n}') = \sum_s \frac{\varphi_s^*(\boldsymbol{n})\varphi_s(\boldsymbol{n}')}{\lambda - \lambda_s}, \tag{4.7.6}$$

where $\varphi_s(\boldsymbol{n})$ and λ_s are the normalized eigenfunctions and eigenvalues of the operator $\hat{\mathcal{L}}$,

$$\hat{\mathcal{L}}\varphi_s - \lambda\varphi_s = 0; \quad \sum_n |\varphi_s(\boldsymbol{n})|^2 = 1.$$

As the eigenfunctions for the ideal crystal vibrations are determined by (1.3.3), it is easy to see that (4.7.6) leads directly to (4.5.12).

We set in (4.7.6) $\boldsymbol{n} = \boldsymbol{n}'$ and sum the relation obtained over all lattice sites:

$$\sum G_\lambda(\boldsymbol{n},\boldsymbol{n}) = \sum_s \frac{1}{\lambda - \lambda_s}. \tag{4.7.7}$$

We introduce the density of eigenvalues $g(\lambda)$ for the operator $\hat{\mathcal{L}}$ and, displacing the parameter λ from the real axis, rewrite (4.7.7) as

$$\frac{1}{N}\sum G_{\lambda-i\gamma}(\boldsymbol{n},\boldsymbol{n}) = \int \frac{g(z)\,dz}{\lambda - i\gamma - z}. \tag{4.7.8}$$

Finally, we use the identity (4.7.4) and apply it to the r.h.s. of (4.6.7) to obtain

$$g(\lambda) = \frac{1}{\pi}\lim_{\gamma\to+0} \operatorname{Im}\frac{1}{N}\sum_n G_{\lambda-i\gamma}(\boldsymbol{n},\boldsymbol{n}). \tag{4.7.9}$$

The formula (4.7.9) is the desired generalization of (4.7.1) to the case of a spatially inhomogeneous system.

The relation between the density of eigenvalues of a certain Hermitian operator and the corresponding Green function (4.7.9) can be written in a more invariant form, if one uses the Green operator concept. The Green function (4.7.6) can be regarded as the site representation of the Green operator (resolvent)

$$\hat{G}_\lambda = (\lambda - \hat{\mathcal{L}})^{-1}, \tag{4.7.10}$$

given in the space of the eigenfunctions of the Hermitian operator $\hat{\mathcal{L}}$.

The sum involved on the r.h.s. of (4.7.9) should then be regarded as the trace of the Green operator $\sum_n G_\lambda(\boldsymbol{n},\boldsymbol{n}) = \operatorname{Tr}\hat{G}_\lambda$.

Thus,

$$g(\lambda) = \frac{1}{\pi}\lim_{\gamma\to+0} \operatorname{Im}\frac{1}{N}\operatorname{Tr}\hat{G}_{\lambda-i\gamma}. \tag{4.7.11}$$

Since the trace operation is invariant with respect to choice of representation, (4.7.11) permits us to calculate the density of eigenvalues $g(\lambda)$ in any representation of the Green operator.

4.8 The Spectrum of Eigenfrequencies and the Green Function of a Deformed Crystal

Generalizing (4.7.1) to the case of inhomogeneous crystal vibrations, we go over to the relation (4.7.9) or its abstract formulation (4.7.11).

If the matrix $\hat{\mathcal{L}}$ in (4.7.10) describes the dynamic force matrix of a deformed crystal, and λ coincides with the vibration eigenfrequency square, then (4.7.11) directly describes the squared frequency distribution in this crystal. However, (4.7.11) is applicable to a nonideal crystal only in the case when the crystal potential energy experiences deformation (perturbation). However, situations when the crystal kinetic energy is also perturbed are possible. If we include the kinetic energy perturbation of stationary vibrations in the matrix $\hat{\mathcal{L}}$, the kinetic energy will be a function of the squared frequency: $\hat{\mathcal{L}} = \hat{\mathcal{L}}(\lambda)$. In this case, (4.7.11) is invalid and it should be generalized.

We consider the linear equation

$$\hat{\mathcal{L}}\varphi - \lambda\varphi = 0, \tag{4.8.1}$$

whose Green operator (resolvent) is determined by (4.7.10)

$$\mathbf{G}(\lambda) = \left(\lambda - \hat{\mathcal{L}}(\lambda)\right)^{-1}. \tag{4.8.2}$$

Let φ_s be the eigenfunctions and α_s the eigenvalues of $\hat{\mathcal{L}}$ (s is the number of a corresponding state): $\hat{\mathcal{L}}(\lambda)\varphi_s = \alpha_s(\lambda)\varphi_s$.

The value of λ corresponding to the eigenvalues of the problem (4.8.1) is found as a solution to the equation

$$\alpha_s(\lambda) = \lambda.$$

We denote the roots of this equation by λ_{sr}, where r is the root number. Then the density of eigenvalues of the problem (4.8.1) reads

$$\nu(\lambda) = \sum_{sr} \delta(\lambda - \lambda_{sr}). \tag{4.8.3}$$

We use the notation $\nu(\lambda)$ for the eigenfrequency distribution function in the general case, leaving the notation $g(\lambda)$ for the density of eigenfrequencies of crystal vibrations ($\lambda = \omega^2$) normalized to unity.

Using the formula (4.7.11) we consider the Green matrix (4.8.2) as a function of the eigenvalue $G(\lambda - i\gamma)$ at $\gamma \to 0$:

$$\begin{aligned}\mathbf{G}(\lambda - i\gamma) &= \left\{\lambda - i\gamma - \hat{\mathcal{L}}(\lambda) + i\gamma\frac{d\hat{\mathcal{L}}}{d\lambda}\right\}^{-1} \\ &= \left\{\left([\lambda - \hat{\mathcal{L}}(\lambda)]\left(1 - \frac{d\hat{\mathcal{L}}}{d\lambda}\right)^{-1} - i\gamma\right)\left(1 - \frac{d\hat{\mathcal{L}}}{d\lambda}\right)\right\}^{-1} \\ &= \left(1 - \frac{d\hat{\mathcal{L}}}{d\lambda}\right)^{-1}\left\{[\lambda - \hat{\mathcal{L}}(\lambda)]\left(1 - \frac{d\hat{\mathcal{L}}}{d\lambda}\right)^{-1} - i\gamma\right\}^{-1}.\end{aligned} \tag{4.8.4}$$

We rewrite (4.8.4) in a somewhat different form:

$$\left(1 - \frac{d\hat{\mathcal{L}}}{d\lambda}\right)\mathbf{G}(\lambda - i\gamma) = \left\{[\lambda - \hat{\mathcal{L}}(\lambda)]\left(1 - \frac{d\hat{\mathcal{L}}}{d\lambda}\right)^{-1} - i\gamma\right\}^{-1}. \tag{4.8.5}$$

We calculate the trace of the operator (4.8.5) by the means of the eigenfunctions $\hat{\mathcal{L}}$:

$$\mathrm{Tr}\left[\left(1-\frac{d\hat{\mathcal{L}}}{d\lambda}\right)\mathbf{G}(\lambda-i\gamma)\right]=\sum_s\left\{\frac{\lambda-\alpha_s(\lambda)}{1-\dfrac{d\alpha_s}{d\lambda}}-i\gamma\right\}^{-1}, \tag{4.8.6}$$

and make the necessary limiting transition:

$$\lim_{\gamma\to+0}\mathrm{Im}\,\mathrm{Tr}\left[\left(1-\frac{d\hat{\mathcal{L}}}{d\lambda}\right)\mathbf{G}(\lambda-i\gamma)\right]=\pi\sum_s\left|1-\frac{d\alpha_s}{d\lambda}\right|\delta(\lambda-\alpha_s(\lambda)). \tag{4.8.7}$$

Finally, we make use of the expression for the δ-function of a complex argument:

$$\left|1-\frac{d\alpha_s}{d\lambda}\right|\delta(\lambda-\alpha_s(\lambda))=\sum_r\delta(\lambda-\lambda_{sr}). \tag{4.8.8}$$

Thus, returning to the definition (4.8.3) and comparing it with (4.8.7) and (4.8.8), we find the final relation:

$$\nu(\lambda)=\frac{1}{\pi}\lim_{\gamma\to+0}\mathrm{Im}\,\mathrm{Tr}\left[\left(1-\frac{d\hat{\mathcal{L}}}{d\lambda}\right)\mathbf{G}(\lambda-i\gamma)\right]. \tag{4.8.9}$$

Equation (4.8.9), suggested by Chebotarev (1980), generalized (4.7.11) to the case of an arbitrary matrix $\hat{\mathcal{L}}$ and is the basis for describing the spectrum of deformed crystal vibrations.

4.8.1 Problems

1. Find Green function for stationary vibrations of a one-dimensional crystal with nearest-neighbor interactions.

Solution.

$$G_\varepsilon(n)=\frac{i}{\varepsilon\sqrt{\omega_m^2-\varepsilon}}\begin{cases}e^{-iakn}, & n<0;\\ e^{iakn}, & n>0,\end{cases}$$

where $ak=2\arcsin(\sqrt{\varepsilon}/\omega_m)$.

2. Find the Green function for stationary vibrations of a 1D crystal, accounting for the interactions of not only the nearest neighbors, when $\omega=\omega(k)$ is a nonmonotonic function in the interval $0<k<\pi/a$.

5
Acoustics of Elastic Superlattices: Phonon Crystals

5.1
Forbidden Areas of Frequencies and Specific Dynamic States in such Areas

While studying the vibration spectrum of a crystal lattice we have seen that the vibration eigenfrequencies always occupy a finite interval of possible frequencies or several finite intervals of frequencies. In the case of a monatomic crystal lattice one speaks of the bands of acoustic vibrations. Every one of these bands has got some upper limit. Denote the largest of them $\omega_{\max}$. The frequencies of mechanical vibrations of the crystal with such a lattice can not be higher than $\omega_{\max}$. Values of frequencies $\omega > \omega_{\max}$ are *forbidden.*

In a polyatomic lattice, vibrations of the optical type exist always and their frequencies are as a rule higher than the frequencies of acoustic vibrations. In any case, the optical vibrations are higher than the acoustic vibrations at least at small enough values of the vector $\boldsymbol{k}$. Therefore, there is a forbidden band (gap) in the region of small k between the acoustic and optical frequencies. This gap disappears in many cases at the values of $\boldsymbol{k}$ close to boundaries of the Brillouin zones (see Fig. 3.2).

However, there are crystals whose vibration spectrum has got a frequency gap separating the high-frequency and low-frequency bands at any $\boldsymbol{k}$. Such a situation can be found in many molecular crystals where the frequencies close to intermolecular vibrations of the single molecule lie higher than the frequencies of usual mechanical vibrations (acoustic and optical types). The forbidden gap can exist even in "convenient" crystals of NiO type see Fig. 5.1.

Thus, crystal eigenvibrations can not possess frequencies within forbidden gaps. But what response of the crystal would be expected to an external periodical perturbation with the frequency belonging to such a band? Consider a primitive cubic lattice occupying the half space $z < 0$ and having a plane boundary surface $z = 0$ where the z-axis is chosen along the edge of the elementary cube (parallel to the axis of symmetry C_4). We confine ourselves to the scalar model and interactions of the nearest neighbors. Then, both the equation of motion (2.1.13) and the dispersion relation take

The Crystal Lattice: Phonons, Solitons, Dislocations, Superlattices, Second Edition. Arnold M. Kosevich

ISBN: 3-527-40508-9

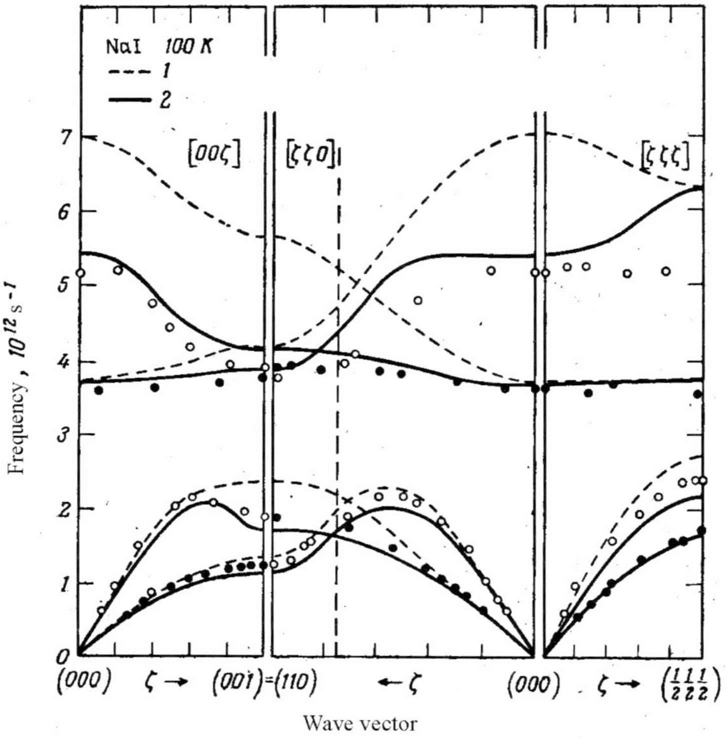

Fig. 5.1 Experimental points and dispersion curves calculated for the crystal NaI (Woods et al. 1960).

the following forms

$$m\omega^2 u(\boldsymbol{n}) = \alpha \sum_{\boldsymbol{n}_0} [u(\boldsymbol{n}) - u(\boldsymbol{n} + \boldsymbol{n}_0)], \tag{5.1.1}$$

$$\omega^2 = \frac{1}{3}\omega_m^2 \left(\sin^2 \frac{ak_x}{2} + \sin^2 \frac{ak_y}{2} + \sin^2 \frac{ak_z}{2} \right); \tag{5.1.2}$$

here the number-vector $\boldsymbol{n}_0$ connects the site $\boldsymbol{n}$ with the six nearest neighbors and ω_m is the maximum frequency of the eigenvibrations : $\omega_m^2 = 4\alpha/m$.

Suppose that the lattice under consideration contacts along the plane $z = 0$ some medium in which bulk mechanical vibrations with high frequencies can be excited. This may be the same cubic lattice with another elastic coefficients $\alpha(\boldsymbol{n})$ or a polyatomic lattice where the high-frequency vibrations of the optical type exist. In electrically active crystals (like ionic or ferroelectric crystals) similar vibrations can be excited by the electromagnetic waves that do not act directly on the lattice under consideration. In any case, suppose that the mechanical vibrations in the half-space $z > 0$ possess the frequencies exceeding all possible frequencies of the lattice under consideration. These vibrations create on the interface boundary $z = 0$ some distribution of forces like a traveling wave. Let the perturbation wave propagate along the x-axis and have the form

$$f(\boldsymbol{n},t) = f_0 e^{ikan_x - i\omega t}. \tag{5.1.3}$$

Perturbation (5.1.3) on the plane $z = 0$ is perceived by the lattice considered as a definite boundary condition. Under any condition of the type (5.1.3) a solution to the linear dynamic equation of the motion (5.1.1) takes the form

$$u = w(n_z)e^{ikan_x - i\omega t}, \tag{5.1.4}$$

where the function $w(n_z)$ obeys the equation

$$\left[\omega^2 - \frac{1}{3}\omega_m^2 \sin^2\left(\frac{ak}{2}\right)\right] w(n) = \frac{1}{12}\omega_m^2[2w(n) - w(n-1) - w(n+1)]. \tag{5.1.5}$$

Since the maximum frequency corresponds to $ak_z = \pi$ and we are interested in the vibrations with $\omega > \omega_m$ the solution to (5.1.5) in the half-space $z < 0$ should be taken in the form

$$w(n) = (-1)^n e^{\kappa a n}, \quad n < 0. \tag{5.1.6}$$

We substitute (5.1.6) into (5.1.5) and come to the expression

$$\omega^2 = \frac{1}{3}\omega_m^2\left(\sin^2\frac{ak}{2} + \cosh^2\frac{a\kappa}{2}\right). \tag{5.1.7}$$

At the frequencies $\omega^2 > \dfrac{2}{3}\omega_m^2$ the parameter κ is real ($\cosh^2\dfrac{a\kappa}{2} > 1$) and we must choose $\kappa > 0$ in (5.1.7).

Solution (5.1.7) describes crystal vibrations with an amplitude decaying exponentially when the distance from the boundary surface increases. This property of the solution discussed is its principal difference from the eigenvibrations of an infinite crystal lattice. The vibrations under consideration represent an example of *localized vibrations*. If the localization takes place near a boundary surface of the crystal one calls them *surface vibrations*. We consider another example of localized vibrations later. Frequencies of the localized vibrations in all such examples lie outside of the continuous spectrum of the eigenvibrations of the crystal lattice (above the frequency spectrum or inside the gaps).

5.2
Acoustics of Elastic Superlattices

We have paid attention to a possibility of existence of the forbidden bands (gaps) in the vibration spectrum of a crystal lattice. Now it is natural to raise a question whether such a possibility is connected with microscopic (atomic) periodicity of the crystal or it is the general consequence of any periodicity independently of a value of the period. In dynamics of a crystal lattice the period of the crystal structure a takes part in the analysis as a parameter and its value arises only at the estimation of the elastic moduli of crystals. As to peculiarities of the vibrational spectrum they are determined by the dimensionless parameter ak, where k is a value of the wave vector. Therefore, the peculiarities of the frequency spectrum, in particular the problem of the existence of gaps, depend not on a concrete value of a, but on its relation to the wavelength. Consider this fact and consider longwave and low-frequency (acoustic) vibrations putting $ak \ll 1$.

The dynamics of a crystal in the (k, ω)-area $ak \ll 1$ and $\omega \ll \omega_D$ does not differ from the dynamics of a continuum media and can be described by the theory of elasticity. Consequently, the vibration spectrum of a homogeneous crystal coincides with the spectrum of sound waves; it is continuous and without any gaps. However, if the homogeneity of the crystal is broken and a macroscopic periodicity comes into existence the situation changes markedly.

Consider a one-dimensional structure consisting of periodically arranged (along the x-axis) layers of elastic isotropic materials of two types [d_α are the thickness of layers ($\alpha = 1, 2$), s_α is the velocity of the sound wave in the α layer, and the structure period is $d = d_1 + d_2$]. The periodic structure under consideration has a macroscopic period d, which, by definition, greatly exceeds the interatomic distance a. Such a periodic structure will be called an *elastic superlattice* (SL). Sometimes a macroscopic periodic structure consisting of alternating elastic materials that differ in their elastic moduli and sound speeds is called a *phonon crystal*.

The field of the elastic wave $\boldsymbol{u}(\boldsymbol{r}, t)$ propagating perpendicular to the layer plane is determined by a standard wave equation. In a system of isotropic blocks, the waves of two possible polarizations are independent, and we can restrict ourselves to analysis of dynamic equations for the scalar fields u^α

$$\frac{\partial^2 u^\alpha}{\partial t^2} - s_\alpha^2 \frac{\partial^2 u^\alpha}{\partial x^2} = 0, \qquad \alpha = 1, 2. \tag{5.2.1}$$

The velocity of a wave is $s_\alpha = \sqrt{\mu_\alpha / \rho_\alpha}$ (μ_α and ρ_α are the elastic moduli and mass densities, respectively).

Equation (5.2.1) should be solved using the boundary conditions according to which the displacements u^α and stresses $\sigma^\alpha = \mu_\alpha \frac{\partial u^\alpha}{\partial x}$ are continuous at all boundaries of the blocks.

Suppose that the elastic properties of two materials differ slightly and introduce the notation $\delta s_1^2 = s_1^2 - s^2$ and $\delta s_2^2 = s_2^2 - s^2$ where the average speed squared s^2 is determined by the condition

$$s^2 d = s^2 d_1 + s_2^2 d_2. \tag{5.2.2}$$

Take into account $\delta s_\alpha^2 \ll s^2$ and rewrite (5.2.1) in the form

$$\frac{\partial^2 u^\alpha}{\partial t^2} - s^2 \frac{\partial^2 u^\alpha}{\partial x^2} = \delta s_\alpha^2 \frac{\partial^2 u^\alpha}{\partial x^2}. \tag{5.2.3}$$

Each of the vibration eigenmodes appearing in a periodic structure with a period d is characterized by the quasi-wave number k. In the zero approximation with respect to the small parameter $\delta s/s \ll 1$ the solution to (5.2.3) has the form of a plane wave

$$u(x,t) = \frac{1}{\sqrt{L}} e^{ikx - i\omega t}, \quad L = Nd, \tag{5.2.4}$$

where N is the number of layers in the SL, and the frequency is proportional to the wave number: $\omega = sk$.

However, one should remember that a periodical perturbation (boundary conditions) exists even in this approximation. This means that a frequency of solution (5.2.4) is a periodical function of k with the period of the reciprocal lattice $2\pi/d$. In order to combine the sound dispersion relation $\omega = sk$ with such a periodicity let us analyze a set of diagrams of this dispersion relation shifted by the value $2\pi n/d (n = 1, 2.3, \ldots)$ along the x-axis (see Fig. 5.2). Figure 5.3a represents the diagrams of the same dispersion relations inside one Brillouin zone.

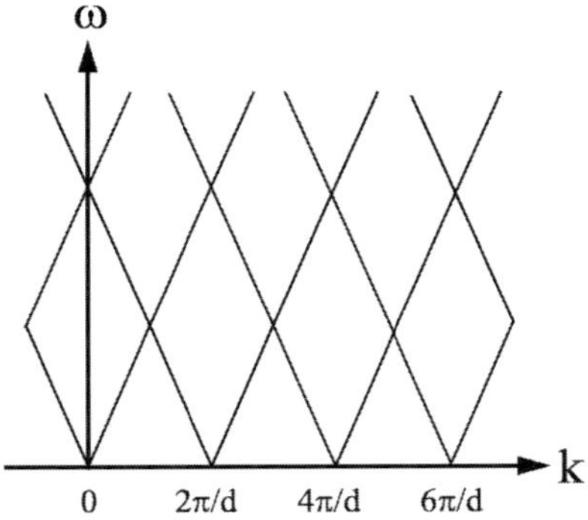

Fig. 5.2 ω-k relations in the periodic-zone scheme.

One can see that a crossover situation that was met earlier in Chapter 3 appears in the center of the zone ($k = 0$) and at its boundaries ($k = \pm\pi/d$). There is a degeneracy of frequencies at the k mentioned. The degeneracy is a consequence of the supposition neglecting differences between parameters of two layers. The term in the r.h.p. of (5.2.3) plays the role of a small perturbation that can remove the degeneracy. It is convenient to combine two equations ($\alpha = 1, 2$) introducing the discontinuous function $U(x) = U(x + d)$ that inside the interval of one period equals ($x = nd + \xi$):

$$U(\xi) = \begin{cases} \delta s_1^2, & 0 < \xi < d_1; \\ \delta s_2^2, & d_1 < \xi < d. \end{cases} \tag{5.2.5}$$

Then, the equations of SL vibrations with the frequency ω can be represented as

$$\omega^2 u = -s^2 \frac{d^2 u_s}{dx^2} + U(x) \frac{d^2 u_s}{dx^2}. \tag{5.2.6}$$

The periodic function $U(x)$ should be expanded in the following Fourier series

$$U(x) = \sum_m U_m e^{2\pi m x/d}, \tag{5.2.7}$$

where

$$U_m = \frac{1}{L}\int U(x)e^{-2\pi i m x/d}dx = \frac{1}{d}\int_0^d U(\xi)e^{-2\pi i m \xi/d}d\xi, \tag{5.2.8}$$

and $U_0 = 0$ as it follows from (5.2.2). The lowest degenerated frequency corresponds to the wave vectors $k = \pm\pi/d$. We seek a solution to (5.2.6) in the form of a linear combination

$$u = (u_+^0 e^{i\pi m x/d} + u_-^0 e^{-i\pi m x/d})e^{iqx}, \tag{5.2.9}$$

with a frequency ω slightly differing from $s\pi/d$ at small q (further $q \ll \pi/d$). We substitute (5.2.9) into (5.2.6), multiply in turn by $\exp -i(\pi/d+q)x$ and $\exp -i(\pi/d-q)x$ and perform the integration with respect to x. Then we obtain in the main approximation the following set of equations for coefficients u_+^0 and u_-^0:

$$\begin{aligned}
&\left[\omega^2 - \left(\frac{\pi s}{d}\right)^2 - \frac{2\pi s^2}{d}q\right]u_+^0 + \left(\frac{\pi}{d}\right)^2 U_1 u_-^0 = 0,\\
&\left[\omega^2 - \left(\frac{\pi s}{d}\right)^2 + \frac{2\pi s^2}{d}q\right]u_-^0 + \left(\frac{\pi}{d}\right)^2 U_1^* u_+^0 = 0.
\end{aligned} \tag{5.2.10}$$

The condition of solvability of set (5.2.10) gives a dispersion relation removing the degeneration at the boundaries of the Brillouin zone:

$$\omega^2 = \left(\frac{\pi}{d}\right)^2\left\{s^2 \pm \left[s^4\left(\frac{2qd}{\pi}\right)^2 + |U_1|^2\right]^{1/2}\right\}. \tag{5.2.11}$$

A gap is opening at the Brillouin zone ($q = 0$):

$$\delta\omega = \frac{\pi\,|U_1|}{cd}, \tag{5.2.12}$$

where ($d_1 < d$)

$$|U_1| = \frac{1}{\pi d}\left|\delta s_1^2 - \delta s_2^2\right|\sin\left(\frac{\pi d_1}{d}\right) \approx \frac{2s}{\pi d}|s_1 - s_2|\sin\left(\frac{\pi d_1}{d}\right). \tag{5.2.13}$$

The forbidden gap appears only at different sound velocities in two neighboring layers.

The higher gaps at the degeneration point can be calculated analogously if one takes pairs of waves with a difference of the wave vectors $k - k' = 2\pi(n - n')$, where n and

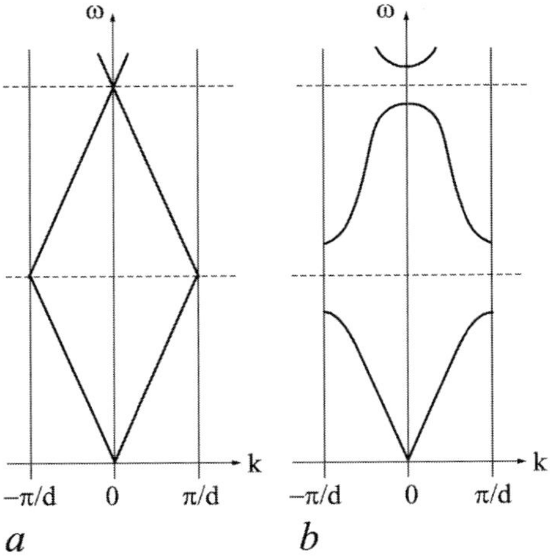

Fig. 5.3 Dispersion curves inside one Brillouin zone: (*a*) in the zero approximation (without gaps), (*b*) deformation of curves by small perturbations.

n' are integers. A set of allowed bands and gaps appears in a SL. Two lowest bands of frequencies allowed are schematically shown in Fig. 5.3b. A frequency spectrum of the SL differs essentially both from the usual sound spectrum $\omega = sk$, where s is the sound velocity and from the vibration spectrum of a crystal lattice. The number of vibration branches in the crystal is determined by the number of atoms in the crystal unit cell, but in a SL a number of vibration branches appears. The total number of vibration branches in a SL is limited only to the value d/a, that is to the number of atoms in one period of the SL (inside one unit cell of the SL).

5.3 Dispersion Relation for a Simple Superlattice Model

We return to (5.2.1) and note that each of the eigensolutions to (5.2.1) in a periodic structure with a period d is characterized by the quasi-wave number k. Natural oscillations of the field in the unit cell with the number n can be written in the form

$$u_n(x) = u_0(x)e^{iknd}, \qquad u_0(x+d) = u_0(x). \tag{5.3.1}$$

We write down solutions to the pair of equations (5.2.1) in the interval of one unit cell with the number n ($nd < x < (n+1)d$)

$$\begin{aligned} u^n(x) &= a_1^{(n)} e^{ik_1\xi} + a_2^{(n)} e^{-ik_1\xi}, \quad 0 < \xi < d_1; \\ u^n(x) &= b_1^{(n)} e^{ik_2\xi} + b_2^{(n)} e^{-ik_2\xi}, \quad d_1 < \xi < d, \end{aligned} \tag{5.3.2}$$

where $\xi = x - nd$ and $k_1 = \omega/s_1$, $k_2 = \omega/s_2$, and ω is the frequency. Amplitudes in the neighboring cells are connected by the conditions

$$a^{(n+1)} = a^{(n)} e^{ikd}, \qquad b^{(n+1)} = b^{(n)} e^{ikd}. \tag{5.3.3}$$

Boundary conditions at the points $\xi = 0$ and $\xi = d$ lead to the set of four homogeneous algebraic equations for the amplitudes a_1, a_2, b_1, and b_2. Equality of the determinant of this set to zero gives the following dispersion relation (Rytov, 1955)

$$\cos kd = \cos k_1 d_1 \cos k_2 d_2 - \frac{1}{2}\left(\frac{k_1}{k_2} + \frac{k_2}{k_1}\right) \sin k_1 d_1 \sin k_2 d_2. \tag{5.3.4}$$

A derivation of (5.3.4) using (5.3.2) and (5.3.3) can be considered as a problem exercise for this section.

Equation (5.3.4) determines in a complicated form the dependence of the frequency ω on the wave number k: $\omega(k)$. This relation coincides with an accuracy to notation with that obtained by Kronig and Penney for a quantum particle in a one-dimensional periodic potential (Kronig and Penney, 1930).

Expression (5.3.4) gives the implicit dependence of the frequency on the quasi-wave number and allows us to describe readily the spectrum of long-wave vibrations ($kd \ll 1$), for which the sound spectrum is naturally obtained with average elastic modulus $\langle\mu\rangle$ and the density $\langle\rho\rangle$: $\langle\rho\rangle d = \rho_1 d_1 + \rho_2 d_2$ and $d/\langle\mu\rangle = d_1/\mu_1 + d_{2/\mu_2}$. Based on such a representation of μ, which contains only ratios d_α/μ_α, it is interesting to consider a limiting case, which can demonstrate the most characteristic properties of the superlattice spectrum, when $d_2 \to 0$ and $\mu_2 \to 0$ for $d_2/\mu_2 = M = \text{const}$. In this case, $d_1 \to d$, $k_2 d_2 = \omega d_2/s_2 = \sqrt{\rho_2 d_2 \omega}\sqrt{d_2/\mu_2} \to 0$. Then the dispersion relation for the system is described by the equation

$$\cos kd = \cos z - Qz \sin z, \tag{5.3.5}$$

where $z = q_1 d = \omega d/s_1$ and $Q = \rho_2 \mu_1 M/(2\rho_{1d})$. Note that (5.3.5) gives the dispersion relation for an elastic SL consisting of periodic elastic blocks of length d with the parameters μ_1 and s_1 under special boundary conditions. If the parameter Q is small, then the system under study represents a periodic sequence of elastic regions that are weakly connected with each other.

The allowed vibrational frequencies of a continuous spectrum of the system under consideration can be qualitatively found by analyzing graphically (5.3.5), as shown in Fig. 5.4. For the beginning we repeat our analysis concerning Fig. 1.15 in Chapter 1. If the r.h.p. of (5.3.5) runs the values between ± 1, the roots of the equation have the values in the intervals shown on the abscissa.

Note that, as z increases, the allowed frequencies are localized within the narrowing intervals near the values $k_1 d = \pm m\pi$, where m is a large integer. For the condition $m^2 Q \gg 1$, the dispersion relation for the m-th band can be readily found.

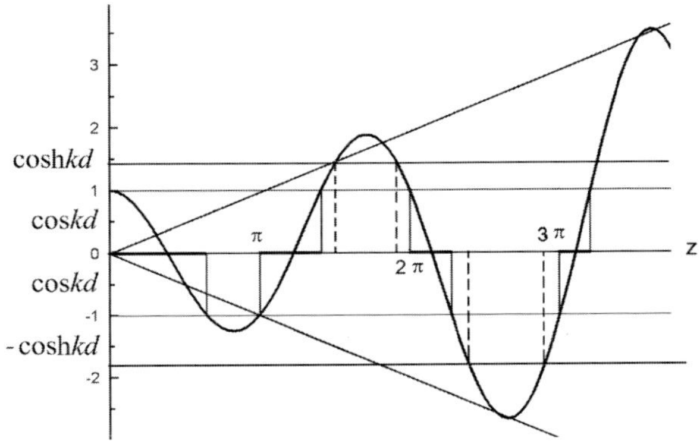

Fig. 5.4 Graphical solution of (5.3.5) in the cases $\cosh kd > 1$.

Indeed, near odd $m = 2p + 1$ (see the vicinity of $z = 3\pi$) in Fig. 5.4 , we can write with sufficient accuracy ($s = s_1$)

$$\cos kd = -1 + Qm\pi(z - m\pi) = -1 + \frac{m\pi d}{s} Q\left(\omega - \frac{m\pi s}{d}\right),$$

which yields

$$\omega = m\omega_0 + \frac{s}{m\pi Qd}(1 + \cos kd), \tag{5.3.6}$$

where $\omega_0 = \pi s/d$. Similarly, near even $m = 2p$ (see the vicinity of $z = 4\pi$) in Fig. 5.4, we can write

$$\cos kd = 1 - Qm\pi(z - m\pi) = 1 - \frac{m\pi d}{s} Q\left(\omega - \frac{m\pi s}{d}\right),$$

which gives

$$\omega = m\omega_0 + \frac{s}{m\pi Qd}(1 - \cos kd). \tag{5.3.7}$$

By combining (5.3.6) and (5.3.7), we can obtain the dispersion relations for the m-th band:

$$\omega = m\omega_0 + \frac{2s}{m\pi Qd}\begin{cases} \sin^2\left(\frac{kd}{2}\right), & m = 2p; \\ \cos^2\left(\frac{kd}{2}\right), & m = 2p + 1. \end{cases} \tag{5.3.8}$$

One can easily see that expressions (5.3.8) represent the size-quantization spectrum of phonons in a layer of thickness d, whose levels split into minibands due to a low "transparency" of the interface between layers, that is due to a weak interaction between adjacent blocks.

The frequency spectrum that we obtained is of interest because its high-frequency part has a set of narrowing allowed frequency bands in which the dispersion relation can be calculated analytically with good accuracy. The spectrum has a number of forbidden bands (gaps in a continuous spectrum).

Consider now the possibility of appearance of vibrational states in forbidden bands. Such vibrations correspond to solutions of the type

$$u_n \propto \exp(\mp \kappa n d) \quad \text{for} \quad k = i\kappa,$$

or

$$u_n \propto (-1)^n \exp(\mp \kappa n d) \quad \text{for} \quad k = i\kappa + \pi,$$

exponentially decreasing (increasing) with the number n. It is obvious that such states can have a physical meaning only on the coordinate semi-axis under the condition that a solution vanishing at infinity is chosen, which reflect some boundary conditions at the coordinate origin.

For the solution of the first type ($k = i\kappa$), the frequency dependence of the parameter κ can be found from the relation

$$\cosh(\kappa d) = \cos z - Qz \sin z > 1, \tag{5.3.9}$$

while, for the solution of the second type ($k = i\kappa + \pi$), it can be found from the relation

$$-\cosh(\kappa d) = \cos z - Qz \sin z > 1 < -1. \tag{5.3.10}$$

The solutions of the first type correspond to frequencies in the intervals $(2p-1)\pi < z < 2p\pi$, and those of the second type, in intervals $2p\pi < z < (2p+1)\pi$ (see Fig. 5.4). Note that such situations appear on the semi-axis, for example, at the ends of the SL.

5.3.1
Problem

Show that (5.3.5) describes the dispersion relation for an elastic SL consisting of periodic blocks of length d under the following boundary conditions on the interfaces: the normal stresses are continuous ($[\sigma]_-^+ = 0$, i.e., $[\partial u/\partial x]_-^+ = 0$), while elastic displacements exhibit the jump $[u]_-^+ = Q(\rho_1/\rho_2)\sigma$.

Part 3 Quantum Mechanics of Crystals

The Crystal Lattice: Phonons, Solitons, Dislocations, Superlattices, Second Edition. Arnold M. Kosevich

ISBN: 3-527-40508-9

6
Quantization of Crystal Vibrations

6.1
Occupation-Number Representation

To describe crystal vibrations one uses the classical equations of motion of atoms (or molecules) positioned at the lattice sites. A classical description of crystal vibrations is only a rough approximation, and from the beginning one should proceed from quantum laws. Small-amplitude vibrations of an ideal crystal, however, represent the rare case of a physical system where a quasi-classical treatment leads to the same results as those obtained in a rigorous quantum-mechanical approach. In this approximation the system of quantized vibration in the crystal is assumed to be equivalent to a system of independent harmonic oscillators. The classification of states and the calculation of the energy spectrum of a harmonic oscillator at a quasi-classical level are known to be accurate quantum mechanically.

Thus, for the majority of crystals the vibrations can be quantized at a late stage in the calculations when the vibration dispersion law is found, the vibration field is represented as a set of harmonic oscillators and the harmonic oscillator frequencies are determined. In particular, for the initial stage of quantization one may take the Hamilton function (2.11.14) written in terms of the real canonically conjugated generalizes coordinates $X(\mathbf{k})$ and momenta $Y(\mathbf{k})$. Since the quantum treatment is not dependent on the vector character of the displacements and the momenta corresponding to them, we begin by using a scalar model based on the Hamilton function

$$\mathcal{H} = \sum_{k} H(\mathbf{k}), \quad H(\mathbf{k}) = \frac{1}{2}\left[Y_k^2 + \omega^2(\mathbf{k})X_k^2\right]. \tag{6.1.1}$$

In quantum mechanics the Hamilton function is regarded as an operator (Hamiltonian) whose dynamic variables $X(\mathbf{k})$ and $Y(\mathbf{k})$ in (6.1.1) are replaced by the operators

The Crystal Lattice: Phonons, Solitons, Dislocations, Superlattices, Second Edition. Arnold M. Kosevich

ISBN: 3-527-40508-9

with commutative relations

$$[X(\boldsymbol{k}), Y(\boldsymbol{k}')] = i\hbar\delta_{\boldsymbol{k}\boldsymbol{k}'};$$
$$[X(\boldsymbol{k}), X(\boldsymbol{k}')] = [Y(\boldsymbol{k}), Y(\boldsymbol{k}')] = 0, \tag{6.1.2}$$

where $[A, B] = AB - BA$. We do not use special notation for the operators of physical quantities, but have to allow for the noncommutativity of these quantities in all calculations.

The simplest and most widely used way of obtaining the quantum spectrum of a multiparticle system is by writing and diagonalizing its Hamiltonian in the *occupation number representation*. Since the Hamiltonian (6.1.1) is already diagonalized in the $\boldsymbol{k}$-states, the choice of new operators can be performed by following linear transformations for a given value of $\boldsymbol{k}$:

$$X(\boldsymbol{k}) = u(\boldsymbol{k})a_{\boldsymbol{k}} + u^*(\boldsymbol{k})a^\dagger_{\boldsymbol{k}}; \qquad Y(\boldsymbol{k}) = v(\boldsymbol{k})a_{\boldsymbol{k}} + v^*(\boldsymbol{k})a^\dagger_{\boldsymbol{k}}. \tag{6.1.3}$$

The operators $a_{\boldsymbol{k}}$ and $a^\dagger_{\boldsymbol{k}}$ are the Hermitian conjugated operators, $u(\boldsymbol{k})$ and $v(\boldsymbol{k})$ are complex functions of the vector $\boldsymbol{k}$ whose choice should satisfy the requirements (6.1.2) and reduce the Hamiltonian $H(\boldsymbol{k})$ to a product of operators $a_{\boldsymbol{k}}$ and $a^\dagger_{\boldsymbol{k}}$.

We transform an individual term in the Hamiltonian (6.1.1)

$$\begin{aligned} Y^2(\boldsymbol{k}) + \omega^2(\boldsymbol{k})X^2(\boldsymbol{k}) &= (|v|^2 + \omega^2|u|^2)(a_{\boldsymbol{k}}a^\dagger_{\boldsymbol{k}} + a^\dagger_{\boldsymbol{k}}a_{\boldsymbol{k}}) \\ &+ (v^2 + \omega^2u^2)a_{\boldsymbol{k}}a_{\boldsymbol{k}} + [(v^*)^2 + \omega^2(u^*)^2]a^\dagger_{\boldsymbol{k}}a^\dagger_{\boldsymbol{k}}, \end{aligned} \tag{6.1.4}$$

and the first of the conditions (6.1.2)

$$\begin{aligned} [X(\boldsymbol{k}), Y(\boldsymbol{k}')] &= u(\boldsymbol{k})v^*(\boldsymbol{k}')[a_{\boldsymbol{k}}a^\dagger_{\boldsymbol{k}'} + u^*(\boldsymbol{k})v(\boldsymbol{k}')[a^\dagger_{\boldsymbol{k}}a_{\boldsymbol{k}'}] \\ &+ u(\boldsymbol{k})v(\boldsymbol{k}')[a_{\boldsymbol{k}}, a_{\boldsymbol{k}'}] + u^*(\boldsymbol{k})v^*(\boldsymbol{k}')[a^\dagger_{\boldsymbol{k}'}, a^\dagger_{\boldsymbol{k}'}] = i\hbar\delta_{\boldsymbol{k}\boldsymbol{k}'}. \end{aligned} \tag{6.1.5}$$

Analyzing (6.1.4), it is easy to come to a conclusion that if we choose the functions $u(\boldsymbol{k})$ and $v(\boldsymbol{k})$ as

$$u(\boldsymbol{k}) = \sqrt{\frac{\hbar}{2|\omega(\boldsymbol{k})|}}, \qquad v(\boldsymbol{k}) = -i|\omega(\boldsymbol{k})|\,u(\boldsymbol{k}), \tag{6.1.6}$$

the Hamiltonian of vibrations (6.1.1) will reduce to

$$\mathcal{H} = \frac{1}{2}\sum_{\boldsymbol{k}} \hbar|\omega(\boldsymbol{k})|\,(a^\dagger_{\boldsymbol{k}}a_{\boldsymbol{k}} + a_{\boldsymbol{k}}a^\dagger_{\boldsymbol{k}}). \tag{6.1.7}$$

According to (6.1.5), the conditions (6.1.2) will then also be satisfied, if the operators $a_{\boldsymbol{k}}$ and $a^\dagger_{\boldsymbol{k}}$ obey the following commutation rules

$$[a_{\boldsymbol{k}}, a^\dagger_{\boldsymbol{k}}] = \delta_{\boldsymbol{k}\boldsymbol{k}'}, \quad [a_{\boldsymbol{k}}, a_{\boldsymbol{k}'}] = [a^\dagger_{\boldsymbol{k}'}, a^\dagger_{\boldsymbol{k}'}] = 0. \tag{6.1.8}$$

Using the first rule from (6.1.8) we simplify the Hamiltonian (6.1.7)

$$\mathcal{H} = \sum_k \hbar|\omega(\boldsymbol{k})| \left(a_k^\dagger a_k + \frac{1}{2} \right) . \tag{6.1.9}$$

Since the Hamiltonian (6.1.9) involves only the absolute frequency value, we can consider further only positive vibration frequencies.

Using the operators a_k and $a_k^\dagger$ is the most efficient way in the representation, where the operator $a_k^\dagger a_k$ included in the Hamiltonian is diagonal. It turns out that if a_k and $a_k^\dagger$ satisfy the relations (6.1.8), the eigenvalues of the operator $a_k^\dagger a_k$ are non-negative integer numbers of a natural series

$$a_k^\dagger a_k = N_k, \quad N_k = 0, 1, 2, \ldots . \tag{6.1.10}$$

This property of the operator $a_k^\dagger a_k$ is proved in quantum mechanics, however, in our case it follows directly from (6.1.9). Indeed, the energy levels of a harmonic oscillator with frequency ω are known

$$E_n = \left(n + \frac{1}{2} \right) \hbar\omega, \quad n = 0, 1, 2, \ldots , \tag{6.1.11}$$

thus, the eigenvalues of the Hamiltonian (6.1.1) being the sum of the energies of independent harmonic oscillators can be represented as

$$\mathcal{E} = \sum_k \left(N_k + \frac{1}{2} \right) \hbar\omega(\boldsymbol{k}), \tag{6.1.12}$$

where N_k are non-negative integers. Comparing the expressions (6.1.12) and (6.1.9) we are convinced of the validity of (6.1.10).

The numbers N_k are called the *occupation numbers* of the states $\boldsymbol{k}$. When the systems consisting of many identical particles are studied from the point of view of quantum mechanics, it is convenient to use a mathematical method where in the occupation number representation various vibrational crystal states are characterized by different sets of numbers N_k, and the action of the operators a_k and $a_k^\dagger$ changes these numbers.

In applications it is important to know not only how the Hamiltonian is written in terms of the operators a_k and $a_k^\dagger$, but also the form of the displacement operator, which is always initial.

We note that the linear transformation (6.1.3) taking into account (6.1.6) is written in the form

$$X(\boldsymbol{k}) = \sqrt{\frac{\hbar}{2\omega(\boldsymbol{k})}} (a_k^\dagger + a_k), \qquad Y(\boldsymbol{k}) = i\sqrt{\frac{1}{2}\hbar\omega(\boldsymbol{k})} (a_k^\dagger - a_k). \tag{6.1.13}$$

Thus, the sum of the operators $a_k^\dagger$ and a_k is the coordinate, and their difference is the momentum of the corresponding harmonic oscillator.

Using now the chain of transformations (6.1.13) and (2.11.15) to find a complex expression for the operators of normal coordinates, we obtain

$$Q(\mathbf{k}) = u_k(a_k - a^\dagger_{-k}), \qquad Q^\dagger(\mathbf{k}) = u_k(a^\dagger_k + a_{-k}), \tag{6.1.14}$$

where the multiplier u_k is determined in (6.1.6). Generally, the complex conjugate normal coordinate $Q^*(\mathbf{k})$ corresponds to a Hermitian conjugate operator $Q^\dagger(\mathbf{k})$.

Using (6.1.14) and (1.6.3), we obtain the atomic displacement operator

$$\begin{aligned} u(\mathbf{n}) &= \left(\frac{\hbar}{2mN}\right)^{1/2} \sum_k \frac{1}{\sqrt{\omega(\mathbf{k})}} \left[a_k e^{i\mathbf{k}\mathbf{r}(\mathbf{n})} + a^\dagger_k e^{-i\mathbf{k}\mathbf{r}(\mathbf{n})}\right] \\ &= \left(\frac{\hbar}{2mN}\right)^{1/2} \sum_k \frac{1}{\sqrt{\omega(\mathbf{k})}} (a_k + a^\dagger_{-k}) e^{i\mathbf{k}\mathbf{r}(\mathbf{n})}. \end{aligned} \tag{6.1.15}$$

We note that in replacing the displacements with the operators by (6.1.15) the denominator always includes the multiplier $\sqrt{\omega(\mathbf{k})}$ that accompanies the operators a_k or $a^\dagger_k$.

For a real crystal lattice when the operators $a_\alpha(\mathbf{k})$ and $a^\dagger_\alpha(\mathbf{k})$ belonging to different branches of the dispersion law ($\alpha = 1, 2, \ldots, 3q$) are introduced, the commutation relations (6.1.8) are generalized to

$$\left[a_\alpha(\mathbf{k}), a^\dagger_{\alpha'}(\mathbf{k}')\right] = \delta_{\alpha\alpha'}\delta_{\mathbf{k}\mathbf{k}'},$$

$$[a_\alpha(\mathbf{k}), a_{\alpha'}(\mathbf{k}')] = \left[a^\dagger_\alpha(\mathbf{k}), a^\dagger_{\alpha'}(\mathbf{k}')\right] = 0.$$

The displacement operator of the $(\mathbf{n}, s)$-th atom is associated with the operators $a_\alpha(\mathbf{k})$ and $a^\dagger_\alpha(\mathbf{k})$ by relations

$$\mathbf{u}_s(\mathbf{n}) = \left(\frac{\hbar}{2m_s N}\right)^{1/2} \sum_{k\alpha} \frac{\mathbf{e}_s(\mathbf{k}, \alpha)}{\sqrt{\omega_\alpha(\mathbf{k})}} [a_\alpha(\mathbf{k}) + a^\dagger_\alpha(-\mathbf{k})] e^{i\mathbf{k}\mathbf{r}(\mathbf{n})}. \tag{6.1.16}$$

Finally, the Hamiltonian of crystal vibrations in a harmonic approximation is reduced to

$$\mathcal{H} = \sum_{\alpha k} \hbar\omega_\alpha(\mathbf{k}) \left[a^\dagger_\alpha(\mathbf{k}) a_\alpha(\mathbf{k}) + \frac{1}{2}\right], \tag{6.1.17}$$

generalizing (6.1.9).

It is clear that the operators of the physical quantities can be expressed directly through the operators a_k and $a^\dagger_k$. We discuss their properties based on the Heisenberg representation when the dynamic processes are described by the time dependence of the operators of physical quantities whose equations of motion are analogous to classical Hamilton equations.

We omit the index α, i. e., return to a scalar model. From the Hamiltonian (6.1.9) there follows a very simple "equations of motion" for the operator a_k and $a^\dagger_k$. Indeed, according to the definition of a time derivative of the operator

$$\begin{aligned}\frac{da(\boldsymbol{k})}{dt} &= \frac{i}{\hbar}[\mathcal{H}, a_{\boldsymbol{k}}] = \frac{i}{\hbar}\hbar\omega(\boldsymbol{k})[a^{\dagger}_{\boldsymbol{k}}a_{\boldsymbol{k}}, a_{\boldsymbol{k}}]\\ &= i\omega(\boldsymbol{k})[a^{\dagger}_{\boldsymbol{k}}, a_{\boldsymbol{k}}]a_{\boldsymbol{k}} = -i\omega(\boldsymbol{k})a_{\boldsymbol{k}}.\end{aligned}$$

Similarly, we obtain "the equation of motion" for the operator $a^{\dagger}_{\boldsymbol{k}}$. Writing these equations simultaneously we obtain

$$\frac{da(\boldsymbol{k})}{dt} = -i\omega(\boldsymbol{k})a(\boldsymbol{k}), \qquad \frac{da^{\dagger}(\boldsymbol{k})}{dt} = i\omega(\boldsymbol{k})a^{\dagger}(\boldsymbol{k}). \tag{6.1.18}$$

We note that the equations of motion (6.1.18) for the operators $a_{\boldsymbol{k}}$ and $a^{\dagger}_{\boldsymbol{k}}$ are first-order differential equations, whereas (1.6.6) for the normal coordinates are equations of the second order.

The explicit time dependence of the operators follows from (6.1.18)

$$a_{\boldsymbol{k}}(t) = e^{-i\omega(\boldsymbol{k})t}a_{\boldsymbol{k}}(0), \qquad a^{\dagger}_{\boldsymbol{k}}(t) = e^{i\omega(\boldsymbol{k})t}a^{\dagger}_{\boldsymbol{k}}(0). \tag{6.1.19}$$

To determine the action of the operators $a_{\boldsymbol{k}}$ and $a^{\dagger}_{\boldsymbol{k}}$ on the occupation numbers, we use the matrix of the coordinate and the momentum of a 1D harmonic oscillator. If M is the mass, ω the frequency, and n is the state number or the energy level number (6.1.11) of a harmonic oscillator, the nonzero matrix elements of its coordinate X and the momentum Y are[1]

$$\begin{aligned}X_{n-1,n} &= \left(\frac{\hbar n}{2M\omega}\right)^{1/2} e^{-i\omega t}, \quad Y_{n-1,n} = -i\omega M X_{n-1,n},\\ Xn, n-1 &= \left(\frac{\hbar n}{2M\omega}\right)^{1/2} e^{i\omega t}, \quad Y_{n,n-1} = i\omega M X_{n,n-1}.\end{aligned} \tag{6.1.20}$$

Using (6.1.20) in the case $M = 1$ and the linear relations (6.1.13), it is easily seen that the matrix elements of the operators $a_{\boldsymbol{k}}$ and $a^{\dagger}_{\boldsymbol{k}}$ are nonzero only for transitions in which the corresponding occupation numbers $N_{\boldsymbol{k}}$ (with the same value of $\boldsymbol{k}$) change by unity.

We denote by

$$|\ldots N_{\boldsymbol{k}} \ldots\rangle \equiv | N_{\boldsymbol{k}}\rangle = \Psi_{\{N\}_{\boldsymbol{k}}}(Q)$$

the wave function of some crystal state that is the eigenfunction of the operator $a^{\dagger}_{\boldsymbol{k}}a_{\boldsymbol{k}}$ and is characterized by a set of occupation numbers $N_{\boldsymbol{k}}$. Then, only the following elements of these operators are nonzero:

$$\begin{aligned}\langle N_{\boldsymbol{k}-1} \left| a_{\boldsymbol{k}} \right| N_{\boldsymbol{k}}\rangle &= \sqrt{N_{\boldsymbol{k}}}e^{-i\omega(\boldsymbol{k})t},\\ \langle N_{\boldsymbol{k}}+1 \left| a^{\dagger}_{\boldsymbol{k}} \right| N_{\boldsymbol{k}}\rangle &= \sqrt{N_{\boldsymbol{k}}+1}e^{i\omega(\boldsymbol{k})t}.\end{aligned} \tag{6.1.21}$$

1) We use the ordinary matrix elements representation $A_{mn} = \langle m \left| A \right| n\rangle = \int \psi^{*}_{m}(Q)A\psi_{n}(Q)dQ$, where $\psi_{n}(Q) \equiv |n\rangle$ are the corresponding normalized wave functions (Q is a set of coordinates).

Using (6.1.21), we get

$$\begin{aligned} a_k|N_k\rangle &= \sqrt{N_k}e^{-i\omega(k)t}|N_k - 1\rangle, \\ a_k^\dagger|N_k\rangle &= \sqrt{N_k + 1}e^{i\omega(k)t}|N_k + 1\rangle. \end{aligned} \tag{6.1.22}$$

Thus, the operator a_k transforms the function with the occupation number N_k into the function with the occupation number $N_k - 1$, i. e., reduces the occupation number by unity, and the operator $a_k^\dagger$ increases it by unity.

We denote by $\langle A\rangle$ the mean value of the operator A in a certain quantum state $|N_k\rangle$:

$$\langle A\rangle = \langle N_k\,|A|\,N_k\rangle. \tag{6.1.23}$$

On the basis of (6.1.21) or (6.1.22) we can conclude how to calculate the quantum-mechanical average for the operator A that has the form of the sum of products of any number of the operators a_k and $a_k^\dagger$. If the number of operators a_k in this product does not equal the number of operators $a_k^\dagger$ (with the same value of $\boldsymbol{k}$), the mean value (6.1.23) vanishes automatically. In particular,

$$\langle a_k\rangle = \langle a_k^\dagger\rangle = 0, \qquad \langle a_k a_{k'}\rangle = \langle a_k^\dagger a_{k'}^\dagger\rangle = 0. \tag{6.1.24}$$

At the same time, (6.1.10) follows naturally from (6.1.21) and its generalization

$$\langle a_k^\dagger a_{k'}\rangle = N_k\delta_{kk'}, \qquad \langle a_k a_{k'}^\dagger\rangle = (N_k + 1)\delta_{kk'}, \tag{6.1.25}$$

is consistent with commutation relations (6.1.8).

6.2 Phonons

Let us consider the ground state of a crystal corresponding to the least vibration energy. The obvious definition of the ground state: $N_k = 0$ for all $\boldsymbol{k}$ follows from the form of energy levels (6.1.12) and the fact that N_k are non-negative numbers. The energy of the ground state

$$\mathcal{E}_0 = \frac{1}{2}\sum_k \hbar\omega(\boldsymbol{k}) = \frac{\hbar}{2}\int_0^{\omega_m} \omega\nu(\omega)\,d\omega, \tag{6.2.1}$$

is called the zero-point energy and the vibrations with $N_k = 0$ are called *zero lattice vibrations*. Let $|0\rangle$ be the wave function of the ground state. Then, according to (6.1.22)

$$a_k|0\rangle = 0. \tag{6.2.2}$$

Thus, the wave function of the ground state (the state vector) is the eigenfunction not only of the binary operator $a_k^\dagger a_k$, but also of the operator a_k. In the latter case it corresponds to a zero eigenvalue.

Any excited state corresponds to a certain set of nonzero integers $N_{\mathbf{k}}$. As follows from (6.1.22), the wave function of the excited state $|N_{\mathbf{k}}\rangle$ can be obtained by an $N_{\mathbf{k}}$-fold action of the operator $a_{\mathbf{k}}^{\dagger}$ on the ground-state vector $|0\rangle$.

On writing the vibration energy (6.1.12) in the form

$$\mathcal{E} = \mathcal{E}_0 + \sum_{\mathbf{k}} \hbar\omega(\mathbf{k}) N_{\mathbf{k}}, \tag{6.2.3}$$

it becomes clear that a weakly excited (with small vibrations) crystal state is equivalent to an ideal gas of quasi-particles, the energy of each particle being $\hbar\omega(\mathbf{k})$, and their numbers in different states being given by the set $N_{\mathbf{k}}$. The quasi-particles arise due to the quantization of vibrational collective excitations, which represent the *elementary excitations* of a vibrating crystal.

Using the de Broglie principle and the general statements of mechanics, the motion of an individual quasi-particle can be characterized by the velocity $\mathbf{v} = \partial\omega/\partial\mathbf{k}$ and the quasi-momentum

$$\mathbf{p} = \hbar\mathbf{k}. \tag{6.2.4}$$

We call the quantity (6.2.4) a quasi-momentum, as $\mathbf{k}$ is a quasi-wave vector and its specific properties in a periodic structure are automatically extended to the vector $\mathbf{p}$.

The quasi-particles introduced in this way are called *phonons* and the operator $a_{\mathbf{k}}^{\dagger}a_{\mathbf{k}}$ is the operator of the number of phonons. The names of the operators $a_{\mathbf{k}}$ and $a_{\mathbf{k}}^{\dagger}$ reflect the properties of (6.1.22): the operator $a_{\mathbf{k}}$ decreases by unity the number of these phonons with quasi-wave vector $\mathbf{k}$, and the operator $a_{\mathbf{k}}^{\dagger}$ increases by unity the number of these phonons in the crystal are called the *annihilation* (or absorption) and the *creation* (emission) operators of a phonon.

If the creation and annihilation operators of particles obey the commutation relations (6.1.8), the corresponding particles are described by Bose statistics. In our case this assertion proves the fact represented by (6.1.10), implying that in a state with quasi-wave vector $\mathbf{k}$ there can be any number of phonons.

We note that in terms of thermodynamics a weakly excited state of the crystal is equivalent to an ideal gas of phonons, whose Hamiltonian has the form

$$\mathcal{H} = \mathcal{E}_0 + \sum_{\mathbf{k}} \hbar\omega(\mathbf{k}) a_{\alpha}^{\dagger}(\mathbf{k}) a_{\alpha}(\mathbf{k}). \tag{6.2.5}$$

The ground state of a crystal is a *phonon vacuum* and its physical properties are manifest in the existence of zero vibrations. The intensity of zero vibrations is characterized by the squared amplitude of each normal vibration that is determined by the same formula as for a 1D harmonic oscillator. Let us consider the square of the shift at the $\mathbf{n}$-th site, i. e., the squared operator (6.1.15) and find its mean value in the ground state. Taking into account (6.1.24), (6.1.25) we obtain

$$\langle u^2 \rangle_0 = \langle 0|u^2(\mathbf{n})|0\rangle = \frac{\hbar}{2mN} \sum_{\mathbf{k}} \frac{1}{\omega(\mathbf{k})}. \tag{6.2.6}$$

When a crystal is excited, phonons appear and their number depends on a specific choice of N_k, but phonon gas characteristics involve only the mean occupation numbers. We denote the phonon mean number in the corresponding state through $f_\alpha(\boldsymbol{k})$ and call it the $f_\alpha(\boldsymbol{k})$ *phonon distribution function.*

The function of the equilibrium boson gas distribution is known to be given by the Bose–Einstein distribution. Using this function one should remember that the total number of phonons characterizing the intensity of the lattice mechanical vibrations is not conserved and depends on the crystal excitation degree. Thus, the chemical potential of the phonon gas is zero and the mean thermodynamic values of the occupation numbers are determined by

$$\langle\langle N_\alpha(\boldsymbol{k})\rangle\rangle = f_0\left[\omega(\boldsymbol{k})\right]; \qquad f_0(\omega) = \left[\exp\left(\frac{\hbar\omega}{T} - 1\right)\right]^{-1}, \tag{6.2.7}$$

where the brackets $\langle\langle\ldots\rangle\rangle$ denote averaging in the equilibrium thermodynamic state and the temperature T here and below is given in units of energy.

A simple form of the Hamiltonian (6.2.5) and the Bose-type distribution (6.2.7) allow one easily to construct the thermodynamics of a weakly excited crystal. If mechanical atomic vibrations exhaust all possible forms of internal motions in a crystal (6.2.3) determines the total crystal energy whose mean value coincides with the internal energy. If there are other forms of motion in a crystal (electron motion, change of spin magnetic moment or similar) (6.2.3) gives only a so-called *lattice* part of the crystal energy. In the latter case all results below listed results can be applied only when the phonon interaction with elementary excitations of the other types is very weak.

6.3 Quantum-Mechanical Definition of the Green Function

The initial definition of the Green function used in quantum theory differs at first sight from that accepted in Chapter 4. Nevertheless, it leads to the same properties of the Green function. We illustrate this by considering an example of the Green retardation function and restrict ourselves to a scalar model.

The quantum-mechanical expression for the Green retardation function is

$$G_R(\boldsymbol{n} - \boldsymbol{n}') = -\frac{im}{\hbar}\Theta(t)\left\langle\left[u(\boldsymbol{n},t), u(\boldsymbol{n}',t)\right]\right\rangle, \tag{6.3.1}$$

where $u(\boldsymbol{n},t)$ is the atomic displacement operator at time t; $\Theta(t)$ is the discontinuous unit Heaviside function

$$\Theta(t) = \begin{cases} 0, & t < 0 \\ 1, & t > 0. \end{cases}$$

The presence of the function $\Theta(t)$ in (6.3.1) is connected with the singularity of the Green retarding function shown, e. g., in (4.6.5). It is necessary to recall the procedures used to determine the operators $u(\boldsymbol{n},t)$ in (6.3.1). Since the time dependence of the operator $u(\boldsymbol{n})$ is described by

$$-i\hbar\frac{du}{dt} = [\mathcal{H}, u] \,, \tag{6.3.2}$$

where $\mathcal{H}$ is the Hamiltonian of crystal vibrations, one can, generally, write

$$u(t) = \exp\left(\frac{i}{\hbar}\mathcal{H}t\right) u(0) \exp\left(-\frac{i}{\hbar}\mathcal{H}t\right).$$

Since the operator $u(0)$ does not commute with $\mathcal{H}$, it follows that for $t \neq 0$ it does not commute with the operator $u(t)$. However, in general

$$[u(\boldsymbol{n},t), u(\boldsymbol{n}',t)] = [u(\boldsymbol{n},0), u(\boldsymbol{n}',0)] = 0\,. \tag{6.3.3}$$

We take the time derivative of (6.3.1):

$$\begin{aligned} &i\hbar\frac{d}{dt}G_R(\boldsymbol{n}-\boldsymbol{n}',t) = m\delta(t)\left\langle\left[u(\boldsymbol{n},t),u(\boldsymbol{n}',0)\right]\right\rangle \\ &+\Theta(t)\left\langle\left[m\frac{du(\boldsymbol{n},t)}{dt},u(\boldsymbol{n}',0)\right]\right\rangle = \Theta(t)\left\langle[p(\boldsymbol{n},t),u(\boldsymbol{n}',0)]\right\rangle. \end{aligned} \tag{6.3.4}$$

We have used (6.3.3) when writing (6.3.4) and introduced the momentum operator $p(\boldsymbol{n}) = m\,du(\boldsymbol{n})/dt$. Let us differentiate the obtained relation with respect to time

$$\begin{aligned} &i\hbar\frac{d^2}{dt^2}G_R(\boldsymbol{n}-\boldsymbol{n}',t) \\ &\quad = \delta(t)\left\langle[p(\boldsymbol{n},0),u(\boldsymbol{n}',0)]\right\rangle + \Theta(t)\left\langle\left[m\frac{d^2u(\boldsymbol{n},t)}{dt^2},u(\boldsymbol{n}',0)\right]\right\rangle. \end{aligned}$$

As the operators $u(\boldsymbol{n})$ and $p(\boldsymbol{n})$ taken at the same time are canonically conjugated, then

$$[p(\boldsymbol{n},0), u(\boldsymbol{n}',0)] = -i\hbar\delta_{\boldsymbol{n}\boldsymbol{n}'}\,. \tag{6.3.5}$$

On the other hand, the operator d^2u/dt^2 can be expressed through the operator $u(\boldsymbol{n},t)$ by means of (4.5.1). By using this equation and (6.3.5), we obtain

$$\begin{aligned} &i\hbar\frac{d^2}{dt^2}G_R(\boldsymbol{n}-\boldsymbol{n}',t) \\ &\quad = -i\hbar\delta(t)\delta_{\boldsymbol{n}\boldsymbol{n}'} - \sum_{\boldsymbol{n}''}\alpha(\boldsymbol{n}-\boldsymbol{n}'')\Theta(t)\left\langle\left[u(\boldsymbol{n}'',t),u(\boldsymbol{n}',0)\right]\right\rangle. \end{aligned} \tag{6.3.6}$$

Recalling the definition (6.2.7), we come to the following equation for the function $G_R(\boldsymbol{n},t)$:

$$m\frac{d^2}{dt^2}G_R(\boldsymbol{n},t) + \sum_{\boldsymbol{n}'}\alpha(\boldsymbol{n}-\boldsymbol{n}')G_R(\boldsymbol{n}',t) = -m\delta(t)\delta_{\boldsymbol{n}0}, \tag{6.3.7}$$

which is exactly the same as (4.5.5).

Let us verify by means of a direct calculation that the Green function (6.3.1) actually determines a retarded solution to (6.3.7) with the Fourier components (4.6.5). We expand the displacements in (6.3.1) in normal vibrations, introducing the phonon creation and annihilation operators and taking into account their commutation relations as well as the time dependence (6.1.19). The following formula is then easily obtained

$$\left[u(\boldsymbol{n},t), u(\boldsymbol{n}',0)\right] = -\frac{i\hbar}{mN}\sum_{\boldsymbol{k}} \frac{\sin\omega(\boldsymbol{k})t}{\omega(\boldsymbol{k})} e^{i\boldsymbol{k}\boldsymbol{r}(\boldsymbol{n}-\boldsymbol{n}')}. \tag{6.3.8}$$

Furthermore, (6.3.8) for $t = 0$ justifies the rule (6.3.3).

Substituting (6.3.8) into (6.3.1) we write

$$G_R(\boldsymbol{n},t) = -\frac{1}{N}\sum_{\boldsymbol{k}} \Theta(t) \frac{\sin\omega(\boldsymbol{k})t}{\omega(\boldsymbol{k})} e^{i\boldsymbol{k}\boldsymbol{r}(\boldsymbol{n}-\boldsymbol{n}')}.$$

The latter has the form of a spatial Fourier expansion of the Green function with Fourier components (4.6.5).

We emphasize the fact that (6.3.8) is valid for any vibrational crystal state. Thus, the averaging contained in the definition (6.3.1) can be carried out both in the sense of quantum mechanics and thermodynamics.

6.4
Displacement Correlator and the Mean Square of Atomic Displacement

The quantum definition of the Green function (6.3.1) involves averages such as $\langle u(\boldsymbol{n},t)u(\boldsymbol{n}',0)\rangle$, determining the correlation of atomic displacements at different crystal lattice sites. Some physical properties of the crystal are generated by this correlator. A *pair correlation function* of displacements (or simply a *pair correlator*) will be referred to as the average

$$\Phi^{jl}_{ss'}(\boldsymbol{n}-\boldsymbol{n}',t) = \langle u^j_s(\boldsymbol{n},t)u^l_{s'}(\boldsymbol{n}',0)\rangle, \tag{6.4.1}$$

where both the spatial homogeneity of an unbounded crystal and time uniformity are taken into account.

Using (6.1.16), we go over from the displacements to the operators $a_\alpha(\boldsymbol{k})$, $a^\dagger_\alpha(\boldsymbol{k})$ and use the properties of their mean values and also of the polarization vectors

$$\begin{aligned}\Phi^{jl}_{ss'}(\boldsymbol{n},t) = \frac{\hbar}{2N\sqrt{m_s m_{s'}}}\sum_{\alpha\boldsymbol{k}} \frac{e^j_s(\alpha)e^j_{s'}(\alpha)}{\omega_\alpha}&\Big[N_\alpha(\boldsymbol{k})e^{i\omega_\alpha t - i\boldsymbol{k}\boldsymbol{r}(\boldsymbol{n})}\\ &+ \left(N_\alpha(\boldsymbol{k})+1\right)e^{-i\omega_\alpha t + i\boldsymbol{k}\boldsymbol{r}(\boldsymbol{n})}\Big],\end{aligned} \tag{6.4.2}$$

where $\omega_\alpha = \omega_\alpha(\boldsymbol{k})$ and $e(\alpha) = e(\boldsymbol{k},\alpha) = e(-\boldsymbol{k},\alpha)$.

After thermodynamic averaging when $\langle\langle N(\boldsymbol{k})\rangle\rangle$ is determined by (6.2.7), we obtain

$$\langle\langle \Phi^{jl}_{ss'}(\boldsymbol{n},t)\rangle\rangle = \frac{\hbar}{2N\sqrt{m_s m_{s'}}}\sum_{\alpha k}\frac{e^j_s(\alpha)e^l_{s'}(\alpha)}{\omega_\alpha} \times \left\{\coth\left(\frac{\hbar\omega_\alpha}{2T}\right)\cos\left[\omega_\alpha t - \boldsymbol{kr}(\boldsymbol{n})\right] - i\sin\left[\omega_\alpha t - \boldsymbol{kr}(\boldsymbol{n})\right]\right\}. \tag{6.4.3}$$

It is clear that the correlation function is complex. The availability term in (6.4.3) is of quantization origin. Indeed, in the classical limit $\hbar \to 0$ when $\hbar\coth(\hbar\omega/2T) \to (2T/\omega)$, there remains

$$\langle\langle \Phi^{jl}_{ss'}(\boldsymbol{n},t)\rangle\rangle = \frac{T}{N\sqrt{m_s m_{s'}}}\sum_{\alpha k}\frac{e^j_s(\alpha)e^l_{s'}(\alpha)}{\omega^2_\alpha}\cos[\omega_\alpha t - \boldsymbol{kr}(\boldsymbol{n})].$$

It follows from the definition (6.4.1) and the relation (6.4.2) that the mean square of atomic displacement at any site (independent, naturally, of the site number) equals $\Phi^{ll}{}_{ss}(0,0)$:

$$\langle u^2_s(\boldsymbol{n})\rangle = \langle u^2_s\rangle = \frac{\hbar}{m_s N}\sum_{\alpha k}\frac{|e_s(\boldsymbol{k},\alpha)|^2}{\omega_\alpha(\boldsymbol{k})}\left(N_\alpha(\boldsymbol{k}) + \frac{1}{2}\right). \tag{6.4.4}$$

In a monatomic crystal lattice there is no index s, so that one can use the relation (1.3.2) leading to

$$\langle u^2\rangle = \frac{\hbar}{mN}\sum_{\alpha k}\frac{1}{\omega_\alpha(\boldsymbol{k})}\left[N_\alpha(\boldsymbol{k}) + \frac{1}{2}\right]. \tag{6.4.5}$$

To characterize the atomic displacements in the excited crystal states it is reasonable to consider the mean thermodynamic values of the displacement squares at nonzero absolute temperature T of the crystal. The averaging (6.4.5) over the phonon equilibrium distribution and performing an integration over frequencies gives the mean square of thermal atomic displacements in a monatomic lattice

$$\langle\langle u^2\rangle\rangle = \frac{\hbar}{2mN}\int\frac{\nu(\omega)}{\omega}\coth\left(\frac{\hbar\omega}{2T}\right)d\omega. \tag{6.4.6}$$

The expression (6.4.6) is much simplified at high temperatures when the ratio $T/\hbar$ is much higher than all possible frequencies of crystal vibrations, i.e., $T \gg \hbar\omega$. Indeed, in this case

$$\langle\langle u^2\rangle\rangle = \frac{T}{mN}\int\frac{\nu(\omega)}{\omega^2}d\omega. \tag{6.4.7}$$

A simple (linear) temperature dependence of the mean square of atomic displacement at high temperatures remains in a polyatomic lattice, but the expression for $\langle\langle u^2\rangle\rangle$ becomes much more complicated. It follows from (6.4.4) that

$$\langle\langle u^2\rangle\rangle = \frac{V_0}{(2\pi)^3}\frac{\hbar}{2m_s}\int\frac{d\omega}{\omega}\coth\left(\frac{\hbar\omega}{2T}\right)\sum_\alpha\oint_{\omega_\alpha(\boldsymbol{k})=\omega}\frac{dS_\alpha}{v}|e_s(\boldsymbol{k},\alpha)|^2,$$

where the last integral is calculated over the isofrequency surface of the α-th branch of vibrations: $\omega_\alpha(\boldsymbol{k}) = \omega = \text{const}$.

The dependence of the mean square of atomic displacement on the crystal dimension is of interest. In a 3D crystal, for $\omega \to 0$ the frequency distribution function vanishes according to $\nu(\omega) \sim \omega^2$. Thus, the integrals in (6.4.6), (6.4.7) for a 3D crystal are finite.

In a 2D crystal $\nu(\omega) \sim \omega$, and the integral (6.4.7) as well as (6.4.6) for $T \neq 0$ diverge logarithmically at the low limit. Consequently, the value of the mean thermal atomic displacement becomes arbitrarily large. It may be said that the thermal fluctuations destroy the long-range order in an unbounded 2D crystal. We stipulate that the crystal is unbounded for the following reason. If we exclude from our treatment rigid-body translation of the crystal ($\boldsymbol{k} = 0$), the minimum value k_{min} according to (2.5.3) can be estimated to be the order of magnitude $k_{\text{min}} \sim \pi/L$ where L is the crystal dimension. Thus, $\omega_{\text{min}} \sim S\pi/L \sim \omega m a/L$ and the logarithmic divergence of the above integrals for a 2D crystal means that $\langle u^2 \rangle \propto \ln(L/a)$. This is a rather weak dependence on L, and the general condition that the crystal specimen is macroscopic ($L \gg a$) is insufficient to assume large fluctuations. An extremely rapid increase of the fluctuations takes place only for $\ln(L/a) \gg 1$, i. e., in fact for an unbounded crystal.

If $T = 0$ the integral (6.4.6) remains finite. In other words, zero vibrations do not break the long-range order in a 2D crystal.

Finally, for a 1D crystal $\nu(0) \neq 0$. Hence, the integral (6.4.6) diverges at any temperature – the mean atomic displacement value is infinite. Thus, the long-range order in a 1D crystal is broken both by thermal and zero vibrations. The absence of a Plank constant in (6.4.7) makes it possible to conclude that at high temperatures the quantization of vibrations is not essential and to describe the averaged atomic motions in the lattice one can use the classical representations.

6.5
Atomic Localization near the Crystal Lattice Site

At the end of the previous section the mean square of an atomic displacement from equilibrium was calculated. However, a detailed description of localized atomic motion in the crystal is given by the distribution function of its coordinate, i. e., the probability density of the random value $u_s(\boldsymbol{n})$.

Let the function $P(u)du$ determine the probability for u to be in the interval $(u, u + du)$ for $du \to 0$. We consider the Fourier transformation of the function $P(u)$ that is sometimes called the characteristic function

$$\sigma(g) = \int_{-\infty}^{\infty} P(u) e^{igu}\, du, \qquad P(u) = \frac{1}{2\pi} \int_{-\infty}^{\infty} \sigma(g) e^{-igu} dg. \tag{6.5.1}$$

From the definition of $P(u)$, the value $\sigma(q)$ represents the mean value of the function $\exp(iqu)$. If we now assume that $P(u)$ gives the density of a thermodynamic probability in a system with a given temperature then it is possible to write $\sigma(q) = \langle\langle \exp(iqu) \rangle\rangle$.

We begin by analyzing a scalar crystal model and suppose that the random value u is the atomic displacement relative to the site with number $\boldsymbol{n}$:

$$\sigma(g) = \langle\langle \exp[igu(\boldsymbol{n})] \rangle\rangle. \tag{6.5.2}$$

It is clear that the average (6.5.2) is independent of the site number. Therefore, we may set $\boldsymbol{n} = 0$, combining the site chosen with the coordinate origin.

Let us calculate the average (6.5.2) in the occupation-number representation. We go over in (6.5.2) to the phonon creation and annihilation operators, writing

$$gu(0) = \sum_{\boldsymbol{k}} \left[C(\boldsymbol{k}) a_{\boldsymbol{k}} + C^*(\boldsymbol{k}) a_{\boldsymbol{k}}^{\dagger} \right], \tag{6.5.3}$$

where the number function $C(\boldsymbol{k})$ is real in the given case

$$C(\boldsymbol{k}) = g \sqrt{\frac{\hbar}{2mN\omega(\boldsymbol{k})}}. \tag{6.5.4}$$

Since the operators $a_{\boldsymbol{k}}$ and $a_{\boldsymbol{k}}^{\dagger}$ with different $\boldsymbol{k}$ commute, (6.5.2) can be written in a harmonic approximation as a product of the multipliers that contain the averaging over states of one of the normal coordinates. Indeed, the Hamiltonian (6.1.9) is the sum of independent operators with different $\boldsymbol{k}$ and, thus

$$\sigma(g) = \prod_{\boldsymbol{k}} \langle\langle e^{i[C(\boldsymbol{k}) a_{\boldsymbol{k}} + C^*(\boldsymbol{k}) a_{\boldsymbol{k}}^{\dagger}]} \rangle\rangle. \tag{6.5.5}$$

To estimate this value we use a simple method. We take into account that in accordance with (6.5.4), all coefficients $C(\boldsymbol{k})$ are proportional to $1/\sqrt{N}$, and for a macroscopic crystal N is extremely large. Proceeding from this, we expand the exponent in (6.5.5) as a power series of $C(\boldsymbol{k})$, up to terms of the third order of smallness. We then remember that nonzero diagonal elements can be present only in the products of an even number of operators $a_{\boldsymbol{k}}$ and $a_{\boldsymbol{k}}^{\dagger}$

$$\begin{aligned} \sigma(g) = & \prod_{\boldsymbol{k}} \left[1 - \frac{1}{2} |C(\boldsymbol{k})|^2 \langle\langle a_{\boldsymbol{k}}^{\dagger} a_{\boldsymbol{k}} + a_{\boldsymbol{k}} a_{\boldsymbol{k}}^{\dagger} \rangle\rangle \right] \\ = & \prod_{\boldsymbol{k}} \left[1 - \frac{1}{2} |C(\boldsymbol{k})|^2 \langle\langle 2N_{\boldsymbol{k}} + 1 \rangle\rangle \right] \\ = & \prod_{\boldsymbol{k}} \left\{ 1 - \frac{1}{2} |C(\boldsymbol{k})|^2 \coth\left[\frac{\hbar\omega(\boldsymbol{k})}{2T} \right] \right\}. \end{aligned} \tag{6.5.6}$$

Using now the equality

$$\lim_{N\to\infty} \prod_{n=1}^{N} \left(1 - \frac{1}{N} \xi_n \right) = \exp\left(- \lim_{n\to\infty} \frac{1}{N} \sum_{n=1}^{N} \xi_n \right),$$

we transform the product (6.5.6) into the sum

$$\sigma(g) = \exp\left\{-\frac{1}{2}\sum_k |C(\boldsymbol{k})|^2 \coth\left[\frac{\hbar\omega(\boldsymbol{k})}{2T}\right]\right\}. \tag{6.5.7}$$

We substitute into (6.5.7) the explicit expression (6.5.4) and compare the exponent (6.5.7) with (6.4.5) for a scalar model. It is clear that

$$\sigma(g) = \exp\left(-\frac{1}{2}g^2\langle\langle u^2\rangle\rangle\right). \tag{6.5.8}$$

Although we derived (6.5.8) with respect to a scalar model, after obvious generalization it remains valid for any crystal lattice:

$$\sigma(g) = \langle\langle e^{i\boldsymbol{g}\boldsymbol{u}_s(\boldsymbol{n})}\rangle\rangle = \exp\left(-\frac{1}{2}\langle\langle(\boldsymbol{g}\boldsymbol{u}_s)^2\rangle\rangle\right). \tag{6.5.9}$$

Thus, the quantity $u_j(n)$ has the Gaussian probability density

$$P(u) = \frac{1}{\sqrt{2\pi\langle\langle u_j^2\rangle\rangle}} \exp\left(-\frac{1}{2}\frac{u^2}{\langle\langle u_j^2\rangle\rangle}\right). \tag{6.5.10}$$

In a classical theory of the crystal lattice, this result was obtained by Debye (1914) and Wailer (1925); the quantum derivation was first made by Ott (1953).

The probability density of an atomic displacement from its crystal lattice site is strongly temperature dependent, since $\langle\langle u^2\rangle\rangle$ is a function of temperature. The most remarkable fact here is that the atom exhibits an appreciable probability to be displaced from its equilibrium position even at $T = 0$ K due to the zero vibrations. Thus, although the energy of zero vibrations may not be manifest, the zero motion associated with it is accessible to observation.

6.6 Quantization of Elastic Deformation Field

In studying classical mechanics of a crystal lattice, it has been established that in the long-wave limit ($ak \ll 1$) the equations of crystal motion transform into the dynamic equations of elasticity theory. Such a transition corresponds formally to the limit $a \to 0$ and establishes a relation between the mechanics of a (crystal) discrete system and that of a continuous medium (continuum).

On the other hand, the quantum equations for crystal motion transform into the classical dynamic equations by passing to the limit $\hbar \to 0$ (if $\alpha = \text{const}$). We clarify whether the limiting transition $\alpha \to 0$ (if $\hbar = \text{const}$) is meaningful, as the lattice constant a and the Planck constant $\hbar$ are considered in the crystal quantum theory as two independent parameters.

The transition from the quantum mechanics of a discrete crystal lattice to the quantum field theory of elastic deformations can be carried out using a scalar crystal model. This allows one to exclude cumbersome expressions and calculations that refer to a vector quantum field.

A classical field of elastic deformations is the field of the function of coordinates and time $u(\boldsymbol{r}, t)$. The Lagrange function of this field is given by (2.9.20) and corresponds to the following density of the Lagrange function

$$\mathcal{L} = \frac{1}{2}\rho\left(\frac{\partial u}{\partial t}\right)^2 - \frac{1}{2}G\left(\frac{\partial u}{\partial x_i}\right)^2. \tag{6.6.1}$$

It is easy to write the energy of a classical elastic field as

$$E = \int \left(\frac{1}{2}\rho\left(\frac{\partial u}{\partial t}\right)^2 + \frac{1}{2}G\left(\frac{\partial u}{\partial x_i}\right)^2\right) dV. \tag{6.6.2}$$

We also find the momentum of the deformation field. By definition, the field momentum

$$\boldsymbol{P} = -\int \frac{\partial \mathcal{L}}{\partial\left(\frac{\partial u}{\partial t}\right)} \operatorname{grad} u \, dV, \tag{6.6.3}$$

and this definition is independent of a specific form of the Lagrange function density. Using (6.6.1) we obtain

$$\boldsymbol{P} = -\int \rho\frac{\partial u}{\partial t} \operatorname{grad} u \, dV. \tag{6.6.4}$$

We note that the field momentum (6.6.4) does not coincide with the momentum of atoms involved in motion by elastic crystal vibrations. Indeed the momentum (6.6.4), as well as the energy (6.6.2) and the Lagrange function (2.9.20) in the harmonic approximation is quadratic in the derivatives of the function u, and the momentum of the atoms is proportional to the first degree of atomic velocity and is, thus, linear in the time derivatives of the atomic displacement vector.

According to the results of quantizing crystal lattice motion in quantum mechanics, the function $u(\boldsymbol{r})$ is replaced by an operator dependent on coordinates as on parameters. The occupation number representation is the simplest way to write this operator, since we know its expansion (6.1.15) in the lattice.

If we take explicitly into account the finite dimensions of the volume V, then the formula (6.1.15) may also be used in the case of an elastic deformation field by replacing $mN \to \rho V$ and assuming the radius vector $\boldsymbol{r} = \boldsymbol{r}(\boldsymbol{n})$ to be a continuously varying quantity:

$$u(\boldsymbol{r}) = \sqrt{\frac{\hbar}{2\rho V}} \sum_k \frac{1}{\sqrt{\omega(\boldsymbol{k})}} (a_k + a_{-k}^\dagger) e^{ikr}, \tag{6.6.5}$$

where the operators a_k and $a_k^\dagger$ obey the commutation relations (6.1.8).

The expansion (6.6.5) is usually performed not in a finite volume, but in all coordinate space, ($V \to \infty$), which results in a continuous $\boldsymbol{k}$-space. In this case, instead of the sum over discrete k in (6.6.5), it is necessary to introduce an integral, using the rule (2.5.4), and to redefine the phonon creation and annihilation operators

$$a_{\boldsymbol{k}} = \left(\frac{2\pi}{L}\right)^{3/2} b(\boldsymbol{k}), \qquad a^{\dagger}_{\boldsymbol{k}} = \left(\frac{2\pi}{L}\right)^{3/2} b^{\dagger}(\boldsymbol{k}) \tag{6.6.6}$$

(we have denoted $V = L^3$). Simultaneously it is necessary to replace the Kronecker symbol in (6.1.8) by the δ-function

$$\delta_{\boldsymbol{k}\boldsymbol{k}'} \to \frac{(2\pi)^3}{V} \delta(\boldsymbol{k} - \boldsymbol{k}').$$

The operators $b(\boldsymbol{k})$ and $b^{\dagger}(\boldsymbol{k})$ have commutation relations that directly generalize (6.1.8)

$$[b(\boldsymbol{k}), b^{\dagger}(\boldsymbol{k})] = \delta(\boldsymbol{k} - \boldsymbol{k}'), \ \ [b(\boldsymbol{k}), b(\boldsymbol{k}')] = [b^{\dagger}(\boldsymbol{k}), b^{\dagger}(\boldsymbol{k}')] = 0. \tag{6.6.7}$$

On performing the above renormalization, we obtain

$$u(\boldsymbol{r}) = \frac{1}{(2\pi)^{3/2}} \sqrt{\frac{\hbar}{2\rho}} \int \frac{d^3\boldsymbol{r}}{\sqrt{\omega(\boldsymbol{k})}} [b(\boldsymbol{k}) + b^{\dagger}(-\boldsymbol{k})] e^{i\boldsymbol{k}\boldsymbol{r}}. \tag{6.6.8}$$

As a result of the quantization, the field function $u(\boldsymbol{r}, t)$ describing the elastic displacements of atoms and satisfying the dynamic equation (2.9.21) is represented as an integral of linear form in the operators $b(\boldsymbol{k})$ and $b^{\dagger}(\boldsymbol{k})$. Thus, it becomes an operator.

The time evolution of the operator $u(\boldsymbol{r}, t)$ is determined by the time dependence of the operators $b(\boldsymbol{k})$ and $b^{\dagger}(\boldsymbol{k})$, and the latter in a harmonic approximation is based on the relations

$$b_{\boldsymbol{k}}(t) = b_{\boldsymbol{k}}(0) e^{-i\omega(\boldsymbol{k})t}, \qquad b^{\dagger}_{\boldsymbol{k}}(t) = b^{\dagger}_{\boldsymbol{k}}(0) e^{i\omega(\boldsymbol{k})t},$$

where $\omega(\boldsymbol{k}) = sk$.

Using (6.6.8) and the expressions for the energy (6.6.2), it is easy to construct the Hamiltonian of a vibrating continuum

$$\mathcal{H} = E_0 + \int \hbar s k b^{\dagger}(\boldsymbol{k}) b(\boldsymbol{k}) \, d^3\boldsymbol{k}, \tag{6.6.9}$$

where E_0 is the energy of zero vibrations.

The eigenstates of the operator (6.6.9) contain quanta (phonons) with rigorously fixed quasi-momenta. The wave functions of these states in the coordinate representation correspond to plane waves.

We shall clarify ultimately the quantum meaning of the field momentum operator (6.6.3) or (6.6.4). We calculate the spatial and time derivatives of (6.6.5)

$$\begin{aligned} \operatorname{grad} u &= i\sqrt{\frac{\hbar}{2\rho V}}\sum_k \frac{\boldsymbol{k}}{\sqrt{\omega(\boldsymbol{k})}}\left(a_k + a^\dagger_{-k}\right)e^{i\boldsymbol{kr}}; \\ \frac{\partial u}{\partial t} &= i\sqrt{\frac{\hbar}{2\rho V}}\sum_k \sqrt{\omega(\boldsymbol{k})}\left(a^\dagger_{-k} - a_k\right)e^{i\boldsymbol{kr}}. \end{aligned} \tag{6.6.10}$$

We substitute (6.6.10) into (6.6.4)

$$\begin{aligned} \boldsymbol{P} = &\ \frac{1}{2}\sum_{kk'}\hbar\boldsymbol{k}\sqrt{\frac{\omega(\boldsymbol{k})}{\omega(\boldsymbol{k}')}}\left(a^\dagger_{-k} - a_k\right)\left(a_{k'} + a^\dagger_{-k'}\right) \\ &\times\frac{1}{V}\int_{-\infty}^{\infty} e^{i(\boldsymbol{k}+\boldsymbol{k}')\boldsymbol{r}}\,dV \\ = &\ \frac{1}{2}\sum_k \hbar\boldsymbol{k}\left(a_k a_k^\dagger + a_k^\dagger a_k\right) + \frac{1}{2}\sum_k \hbar\boldsymbol{k}\left(a^\dagger_{-k}a^\dagger_k + a_k a_{-k}\right). \end{aligned} \tag{6.6.11}$$

The last sum in (6.6.11) vanishes, as its terms are the odd functions of $\boldsymbol{k}$, and the first sum transforms trivially

$$\boldsymbol{P} = \sum_k \hbar\boldsymbol{k} a_k^\dagger a_k = \int \hbar\boldsymbol{k} b^\dagger(\boldsymbol{k}) b(\boldsymbol{k})\, d^3\boldsymbol{k}. \tag{6.6.12}$$

It follows from (6.6.12) that the operator of the field momentum $\boldsymbol{P}$ is the operator of the total quasi-momentum of a vibrating crystal.

Writing the operators (6.6.8), (6.6.9), (6.6.12) and also the commutation rules (6.6.7) accomplishes the quantization of a field of elastic deformations.

7
Interaction of Excitations in a Crystal

7.1
Anharmonicity of Crystal Vibrations and Phonon Interaction

In a harmonic approximation the phonons are noninteracting quasi-particles. However, the situation is changed if the anharmonicity of crystal vibrations is taken into account – the phonon gas ceases to be ideal.

To clarify the qualitative role of the anharmonicity, we consider an unbounded crystal with a monatomic spatial lattice. We assume the anharmonicity to be small and in the expansion of the potential crystal energy in power of the displacement, we restrict ourselves to cubic terms:

$$
\begin{aligned}
U = U_0 \quad & + \frac{1}{2} \sum \alpha^{ik}(\boldsymbol{n} - \boldsymbol{n}') u^{ik}(\boldsymbol{n}) u^k(\boldsymbol{n}') \\
& + \frac{1}{3} \sum \tilde{S} \left\{ \gamma^{ikl}(\boldsymbol{n} - \boldsymbol{n}', \boldsymbol{n} - \boldsymbol{n}'') \left[u^i(\boldsymbol{n}) - u^i(\boldsymbol{n}') \right] \right. \\
& \left. \times \left[u^k(\boldsymbol{n}) - u^k(\boldsymbol{n}'') \right] \left[u^l(\boldsymbol{n}') - u^l(\boldsymbol{n}'') \right] \right\},
\end{aligned}
\tag{7.1.1}
$$

where $\tilde{S}$ is the symmetrization operation in number vectors $\boldsymbol{n}$, $\boldsymbol{n}'$, $\boldsymbol{n}''$. This takes into account at once the potential energy invariance with respect to crystal motion as a whole. The coefficients $\gamma^{ikl}(\boldsymbol{n}, \boldsymbol{n}')$ characterize the intensity of the crystal vibration anharmonicity. Generally, they are of the order of magnitude $\gamma \sim \alpha / a$.

After introducing normal coordinates and quantizing the vibrations, the quadratic term in (7.1.1) will belong to the Hamiltonian of the phonon ideal gas. The cubic term in (7.1.1) will be quantized by (6.1.16). Denoting the corresponding term in the

The Crystal Lattice: Phonons, Solitons, Dislocations, Superlattices, Second Edition. Arnold M. Kosevich

ISBN: 3-527-40508-9

Hamiltonian by $\mathcal{H}_{\text{int}}$, we obtain

$$\mathcal{H}_{\text{int}} = \left(\frac{\hbar}{2mN}\right)^{3/2} \sum \frac{e_i e'_j e''_l}{\sqrt{\omega\omega'\omega''}} \Gamma_{ijl}(\boldsymbol{k}, \boldsymbol{k}', \boldsymbol{k}'') \left[a_\alpha(\boldsymbol{k}) + a^\dagger_\alpha(-\boldsymbol{k})\right]$$
$$\times \left[a_{\alpha'}(\boldsymbol{k}') + a^\dagger_{\alpha'}(-\boldsymbol{k}')\right] \left[a_{\alpha''}(\boldsymbol{k}'') + a^\dagger_{\alpha''}(-\boldsymbol{k}'')\right] \sum_n e^{i(\boldsymbol{k}+\boldsymbol{k}'+\boldsymbol{k}'')\boldsymbol{r}(\boldsymbol{n})}, \tag{7.1.2}$$

where $\omega = \omega_\alpha(\boldsymbol{k})$, $\omega' = \omega_{\alpha'}(\boldsymbol{k}')$, $\omega'' = \omega_{\alpha''}(\boldsymbol{k}'')$, $\boldsymbol{e} = \boldsymbol{e}_\alpha(\boldsymbol{k})$, $\boldsymbol{e}' = \boldsymbol{e}_{\alpha'}(\boldsymbol{k}')$, $\boldsymbol{e}'' = \boldsymbol{e}_{\alpha''}(\boldsymbol{k}'')$. The tensor function $\Gamma_{ijl}(\boldsymbol{k}, \boldsymbol{k}', \boldsymbol{k}'')$ is defined by the double sum over the sites

$$\Gamma_{ijl}(\boldsymbol{k}, \boldsymbol{k}', \boldsymbol{k}'') = \tilde{S} \sum_{\boldsymbol{n},\boldsymbol{n}'} \gamma_{ijl}(\boldsymbol{n}, \boldsymbol{n}') \left(1 - e^{-i\boldsymbol{k}\boldsymbol{r}}\right) \left(1 - e^{-i\boldsymbol{k}'\boldsymbol{r}'}\right) \left(e^{-i\boldsymbol{k}''\boldsymbol{r}} - e^{-i\boldsymbol{k}''\boldsymbol{r}'}\right),$$

where $\boldsymbol{r}' = \boldsymbol{r}(\boldsymbol{n}')$ and the symmetrization is performed in quasi-wave vectors $\boldsymbol{k}$, $\boldsymbol{k}'$, $\boldsymbol{k}''$.

If one takes the Hamiltonian (7.1.2) into account in deriving the equations of motion for the operators $a_{\boldsymbol{k}}$ and $a^\dagger_{\boldsymbol{k}}$, then instead of a system of separable equations for individual phonons such as (6.1.18), we get a nonlinear system of "coupled" equations that will relate the evolution of the phonon in the state $(\alpha\boldsymbol{k})$, with the evolution of the remaining phonons in the other states. As a result, the operator $\mathcal{H}_{\text{int}}$ can be regarded as the phonon interaction operator. Thus, the phonon interaction displays the anharmonicity of crystal vibrations.

Let us note that the last multiplier in (7.1.2) is nonzero and equals N only for $\boldsymbol{k} + \boldsymbol{k}' + \boldsymbol{k}'' = \boldsymbol{G}$, where $\boldsymbol{G}$ is any reciprocal lattice vector. Thus, (7.1.2) can be written as

$$\mathcal{H}_{\text{int}} = \frac{1}{\sqrt{N}} \sum_{\boldsymbol{k}+\boldsymbol{k}'+\boldsymbol{k}''=\boldsymbol{G}} \frac{V(\boldsymbol{k}, \boldsymbol{k}', \boldsymbol{k}'')}{\sqrt{\omega(\boldsymbol{k})\omega(\boldsymbol{k}')\omega(\boldsymbol{k}'')}}$$
$$\times (a_{\boldsymbol{k}} + a^\dagger_{-\boldsymbol{k}})(a_{\boldsymbol{k}'} + a^\dagger_{-\boldsymbol{k}'})(a_{\boldsymbol{k}''} + a^\dagger_{-\boldsymbol{k}''}), \tag{7.1.3}$$

where for simplicity, we omit the indices that number the branches of vibrations over which the summation should also be performed.

In (7.1.3), a new notation is used

$$V(\boldsymbol{k}, \boldsymbol{k}', \boldsymbol{k}'') \equiv V_{\alpha\alpha'\alpha''}(\boldsymbol{k}, \boldsymbol{k}', \boldsymbol{k}'')$$
$$= \frac{1}{N}\left(\frac{\hbar}{2m}\right)^{3/2} e^i_\alpha(\boldsymbol{k}) e^j_{\alpha'}(\boldsymbol{k}') e^l_{\alpha''}(\boldsymbol{k}'') \Gamma_{ijl}(\boldsymbol{k}, \boldsymbol{k}', \boldsymbol{k}''). \tag{7.1.4}$$

The summation in (7.1.3) is carried out under the condition

$$\boldsymbol{k} + \boldsymbol{k}' + \boldsymbol{k}'' = \boldsymbol{G}, \tag{7.1.5}$$

and since each of the vectors $\boldsymbol{k}$, $\boldsymbol{k}'$, and $\boldsymbol{k}''$ does not go beyond the unit cell then (7.1.5) reduces to the two alternative conditions

$$\boldsymbol{k} + \boldsymbol{k}' + \boldsymbol{k}'' = 0, \tag{7.1.6}$$

$$\boldsymbol{k} + \boldsymbol{k}' + \boldsymbol{k}'' = \boldsymbol{G}_0, \tag{7.1.7}$$

where $\boldsymbol{G}_0$ is a vector that can be only composed of the three fundamental vectors of a reciprocal lattice $\boldsymbol{b}_1$, $\boldsymbol{b}_2$ and $\boldsymbol{b}_3$.

If the phonon interaction takes place under the condition (7.1.6), it is called a *normal* collision (or N-process). If the condition (7.1.7) is satisfied, the interaction is called an anomalous collision or an *umklapp process* (U-process, from the German word *Umklappprozess*).

Let us note a remarkable feature of the coefficients $V(\boldsymbol{k}, \boldsymbol{k}', \boldsymbol{k}'')$ in the interaction Hamiltonian (7.1.3), associated with their behavior at small values of quasi-wave vectors. It follows from the definition (7.1.5) that a function of three independent variables $V(\boldsymbol{k}, \boldsymbol{k}', \boldsymbol{k}'')$ vanishes if at least one of these variables tends to zero. For $ak \ll 1$ this function can be represented in the form

$$V_{\alpha\alpha'\alpha''}(\boldsymbol{k}, \boldsymbol{k}', \boldsymbol{k}'') = e^i_\alpha(\boldsymbol{k}) e^j_{\alpha'}(\boldsymbol{k}') e^l_{\alpha''}(\boldsymbol{k}'') M^{npq}_{ijl} k_n k_p k_q, \tag{7.1.8}$$

where M is some tensor coefficient.

In other words, for $k, k', k'' \longrightarrow 0$ we have

$$V(\boldsymbol{k}, \boldsymbol{k}', \boldsymbol{k}'') \sim kk'k'' \sim \omega\omega'\omega''. \tag{7.1.9}$$

Such a behavior of the corresponding matrix elements in the long-wave limit is quite natural. Indeed, in view of the limit $k \longrightarrow 0$, for estimations one can make use of the results of elasticity theory. However, in elasticity theory the crystal energy is expressed through the strain tensor. The cubic anharmonicity corresponds to terms of third order in deformations in the elastic energy. The deformation tensor for a plane wave of displacements can be estimated as $\varepsilon_{ij} \sim \nabla_i u_j \sim k_i u_j$. Thus, the cubic anharmonicity is characterized by an intensity (7.1.9) at small k. This property of the quantities $V(\boldsymbol{k}, \boldsymbol{k}', \boldsymbol{k}'')$ is used essentially in evaluating the contribution of the vibration anharmonicity to the different macroscopic characteristics of the crystal.

We denote temporarily $a^-(\boldsymbol{k}) = a(\boldsymbol{k})$ and analyze the characteristic product of the phonon creation and annihilation operators that enters into the Hamiltonian (7.1.3), just a product of the type

$$a^\pm_\alpha(\boldsymbol{k}) a^\pm_{\alpha'}(\boldsymbol{k}') a^\pm_{\alpha''}(\boldsymbol{k}''). \tag{7.1.10}$$

It follows from (6.1.21) that the nonzero matrix elements of the operator (7.1.10) are proportional to the multiplier $\exp(\pm i\omega t \pm i\omega' t \pm i\omega'' t)$, whereas before, $\omega = \omega(\alpha, \boldsymbol{k})$; $\omega' = \omega(\alpha', \boldsymbol{k}')$; $\omega'' = \omega(\alpha'', \boldsymbol{k}'')$.

If we now calculate by perturbation theory the probability of a corresponding collision using the matrix element of the operator (7.1.10), then it will be proportional to the δ-function

$$\delta(\pm\omega \pm \omega' \pm \omega''). \tag{7.1.11}$$

The signs in (7.1.11) correspond to the $\pm$ signs in (7.1.10).

Let us consider for instance the term in $\mathcal{H}_{\text{int}}$ proportional to the operator $a^\dagger(\alpha, \mathbf{k})a^\dagger(\alpha_1, \mathbf{k}_1)a(\alpha_2, -\mathbf{k}_2)$. This term describes the two-phonon creation process in the states $(\alpha, \mathbf{k})$ and $(\alpha_1, \mathbf{k}_1)$ and phonon vanishing in the state $(\alpha_2, -\mathbf{k}_2)$. It gives nonzero probability under the condition

$$\omega(\alpha, \mathbf{k}) + \omega(\alpha_1, \mathbf{k}_1) = \omega(\alpha_2, \mathbf{k}_2),$$

which is equivalent to the energy conservation under the collision

$$\hbar\omega(\alpha, \mathbf{k}) = \hbar\omega(\alpha_1, \mathbf{k}_1) + \hbar\omega(\alpha_2, \mathbf{k}_2). \tag{7.1.12}$$

By virtue of the conditions (7.1.4) and relation (6.2.4) the collision occurs either with total quasi-momentum conservation (N-process, Fig. 7.1a)

$$\mathbf{p} + \mathbf{p}_1 = \mathbf{p}_2, \tag{7.1.13}$$

or when the quasi-momentum of "the phonon center of gravity" changes by $\hbar\mathbf{G}_0$ (U-process, Fig. 7.1b)

$$\mathbf{p} + \mathbf{p}_1 = \mathbf{p}_2 + \hbar\mathbf{G}_0. \tag{7.1.14}$$

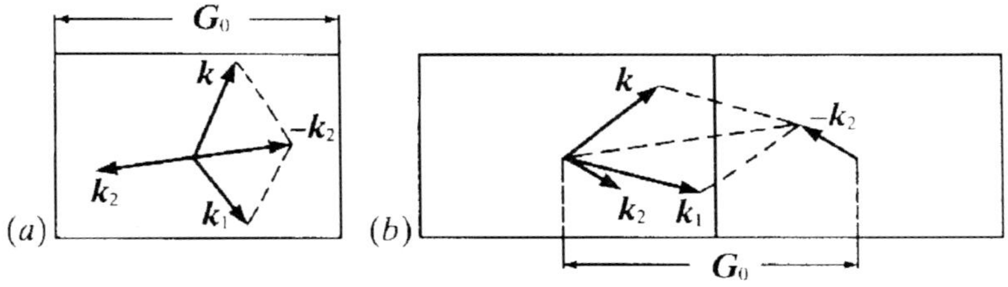

Fig. 7.1 Three-phonon processes: (*a*) is the normal process, (*b*) is the umklapp process.

The term with the operator $a^\dagger_{\mathbf{k}} a_{-\mathbf{k}'} a_{-\mathbf{k}''}$ in the interaction Hamiltonian has an analogous meaning.

7.2
The Effective Hamiltonian for Phonon Interaction and Decay Processes

The term $\mathcal{H}^{(3)}_{\text{int}}$ (cubic in anharmonicity and describing the phonon interaction) present in the crystal Hamiltonian affects both the dynamics of separate quasi-particles and the equilibrium properties of a phonon gas. In particular, the crystal energy cannot now be represented as (6.2.3). However, if the anharmonicity is small, it can be taken into account by perturbation theory in calculating the crystal energy. Small corrections to the energy are generally expressed through the displacements of the eigenfrequencies of crystal vibrations. However, when they are calculated consecutively, it is impossible to confine oneself only to cubic anharmonicity.

Indeed, for the Hamiltonian (7.1.3) we have

$$\left\langle \mathcal{H}_{\text{int}}^{(3)} \right\rangle = 0. \tag{7.2.1}$$

Thus, when taking cubic anharmonicity into account only, the correction to the first order of perturbation theory is absent. The frequency shift can be obtained in second-order perturbation theory only, but the neglected fourth-order terms in the expansion of the displacement energy would lead to the interaction Hamiltonian $\mathcal{H}_{\text{int}}^{(4)}$ involving the products of four operators a_k and $a_k^\dagger$ of the type $a_1^\pm a_2^\pm a_3^\pm a_4^\pm$. Thus, $\left\langle \mathcal{H}_{\text{int}}^{(4)} \right\rangle \neq 0$ and the corresponding anharmonicity would result in the frequency renormalization just in the first order of perturbation theory. It turns out that the corrections to the energy in first-order perturbation theory coming from $\mathcal{H}_{\text{int}}^{(4)}$ are generally of the same second-order corrections coming from $\mathcal{H}_{\text{int}}^{(3)}$.

In this situation, it makes no sense to calculate the eigenfrequency shift with the Hamiltonian (7.1.3). We only note that in calculating a similar shift in the second-order perturbation theory, all terms of the Hamiltonian (7.1.3) make nearly the same contribution to the final result. This is so because the *virtual transitions* in a system for which the phonon energy conservation law should not necessarily be satisfied are taken into account in the second-order perturbation theory. When real collisions of phonons conserving their energy are described the situation appears to be different.

Since we have defined the frequency $\omega = \omega(\alpha, \boldsymbol{k})$ as a positive quantity, the δ-function (7.1.11) may be nonzero only if the terms in its argument have different signs. Hence it follows that to describe real phonon collisions in a crystal with small anharmonicity (7.1.3), one can introduce a simpler effective interaction Hamiltonian. Indeed, if we restrict ourselves to the basic terms of the phonon interaction energy and use the first-order perturbation theory approximation (when real scattering processes are considered), we can omit in the Hamiltonian $\mathcal{H}_{\text{int}}$ the terms involving the products of the operators $a_k a_{k'} a_{k''}$ and $a_k^\dagger a_{k'}^\dagger a_{k''}^\dagger$. The remaining terms, by a simple replacement of the summation indeces, can be written in the form

$$\mathcal{H}_{\text{int}}^{\text{ef}} = \frac{1}{\sqrt{N}} \sum \frac{W(\boldsymbol{k}, \boldsymbol{k}', \boldsymbol{k}'')}{\sqrt{\omega \omega' \omega''}} (a_k^\dagger a_{k'}^\dagger a_{-k''} + a_k^\dagger a_{-k'} a_{k''}), \tag{7.2.2}$$

where the coefficients $W(\boldsymbol{k}, \boldsymbol{k}', \boldsymbol{k}'')$ differ from the $V(\boldsymbol{k}, \boldsymbol{k}', \boldsymbol{k}'')$ in numerical multipliers only, the summation is performed over the quasi-wave vectors under the condition (7.1.4) and the indices that number the vibration branches are omitted.

The permutation of the operators $a_k^\dagger$ and a_k in deriving (7.2.2) gives no additional terms in $\mathcal{H}_{\text{int}}^{\text{ef}}$, since the latter vanish by virtue of the conservation law (7.1.6) and the property (7.1.9) of the function $V(\boldsymbol{k}, \boldsymbol{k}', \boldsymbol{k}'')$.

We analyze the probability of the elementary process described by the Hamiltonian (7.2.2) by focusing on the long-wave acoustic phonons ($ak \ll 1$). We note that if the long-wave phonons only participate in the collision, the process is normal. Thus, we shall first be interested in the probability of normal "three-phonon" processes.

Similar processes take place when the conservation laws (7.1.6), (7.1.12) are obeyed

$$\boldsymbol{k}_1 = \boldsymbol{k}_2 + \boldsymbol{k}_3; \quad \omega(\alpha_1, \boldsymbol{k}_1) = \omega(\alpha_2, \boldsymbol{k}_2) + \omega(\alpha_3, \boldsymbol{k}_3). \tag{7.2.3}$$

To clear up whether all conditions (7.2.3) can be satisfied simultaneously, in particular, whether a quite definite phonon $(\alpha, \boldsymbol{k})$ always vanishes (decays) with an arbitrarily small cubic anharmonicity, we discuss this problem qualitatively. For long-wave phonons, it suffices to consider the approximation where the dispersion laws are almost the same as for sound

$$\omega_\alpha(\boldsymbol{k}) = s_\alpha k, \quad \alpha = 1, 2, 3. \tag{7.2.4}$$

In such an approximation "longitudinal phonons" (l) exist whose dispersion law is close to the isotropic one (s_l =constant) and "transverse" phonons (t) whose velocities satisfy the relation

$$s_t < s_l. \tag{7.2.5}$$

Indeed, the true dispersion law differs insignificantly from (7.2.4) even for small k. For the longitudinal phonons, as a rule, $\omega_l(\boldsymbol{k}) < s_1 k$. Thus, for the process involving only the longitudinal phonons, (7.2.3) cannot be satisfied simultaneously. For phonons of the same type in the isotropic model from (7.2.3), (7.2.4) it follows that $|\boldsymbol{k}_1 + \boldsymbol{k}_2| = k_1 + k_2$. Thus, when the isotropic dispersion law (7.2.4) is obeyed, the process we are interested in would occur only in the case of parallel vectors $\boldsymbol{k}_1, \boldsymbol{k}_2, \boldsymbol{k}_3$. We use this fact to elucidate the possibility of longitudinal phonon decay in a one-dimensional process.

The dispersion law for longitudinal phonons is given by curve 1 in Fig. 7.2. We show the point (k_1, ω_1) corresponding to the state of one of the phonons after the decay. Taking this point as the reference frame origin, starting at this point we construct the curve 2 for the same dispersion law of the second phonon after the decay. To satisfy (7.2.3), curves 1 and 2 should intersect and the intersection point will determine the state of a decaying phonon (k_2, ω_2). These curves, however, do not intersect at small k and hence, the process $l \to l + l$ is impossible. In a scalar model (for the same type of phonons) the dispersion law analyzed would be *nondecaying*.

The conclusion about the nondecaying character of the dispersion law with the property $(\partial^2\omega/\partial k^2) < 0$ is also valid for an anisotropic crystal, if the isofrequency surfaces are convex. These include isofrequency surfaces for the branch of phonons that corresponds mainly to the longitudinal polarization of vibrations. However, the

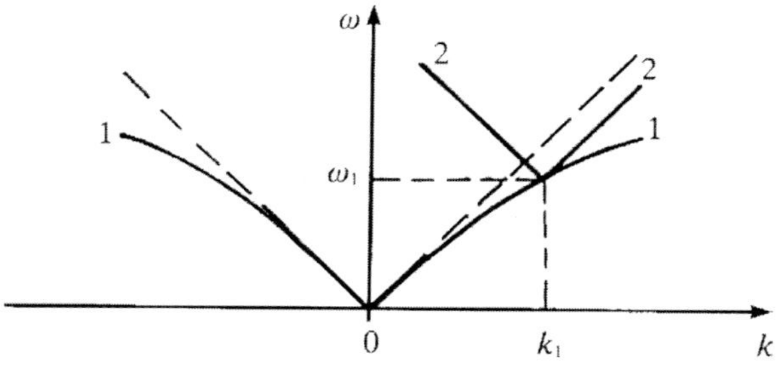

Fig. 7.2 The nondecaying dispersion law.

crystal also involves phonons of another type, namely, transverse ones. By examining the dispersion law of bending vibrations of a layered crystal (Section 1.2.5) it has been established that in certain directions the dispersion law for transverse vibrations has the property $\omega_t(\mathbf{k}) > s_t k$ (Fig. 7.3a, curve 1). Let us check that in this case the $t \rightarrow t + t$ process is possible.

We choose the required direction in reciprocal space along the k_x-axis and repeat the constructions just made above. Curves 1 and 2 may not intersect along the direction Ok_x. We consider a two-dimensional picture on the plane k_xOk_y and construct isofrequency lines corresponding to the frequency ω_2 for the dispersion law 1 in Fig. 7.3a and to the frequencies $\omega_2 - \omega$ for the dispersion law 2. These lines intersect (Fig. 7.3b) and the intersection points determine the wave vectors of a phonon capable of decaying: $\mathbf{k} = \mathbf{k_1} + \mathbf{k_2}$. Thus, the dispersion law running steeper than that of sound dispersion ($\partial^2\omega/\partial k^2 > 0$) is the *decaying* one.

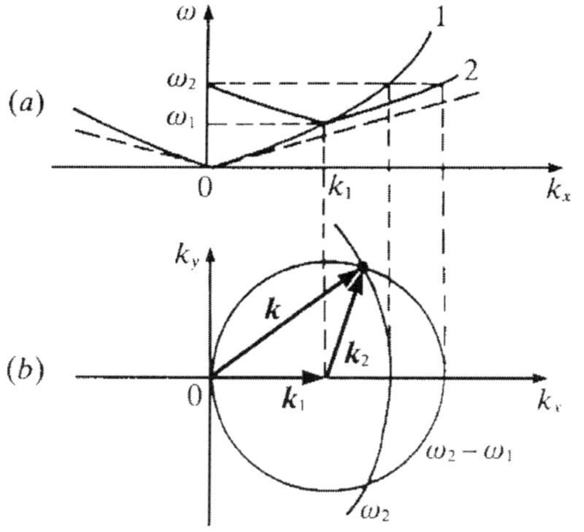

Fig. 7.3 Phonon decay into two phonons of the same branch of vibrations: (*a*) the decaying dispersion law; (*b*) the intersection of isofrequency curves.

The dispersion law may be decaying also due to strong anisotropy of isofrequency surfaces. Even if in each direction of $\mathbf{k}$-space the condition $\partial^2\omega/\partial k^2 < 0$ is satisfied, but the isofrequency surface is not convex (the cross section of such a surface is shown in Fig. 4.1) a decay such as $t \to t + t$ may be allowed.

When both longitudinal (l) and the transverse (t) phonons participate simultaneously in a three-phonon process, the conditions for its realization, i. e., the conditions for satisfying (7.2.3) to be met, are much simplified even if the dispersion laws of each type of phonons are nondecaying. To analyze these processes, we consider the dispersion laws (7.2.4) in view of the condition (7.2.5). One can see that for a one-dimensional model the processes $l \to t + t'$ and $l \to l + t$ (Fig. 7.4) are possible, while processes such as $t \to l + l'$, $t \to t + l$ are forbidden, at least in the isotropic approximation.

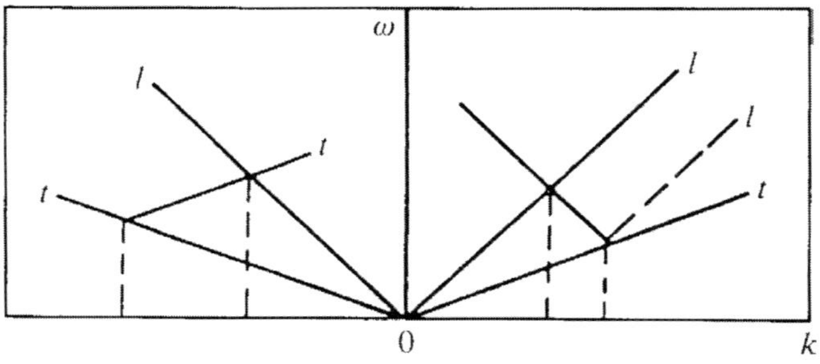

Fig. 7.4 Fulfillment of the dispersion laws for the decays $l \to t + t'$ (on the left) and $l \to l + t$ for a 1D model.

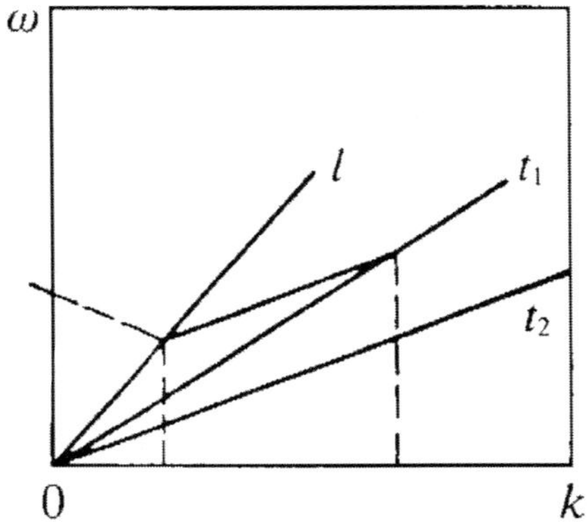

Fig. 7.5 Three-phonon process $t_1 \to l + t_2$.

However, taking into account the existence of two kinds of transverse phonons (t_1 and t_2) with different sound velocities ($s_1 \neq s_2$) the following process is possible (Fig. 7.5):

$$t_1 \to t_2 + l, \qquad s_2 < s_i < s_l.$$

We emphasize that the phonon dispersion law anisotropy assists significantly when the condition (7.2.3) is satisfied, although it does not remove all prohibitions arising in the isotropic approximation.

In conclusion we note that for all processes of single-phonon decay into two phonons there are corresponding inverse processes (fusion of two phonon into a single one). These are allowed or forbidden to the same extent as direct processes.

7.3 Inelastic Diffraction on a Crystal and Reproduction of the Vibration Dispersion Law

In elucidating the conditions for Bragg X-ray (rather soft γ-quantum) reflection, we considered the crystal as an ideal spatial lattice of fixed mathematical points. However, the scattering by a real crystal whose atoms can perform vibrational motion are distinguished from the scattering by fixed centers. First, the incident wave may excite vibrational motion in the crystal, i. e., generate phonons in it. Then, the vibrating lattice atoms are displaced from their equilibrium positions and the scattering by such a system is different from that by an ideal periodic structure (the incident wave interacts with crystal phonons). The second distinction is connected with the root-mean-square value of the displacement.

Let us discuss the first of the phenomena generated by the inelastic interaction of penetrating radiation with a vibrating crystal. In a kinematic theory, the probability of the incident beam scattering with its wave vector changing to $\boldsymbol{q} = \boldsymbol{q}_2 - \boldsymbol{q}_1$, is proportional to the squared modulus of the following integral over the sample volume (Section 0.4.5):

$$U(\boldsymbol{q}_2, \boldsymbol{q}_1) \equiv V(\boldsymbol{q}) = \int U(\boldsymbol{r}) \exp\left(-i\boldsymbol{q}\boldsymbol{r}\right) d^3\boldsymbol{r}, \tag{7.3.1}$$

where $U(\boldsymbol{r})$ is the potential of the crystal interaction with a field or with a particle. In a monatomic lattice

$$U(\boldsymbol{r}) = \sum_n U_0\left[\boldsymbol{r} - \boldsymbol{r}(\boldsymbol{n}) - \boldsymbol{u}(\boldsymbol{n})\right], \tag{7.3.2}$$

and $U_0(\boldsymbol{r} - \boldsymbol{\xi})$ describes the interaction between the penetrating radiation and a single atom at the point $\boldsymbol{\xi}$. We substitute (7.3.2) into (7.3.1)

$$V(\boldsymbol{q}) = F_0(\boldsymbol{q}) \sum_n \exp\left\{-i\boldsymbol{q}\left[\boldsymbol{r}(\boldsymbol{n}) + \boldsymbol{u}(\boldsymbol{n})\right]\right\}. \tag{7.3.3}$$

Here the notation $F_0(\boldsymbol{q})$ is introduced for the *atomic scattering factor*

$$F_0(\boldsymbol{q}) = \int U_0(\boldsymbol{r}) e^{-i\boldsymbol{q}\boldsymbol{r}} d^3\boldsymbol{r}.$$

The sum over the lattice sites that is a multiplier in (7.3.3) will be called the *structure factor* and is denoted as $S(\boldsymbol{q})$. Usually the structure factor is determined as this sum divided into the number of crystal sites N.

We try to calculate the structure factor using the smallness of atomic displacements. As, by assumption, $u \ll a$ and we take interest in the wave vectors $\boldsymbol{q}$ whose modulus does not exceed the order of magnitude of π/a, we expand each term of the structure factor as a power series of $\boldsymbol{u}$, restricting ourselves to the linear terms

$$S(\boldsymbol{q}) = \sum_n e^{-i\boldsymbol{q}[\boldsymbol{r}(\boldsymbol{n})+\boldsymbol{u}(\boldsymbol{n}))]} = \sum_n e^{-i\boldsymbol{q}\boldsymbol{r}(\boldsymbol{n})} - i\boldsymbol{q}\sum_n \boldsymbol{u}(\boldsymbol{n})e^{-i\boldsymbol{q}\boldsymbol{r}(\boldsymbol{n})}. \tag{7.3.4}$$

The first term on the r.h.s. of (7.3.4) describes the diffraction pattern for an ideal periodic structure (Section 0.5). It gives the amplitude of scattering by a crystal for which $\boldsymbol{q} = \boldsymbol{G}$, where $\boldsymbol{G}$ is any reciprocal lattice vector.

It is essential that the crystal state under this scattering does not change.

We now study the second (linear in displacement) term on the r.h.s. of (7.3.4), having denoted it as $M(\boldsymbol{q})$. To understand the physical meaning of this term in the structure factor, we consider $M(\boldsymbol{q})$ as an operator in occupation-number space and express the displacement vector $\boldsymbol{u}(\boldsymbol{n})$ through the phonon creation and annihilation operators (6.1.16)

$$\begin{aligned} M(\boldsymbol{q}) &= -i\boldsymbol{q}\sum_n \boldsymbol{u}(\boldsymbol{n})e^{-i\boldsymbol{q}\boldsymbol{r}(\boldsymbol{n})} = -i\left(\frac{\hbar}{2mN}\right)^{1/2} \\ &\times \sum_{k\alpha} \frac{\boldsymbol{q}\boldsymbol{e}(\alpha,\boldsymbol{k})}{\sqrt{\omega_\alpha(\boldsymbol{k})}} \sum_n \left[a_\alpha(\boldsymbol{k})e^{-i(\boldsymbol{q}-\boldsymbol{k})\boldsymbol{r}(\boldsymbol{n})} + a_\alpha^\dagger(\boldsymbol{k})e^{-i(\boldsymbol{q}+\boldsymbol{k})\boldsymbol{r}(\boldsymbol{n})}\right]. \end{aligned} \tag{7.3.5}$$

On summing $\boldsymbol{n}$ with (0.4.14) taken into account, it is clear that in (7.3.5) there are terms of two kinds. Some of them involve the operator $a_{\boldsymbol{k}}$ as a multiplier and describe the scattering process accompanied with the absorption (annihilation) of a phonon with the quasi-wave vector $\boldsymbol{k}$, so that

$$\boldsymbol{q}_2 - \boldsymbol{q}_1 = \boldsymbol{k} + \boldsymbol{G}. \tag{7.3.6}$$

The other terms involve the operator $a_{\boldsymbol{k}}^\dagger$ as a multiplier and are responsible for the phonon radiation (creation) obeying the dispersion law

$$\boldsymbol{q}_1 - \boldsymbol{q}_2 = \boldsymbol{k} + \boldsymbol{G}. \tag{7.3.7}$$

The matrix elements of the operator $M(\boldsymbol{q})$ thus determine the amplitude of inelastic scattering involving one phonon (the amplitude of a *one-phonon process*). It follows from (7.3.6), (7.3.7) that one-phonon scattering may take place either as an N- or a U-process.

We now obtain fairly straightforward energy conservation laws that govern the processes (7.3.6), (7.3.7) and also look at the one-phonon scattering from another point of view. Since we regard $V(\boldsymbol{q})$ as the energy operator of crystal interaction with an incident beam, the probability of the beam scattering in the first-order perturbation theory is expressed through the squared matrix element $\langle f\,|V(\boldsymbol{q})|\,i\rangle$ for a transition from the initial crystal state i with a set of occupation numbers for phonons $\{N_k\}$ into the final state f with the occupation numbers $\{N'_k\}$. Let E_i and E_f be the crystal energy in the harmonic approximation before and after scattering:

$$E_i = \sum_{k\alpha} \left(N_\alpha(\boldsymbol{k}) + \frac{1}{2} \right) \hbar\omega_\alpha(\boldsymbol{k}), \quad E_f = \sum_{k\alpha} \left(N'_\alpha(\boldsymbol{k}) + \frac{1}{2} \right) \hbar\omega_\alpha(\boldsymbol{k}).$$

Furthermore we introduce the notation where $E_1 = E(\hbar q_1)$ and $E_2 = E(\hbar q_2)$ for the energies of an incident particle (γ-quantum, electron or neutron) before and after scattering, and $E(\boldsymbol{p})$ is the energy of a scattered particle as a function of its momentum.

In the first Born approximation, the probability of the separate scattering by a crystal in the state i is proportional to

$$w(\boldsymbol{q}, E) = \sum_f |\langle f| V(\boldsymbol{q}) |i\rangle|^2 \delta(E + E_f - E_i), \tag{7.3.8}$$

where $E = E_2 - E_1 = \hbar\omega$.

Using the integral representation of the δ-function

$$\delta(z) = \frac{1}{2\pi} \int_{-\infty}^{+\infty} \exp(izt)\, dt,$$

and transforming (7.3.8) we obtain

$$w = \frac{1}{2\pi\hbar} \int_{-\infty}^{+\infty} e^{i\omega t} \sum_f |\langle f| V |i\rangle|^2 \exp \frac{i(E_f - E_i)t}{\hbar}\, dt. \tag{7.3.9}$$

We take into account the identity

$$\sum_f |\langle f| V |i\rangle|^2 \exp \frac{i(E_f - E_i)t}{\hbar} = \sum_f \langle f |V| i\rangle^* \langle f |V| i\rangle \exp \frac{i(E_f - E_i)t}{\hbar}$$

$$= \sum_f \langle f \left| V^\dagger \right| i\rangle \langle f \left| e^{\frac{i}{\hbar}H_t} V e^{-\frac{i}{\hbar}H_t} \right| i\rangle = \sum_f \langle i \left| V^\dagger \right| f\rangle \langle f |V(t)| i\rangle,$$

where we move to the Heisenberg representation of the operators and denote V by $V(0)$. Then,

$$w(\boldsymbol{q}, E) = \frac{1}{2\pi\hbar} \int_{-\infty}^{+\infty} \langle V^*(\boldsymbol{q}, 0) V(\boldsymbol{q}, t) \rangle e^{i\omega t}\, dt. \tag{7.3.10}$$

Here we use the notation $\langle \ldots \rangle$ for the quantum-mechanical average in some crystal state, $\{N_k\}$. We note that by virtue of time homogeneity

$$\langle V^*(\boldsymbol{q}, 0) V(\boldsymbol{q}, t) \rangle = \langle V^*(\boldsymbol{q}, -t) V(\boldsymbol{q}, 0) \rangle. \tag{7.3.11}$$

We come back to the definition (7.3.3), taking into account (7.3.11)

$$w(\boldsymbol{q},E) = \frac{1}{2\pi\hbar}|F_0(\boldsymbol{q})|^2 \sum_{\boldsymbol{n},\boldsymbol{n}'} e^{i\boldsymbol{q}[\boldsymbol{r}(\boldsymbol{n})-\boldsymbol{r}(\boldsymbol{n}')]} \times \int_{-\infty}^{+\infty} \langle e^{i\boldsymbol{q}\boldsymbol{u}(\boldsymbol{n},t)} e^{-i\boldsymbol{q}\boldsymbol{u}(\boldsymbol{n}',0)}\rangle e^{-i\omega t}\,dt. \tag{7.3.12}$$

Thus, in the general case the scattering probability is determined by the Fourier components of the correlation functions

$$\langle e^{i\boldsymbol{q}\boldsymbol{u}(\boldsymbol{n},t)} e^{-i\boldsymbol{q}\boldsymbol{u}(\boldsymbol{n}',0)}\rangle. \tag{7.3.13}$$

For small thermal atomic displacements and insignificant changes in the wave vector in the scattering, we confine ourselves to the expansion (7.3.4). This expansion is sufficient to describe the one-phonon processes of inelastic scattering. Using the definition (7.3.5), we note that

$$\langle M^{\dagger}(\boldsymbol{q},t)\rangle = \langle M(\boldsymbol{q},0)\rangle = 0,$$

and calculate the part quadratic in $M(\boldsymbol{q})$ that contributes to the inelastic scattering probability

$$\begin{aligned}\delta w(\boldsymbol{q},E) &= \frac{1}{2\pi\hbar}|F_0(\boldsymbol{q})|^2 \int_{-\infty}^{+\infty} \langle M^{\dagger}(\boldsymbol{q},t)M(\boldsymbol{q})\rangle e^{-i\omega t}\,dt \\ &= \frac{q_j q_l}{2\pi\hbar}|F_0(\boldsymbol{q})|^2 \sum_{\boldsymbol{n},\boldsymbol{n}'} \int_{-\infty}^{+\infty} \langle u^j(\boldsymbol{n},t)u^l(\boldsymbol{n}',t)\rangle e^{i\boldsymbol{q}[\boldsymbol{r}(\boldsymbol{n})-\boldsymbol{r}(\boldsymbol{n}')]-i\omega t}\,dt.\end{aligned} \tag{7.3.14}$$

The average value coincident with the pair correlator (6.4.1) has arisen in the integrand of (7.3.14). Thus, the probability of one-phonon inelastic processes is proportional to

$$\delta\omega(\boldsymbol{q},E) = N\,|F_0(\boldsymbol{q})|^2\, q_j q_l K_{jl}(\boldsymbol{q},\omega), \tag{7.3.15}$$

where

$$K_{jl}(\boldsymbol{q},\omega) = \frac{1}{2\pi\hbar}\sum_{\boldsymbol{n}} \int_{-\infty}^{+\infty} \Phi_{jl}(\boldsymbol{n},t) e^{i\boldsymbol{q}\boldsymbol{r}(\boldsymbol{n})-i\omega t}\,dt. \tag{7.3.16}$$

Thus, the specificity of the scattering of various particles (or waves) is manifest in the presence of the atomic factor $F_0(\boldsymbol{q})$ only.

The participation of the crystal structure of substances in the generation of one-phonon processes is universal and determined by the pair displacement correlator. The probability of such processes (7.3.15) is proportional to a spatial and a time Fourier transformation of the displacement correlator (6.4.2).

It follows from (6.4.2) and (7.3.16) that

$$K_{jl}(\boldsymbol{q},\omega) = \frac{1}{2mN}\sum_{k\alpha}\frac{e_j(\boldsymbol{k})e_l(\boldsymbol{k})}{\omega_\alpha(\boldsymbol{k})}\{N_\alpha(\boldsymbol{k})\delta\left[\omega-\omega_\alpha(\boldsymbol{k})\right]$$
$$\times\sum_n e^{i(\boldsymbol{q}-\boldsymbol{k})\boldsymbol{r}(\boldsymbol{n})} + \left(N_\alpha(\boldsymbol{k})+\frac{1}{2}\right)\delta\left[\omega+\omega_\alpha(\boldsymbol{k})\right]\sum_n e^{i(\boldsymbol{q}+\boldsymbol{k})\boldsymbol{r}(\boldsymbol{n})}\}, \tag{7.3.17}$$

showing that the Fourier components of the displacement correlation are nonzero only for the frequencies $\omega = \pm\omega_\alpha(\boldsymbol{k})$. This means that one-phonon processes are allowed if one of the conditions

$$E_2 - E_1 = \hbar\omega_\alpha(\boldsymbol{k}), \tag{7.3.18}$$

$$E_1 - E_1 = \hbar\omega_\alpha(\boldsymbol{k}) \tag{7.3.19}$$

is satisfied (we remember that in (7.3.15) $\hbar\omega = E_2 - E_1$).

It is easy to understand from (7.3.17) that the process (7.3.18) is allowed under the condition (7.3.6), and the process (7.3.19) under the condition (7.3.7). The conservation laws (7.3.6), (7.3.7), (7.3.18), (7.3.19) describe the interaction of strange particles (capable of propagating a crystal) with phonons, i. e., the quasi-particles of a crystal.

The inelastic diffraction phenomenon accompanied by one-phonon processes described provides an effective method for describing lattice dynamics. Let us suppose that it is possible to create a narrow beam of particles incident on the crystal, to fix the directions of scattered particles and to measure their energy. Thus, in each scattering process we know $\boldsymbol{q}_1$, $\boldsymbol{q}_2$, E_1 and E_2. If we neglect the umklapp processes whose contribution can be singled out, we determine the values of $\boldsymbol{k}$ and $\hbar\omega$ for a phonon participating in the process. Considering various directions and orientations of the crystal, we can reproduce in principle the function $\omega = \omega(\boldsymbol{k})$.

However, this experiment will be a success only if with the wavelength of the order of a lattice constant ($\lambda \sim a$) one can observe the change in the energy of a beam or a particle by an amount of the order of phonon energy,

$$\delta E \sim \hbar\omega_D \sim \hbar\omega_m \sim 0.01 \text{ eV}.$$

The energy of X-rays with the required wavelength has the order of magnitude

$$E = \hbar\omega = 2\pi\frac{\hbar c}{\lambda} \sim 2\pi\frac{\hbar c}{a} \sim 10^4 \text{ eV}.$$

Thus, using X-rays the indicated energy change δE is impossible to observe. But for neutrons,

$$E = \frac{p^2}{2m} = \frac{(2\pi)^2\hbar^2}{2m\lambda^2} \sim (2\pi)^2\frac{\hbar^2}{ma^2} \sim 0.1 \text{ eV},$$

which corresponds to the energy of thermal neutrons. Phonon absorption or emission changes the neutron energy by δE, a value that can be observed.

Examples of experimentally reproducing dispersion laws for neutron scattering are given in Figs 2.4b, 3.2, 7.6.

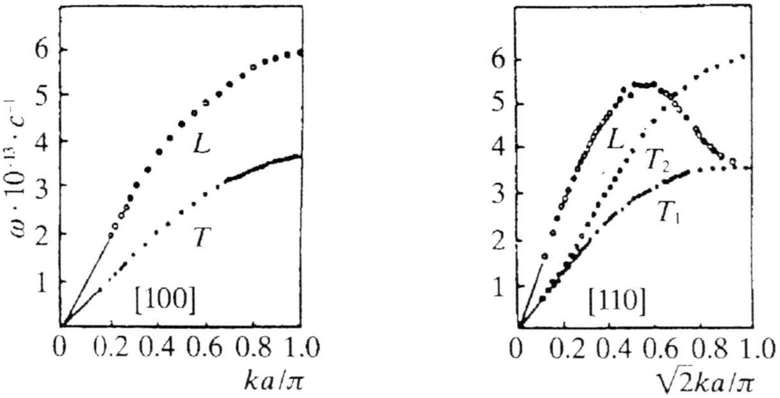

Fig. 7.6 Dispersion curves for the two directions in Al are reproduced by the neutron experiments (Yarnell et al. 1965).

7.4
Effect of Thermal Atomic Motion on Elastic γ-Quantum-Scattering

In distinguishing in (7.3.12) the probability of inelastic one-phonon processes, we have restricted ourselves in (7.3.12), only to the terms containing multipliers $q_i q_k \langle u^i(\boldsymbol{n}, t) u^k(\boldsymbol{n}', 0)\rangle$. We have not taken into account the terms with multipliers such as

$$\langle [\boldsymbol{q}\boldsymbol{u}(\boldsymbol{n}, t)]^2 \rangle = \langle [\boldsymbol{q}\boldsymbol{u}(\boldsymbol{n}, 0)]^2 \rangle, \tag{7.4.1}$$

that also have second order of smallness in $u(n)$. These terms were omitted since they contribute to the elastic scattering of penetrating radiation only (for definiteness, we shall speak of γ-quanta). Indeed, the averages (7.4.1) are time independent and, therefore, after the integration over time they lead to the δ-function $\delta(\omega) = \hbar\delta(E_1 - E_2)$ appearing in (7.3.12).

Thus, terms such as (7.4.1) determine the difference between the elastic γ-quantum diffraction intensity from a vibrating crystal from that on an immobile spatial lattice of atoms. It is easily seen that for a quantitative description of the influence the atomic motion on producing elastic diffraction it is not necessary to restrict ourselves to approximation, which is quadratic in displacements.

We remind ourselves that (6.5.9) is applicable for a monatomic lattice

$$\langle\langle e^{i\boldsymbol{q}\boldsymbol{u}(\boldsymbol{n},t)} \rangle\rangle = \langle\langle e^{i\boldsymbol{q}\boldsymbol{u}(\boldsymbol{n})} \rangle\rangle = e^{-W},$$

where

$$W = \frac{1}{2}\langle\langle (\boldsymbol{qu})^2 \rangle\rangle, \tag{7.4.2}$$

and represent the correlation function (7.3.13) in the form of two terms

$$\begin{aligned} \langle\langle e^{i\boldsymbol{qu}(\boldsymbol{n},t)} e^{-i\boldsymbol{qu}(\boldsymbol{n}',0)} \rangle\rangle &= e^{-2W} \\ + \left\{ \langle\langle e^{i\boldsymbol{qu}(t)} e^{-i\boldsymbol{qu}(0)} \rangle\rangle \right. &- \left. \langle\langle e^{i\boldsymbol{qu}(t)} \rangle\rangle \langle\langle e^{-i\boldsymbol{qu}(0)} \rangle\rangle \right\}. \end{aligned} \tag{7.4.3}$$

We begin to analyze this formula with the second term. In the approximation quadratic in displacements it is equal to the average $\langle\langle M^+(\boldsymbol{q},t)M(\boldsymbol{q},0)\rangle\rangle$ that was considered in calculation of the probability for one-phonon inelastic scattering (7.3.14). It is clear that the next term following the expansion of the second term in (7.4.3) in even powers of the displacement determines the probabilities for multiphonon inelastic scattering processes.

The first term in (7.4.3) is time independent and is thus related to elastic scattering. When substituted in (7.1.3), it contributes to the elastic scattering probability.

The diffraction intensity observed in an experiment is automatically averaged over all the initial crystal states, which is equivalent to a thermodynamic averaging of the scattering probability. On performing a thermodynamic averaging in (7.3.12), we find that the intensity of the elastic γ-quantum diffraction is proportional to the

$$\begin{aligned} w^{el}(\boldsymbol{q},E) &= N e^{-2W} \delta(E) \, |F_0(\boldsymbol{q})|^2 \sum_n e^{i\boldsymbol{qr}(\boldsymbol{n})} \\ &= N\Omega_0 \, e^{-2W} \delta(E) \, |F_0(\boldsymbol{q})|^2 \sum_{\boldsymbol{G}} \delta(\boldsymbol{q}-\boldsymbol{G}), \end{aligned} \tag{7.4.4}$$

where the summation is over the reciprocal lattice vectors; Ω_0 is the unit cell volume of the reciprocal lattice. The factor e^{-2W} describing the weakening of the diffraction maxima intensity due to thermal atomic motion is called the *Debye–Waller factor*.

Since the Debye–Waller factor in (7.4.4) is a multiplying factor, the thermal atomic motion in the crystal does not result in a smearing of the sharp diffraction maxima in γ-quantum scattering it only reduces its intensity.

The generalization of (7.4.4) to the case of a polyatomic lattice is straightforward

$$w^{el}(\boldsymbol{q},E) = N\Omega_0 \delta(E) \left| \sum_s e^{-W_s} F_s(\boldsymbol{q}) e^{i\boldsymbol{q}\boldsymbol{x}_s} \right| \sum_{\boldsymbol{G}} \delta(\boldsymbol{q}-\boldsymbol{G}), \tag{7.4.5}$$

where $\boldsymbol{x}_s$ is the vector of the basis of a polyatomic lattice that gives the s-th atom position in the unit cell; $\boldsymbol{e}_s$ is the polarization vector of the s-th sublattice vibrations,

$$W_s = \frac{1}{2m_s N} \sum_{\boldsymbol{k}\alpha} \frac{\{\boldsymbol{q}\boldsymbol{e}_s(\alpha,\boldsymbol{k})\}^2}{\omega_\alpha(\boldsymbol{k})} \coth \frac{\hbar\omega_\alpha(\boldsymbol{k})}{T}. \tag{7.4.6}$$

Examining the processes of the interaction of the γ-quantum with the crystal has shown that an incident γ-quantum may either give the crystal some momentum to produce a phonon or get from the crystal some momentum generated by the phonon "absorption". In this case the γ-quantum energy will change to a finite value and the scattering will be inelastic.

But another interaction with a crystal is possible when the finite part of the γ-quantum momentum is transferred to a crystal in the process without the phonon emission or absorption. As a result of this interaction, the crystal does not change its state and the γ-quantum conserves its energy. So we have a *process without* recoil characterized by extremely narrow diffraction lines. The proportion of such elastic processes is measured by the Debye–Waller factor.

7.5
Equation of Phonon Motion in a Deformed Crystal

The equations of crystal lattice motion have static inhomogeneous solutions with boundary conditions determined by external loads. When such crystal states are described in terms of long waves, we can consider the inhomogeneous static deformations that deform the crystal lattice (Fig. 7.7). If the deformations are small, they are solutions to the linear equations of elasticity theory. Static deformations do not influence small vibrations (phonon after quantization) due to the linearity of the elastic theory equations, but including the anharmonic interactions transforms a deformed lattice into an inhomogeneous elastic medium.

It is of interest to discuss the possibility to describe the vibrations of such spatial inhomogeneous crystals in terms of phonons. Let us suppose that the macroscopic crystal characteristics vary essentially at distances of the order of δL. How is this inhomogeneity to be taken into account, while preserving the usual concept of phonons?

The phonons were introduced to quantize vibrations of the homogeneous crystal that results in the fact that the states of an individual phonon were characterized by a quasi-wave vector $\boldsymbol{k}$. If the distance δL is large compared to the average phonon wavelength $\delta L \gg \bar{\lambda}$, the phonon concept may be preserved in an inhomogeneous crystal, too. Indeed, to describe the crystal vibrations with the above inhomogeneities instead of normal modes, one should take the wave packets with the interval of wave vectors $\delta\boldsymbol{k}$, where

$$\delta L \gtrsim \frac{1}{\delta k} \gg \bar{\lambda}. \tag{7.5.1}$$

Analyzing (7.5.1) and taking into account the relation between the wavelength and the value of the wave vector we obtain $\delta k \ll k$, implying that the wave packets concerned consist of normal modes whose wave-vector differences much less than their wave vectors. Some conclusions important for further discussion follow from the last assertion.

First the wave packet described may be associated with a vibration with the quasi-wave vector $\boldsymbol{k}$, i. e., with an individual phonon in a state with a given $\boldsymbol{k}$.

Then, the velocity of this new "phonon" is determined by the group velocity of the packet:

$$v = \frac{\partial \omega}{\partial \boldsymbol{k}}. \tag{7.5.2}$$

First, if we measure the spatial positions within δx, where $\bar{\lambda} \ll \delta x \ll \delta L$, then a phonon[1], having a quasi-wave vector $\boldsymbol{k}$ and located at the point $\boldsymbol{r}$ can be introduced. Under such conditions the crystal-state inhomogeneity manifest in the inhomogeneity of a "refraction coefficient" for a phonon, i. e., in the dependence of phonon frequency on the coordinate $\boldsymbol{r}$: $\omega_\alpha = \omega_\alpha(\boldsymbol{k}, \boldsymbol{r})$. The vector $\boldsymbol{k}$ of the corresponding packet will then vary in time

$$\frac{d\boldsymbol{k}}{dt} = -\nabla\omega(\boldsymbol{k}, \boldsymbol{r}) \equiv -\frac{\partial \omega}{\partial \boldsymbol{r}}. \tag{7.5.3}$$

The approximation in which the relation (7.5.3) is meaningful corresponds completely to the Eikonal approximation in geometry optics. By analogy with the description of electromagnetic vibrations in inhomogeneous media, we can describe the equations of motion for a sound beam that follows directly from (7.5.2), (7.5.3)

$$\frac{d\boldsymbol{r}}{dt} = \frac{\partial \omega(\boldsymbol{k}, \boldsymbol{r})}{\partial \boldsymbol{k}}, \qquad \frac{d\boldsymbol{k}}{dt} = -\frac{\partial \omega(\boldsymbol{k}, \boldsymbol{r})}{\partial \boldsymbol{r}}. \tag{7.5.4}$$

The equations of motion (7.5.4) can be derived not only using the wave properties of phonons employed to construct the wave packets. According to the de Broglie principle the phonon is a quasi-particle with energy $\hbar\omega$ and quasi-momentum $\boldsymbol{p} = \hbar\boldsymbol{k}$. Thus, the energy $\hbar\omega(\boldsymbol{k}, \boldsymbol{r})$ can be regarded as the Hamiltonian function and (7.5.4) as the Hamiltonian form of the equations of motion of this quasi-particle.

In the case of small elastic deformations the dependence of $\omega(\boldsymbol{k}, \boldsymbol{r})$ on the strain tensor u_{ik} is linear

$$\omega(\boldsymbol{k}, \boldsymbol{r}) = \omega_0(\boldsymbol{k}) + g_{il}(\boldsymbol{k}) u_{il}(\boldsymbol{r}), \tag{7.5.5}$$

where $\omega_0(\boldsymbol{k})$ is the dependence of frequency on a quasi-wave vector in a nondeformed crystal.

It is interesting to analyze (7.5.5) and the equations of motion (7.5.4) near the edge of the phonon frequency band. We shall the quasi-wave vector from the value of $\boldsymbol{k}_m$ corresponding to the maximum or minimum (if we speak of optical phonons) value of $\omega_0(\boldsymbol{k})$. Then, at small k

$$\omega_0(\boldsymbol{k}) = \omega_m - \frac{1}{2}\gamma_{il} k_i k_l. \tag{7.5.6}$$

Confining ourselves to small k, in (7.5.5) we obtain

$$\omega(\boldsymbol{k}, \boldsymbol{r}) = \omega_m + g_{il}(0) u_{il}(\boldsymbol{r}) - \frac{1}{2}\gamma_{il} k_i k_l, \tag{7.5.7}$$

where ω_m is the frequency band boundary of the continuous spectrum of a nondeformed crystal.

1) The condition $\delta k \delta x \sim 1$ is allowed by the inequality (7.5.1).

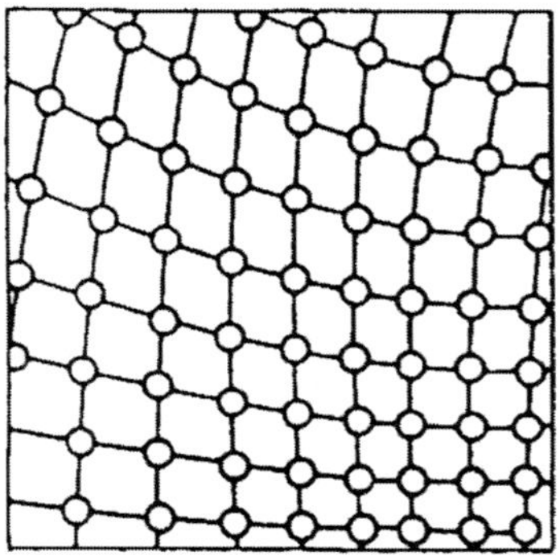

Fig. 7.7 Deformed crystal lattice.

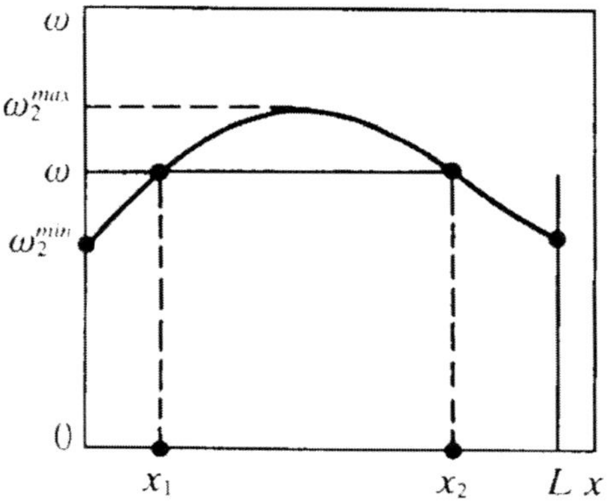

Fig. 7.8 Distortion of an upper edge of the phonon frequency band.

The equations of motion (7.5.4) for this case are

$$\frac{dk_j}{dt} = -g_{il}(0)\frac{\partial u_{il}}{\partial x_j}, \qquad \frac{dx_j}{dt} = -\gamma_{jl}k_l.$$

In a cubic crystal (7.5.7) is much simplified

$$\omega(\boldsymbol{k},\boldsymbol{r}) = \omega_m\left\{1 - \beta u_{ll}(\boldsymbol{r})\right\} - \frac{1}{2}\gamma^2 k^2,$$

$$\gamma = \text{const}, \quad \beta = \text{const} \sim 1.$$

Since the reciprocal-space volume corresponding to the vibrations concerned increases under total compression ($u_{ll} < 0$), it may be concluded that $\beta > 0$. In the simplest case

$$\frac{d\boldsymbol{k}}{dt} = \beta\omega_m\frac{\partial u_{ll}}{\partial \boldsymbol{r}}, \qquad \frac{d\boldsymbol{r}}{dt} = -\gamma^2\boldsymbol{k}. \tag{7.5.8}$$

The phonon motion described by (7.5.8) takes place under the condition

$$\beta u_{ll}(\boldsymbol{r}) + \frac{1}{2}\gamma^2 k^2 = \omega_m - \omega = \text{const.}$$

It is clear that the part of the total crystal compression (at $\beta > 0$) is a "potential well" for a short-wave phonon near the upper spectrum boundary. Thus, getting into the region of forbidden frequencies of an ideal crystal ($\omega > \omega_m$), the phonon cannot go beyond the volume bounded by the surface

$$\beta u_{ll}(\boldsymbol{r}) = \omega_m - \omega .$$

Consider the possible dependence of the upper spectrum boundary ω_2 on the coordinate shown in Fig. 7.8. The short-wave phonon moves in the "potential well" experiencing total internal reflection at the points $x = x_1$ and $x = x_2$.

The phonon motion under such conditions becomes finite, i. e., localized in some volume of a crystal with linear dimensions much exceeding the lattice constant. If the mean free path of the phonon is much larger then its "trajectory", then we can speak of bound states of the phonon concerned. When such phonon motion is quantized, discrete frequencies should appear outside the continuous spectrum band. The number of possible discrete frequencies is determined by the form of the "potential well".

8
Quantum Crystals

8.1
Stability Condition of a Crystal State

The anharmonicities of the crystal lattice vibrations are important in two cases: either at high temperatures when relative displacements of neighboring atoms become significant and the nonlinear character of elastic interatomic forces manifests itself, and even in the low-temperature region when the phenomena generated by the anharmonicity of crystal vibration are examined. The phenomena that are inexplicable from the viewpoint of a harmonic approximation can be exemplified by any process stimulated by an interaction in the phonon gas.

Let us analyze the harmonic displacements of atoms more scrupulously and clear up when they can be considered as small. The necessary condition for the harmonic approximation to be valid is the requirement that the atomic zero vibration amplitude in the crystal be small compared to the lattice period. If we denote in (6.2.6)

$$\frac{1}{\omega_*} = \frac{1}{N}\sum_{\boldsymbol{k}} \frac{1}{\omega(\boldsymbol{k})}, \tag{8.1.1}$$

the ratio of r.m.s. atomic displacement in the ground state of the crystal to the lattice constant a will be of the order of magnitude

$$\Lambda = \frac{\sqrt{\langle u^2 \rangle_0}}{a} \sim \frac{1}{a}\sqrt{\frac{\hbar}{m\omega_*}}.$$

Thus, the harmonic approximation provides a sufficient accuracy in describing the atomic motion in a crystal if $\Lambda \ll 1$.

For almost all crystalline substances the parameter Λ is very small. Moreover, even at high temperatures, when thermal atomic displacements greatly exceed the zero-vibration amplitude, the ratio of the r.m.s. atomic displacement amplitude to the lattice period is generally small. Only at the melting temperature does the value

The Crystal Lattice: Phonons, Solitons, Dislocations, Superlattices, Second Edition. Arnold M. Kosevich

ISBN: 3-527-40508-9

of $\langle\langle u^2 \rangle\rangle$ become comparable with the square of the interatomic distance. The last point allows one to introduce the physical criterion of melting when the temperature rises. The melting temperature can be estimated on the basis of a *Lindemann condition* according to which at the melting point the thermal atomic displacement square (6.4.7) is $\langle\langle u^2 \rangle\rangle = \gamma^2 a^2, \gamma \sim 0.1$.

The crystal melting temperature T_{melt} for an ordinary crystal is always higher than the Debye temperature Θ. At temperatures lower then T_{melt}, in the majority of crystals the ratio of thermal atomic displacement to the lattice period is very small.

However, there exist crystals for which $\Lambda \sim 1$. These are, in the first instance, hydrogen and helium crystals[1]. The standard quasi-classical approach to studying the vibrations of such crystals is already inapplicable. This is not only due to large anharmonicities. The classification of the crystal states based on the concept that atoms in equilibrium are located at certain lattice sites does not correspond to the physical situation.

Since the existence of a crystalline state in this situation is problematic, the dimensionless parameter Λ should be determined without involving the lattice period. It was done by de Bour (1948) in terms of the quantities determining the potential energy of the interaction between the crystal atoms. The interaction energy of the atomic pair can be written with a good accuracy in the form of a Lennard–Jones potential

$$U(r) = 4\epsilon \left\{ \left(\frac{\sigma}{r}\right)^{12} - \left(\frac{\sigma}{r}\right)^{6} \right\}, \tag{8.1.2}$$

where r is the interatomic distance, σ is the spatial interaction distance, the parameter ϵ characterizes the intensity of interaction energy.

Then the characteristic dimensionless *de Bour parameter* is

$$\Lambda_{\text{B}} = \frac{\hbar}{\sqrt{m\epsilon}}. \tag{8.1.3}$$

This parameter Λ_{B} for the crystal coincides, in order of magnitude, with the parameter Λ considered above, so that small zero atomic vibrations in the crystal correspond to $\Lambda_{\text{B}} \ll 1$. This is just the condition for the applicability of the classical approach to the description of the crystal state.

It follows from (8.1.3) that the larger the value of Λ_{B} the smaller is the atomic mass and the intensity of its interaction with its neighbors. For some inert gases the values of Λ_{B} are

^{3}He	^{4}He	H_2	D_2	Ne	Ar	Kr	Xe
0.48	0.43	0.28	0.20	0.09	0.03	0.02	0.01

1) We remind the reader that at normal atmospheric pressure helium remains liquid even at $T = 0$ K, and its ordered ground state has no spatial crystal structure. Helium-4 isotopes form a crystalline state only at pressures higher than 250 Pa, and helium-3 isotopes – at pressures higher than 300 Pa.

Argon, xenon and krypton crystals have comparatively small values of zero-vibration amplitudes and can be considered as ordinary crystals with the atoms localized at the lattice sites.

However, the de Bour parameter for helium and hydrogen crystals is comparable with 1 and, thus, the classical approach to describe the physical properties of these crystals (in particular, their ground state) is inapplicable. Crystals where the zero-vibration amplitude is comparable in order of magnitude with the lattice period are called *quantum crystals*.

The most remarkable representative of substances whose ground state is described in a purely quantum language is helium. The quantum zero vibrations in helium do not allow the formation of a stable crystal lattice at normal pressure. Thus, helium does not crystallize at any temperature (including $T = 0$ K) if the pressure does not exceed a certain limiting value. For pressures lower than this value, at temperatures near absolute zero helium forms a *quantum liquid*.

It is interesting to note that in quantum crystals unusual relations between melting T_{melt} and Debye Θ temperatures ($T_{\text{melt}} \ll \Theta$) are observed. For instance, for H_2 we have $T_{\text{melt}} \approx 14$ K and $\Theta \approx 120$ K; for ^{4}He, $T_{\text{melt}} \approx 3$ K and $\Theta \approx 30$ K. Even for Ne, $T_{\text{melt}} \approx 25$ K, $\Theta \approx 75$ K.

The basic peculiarity of quantum crystal mechanics is that the atoms are not considered as particles vibrating independently near the lattice sites under only a classical force interaction. The reason for this is as follows. A quantum crystal is characterized by a regular spatial structure and has a definite crystal lattice. Thus, on the one hand, the atoms of the quantum crystal make up a spatial lattice and perform vibrational motions near its sites, but, on the other hand, the amplitude of the atomic zero vibrations in the potential field (8.1.2) is of the order of the distance between the sites. To combine these two properties of atomic motion in quantum crystals, we assume the atomic motion is correlated. Thus, if the atomic motion in a quantum crystal is described microscopically it is necessary to take into account the correlation of the motion at small distances (*short range correlation*).

In a quantum crystal the phonons also play the role of weakly excited crystal states. This conclusion is based on studying the low-temperature properties of the hard helium. The phonons prove to be a good approximation to describe the thermal motion in a quantum crystal. Hence it follows that large zero atomic vibrations result only in a renormalization of the long-range correlations (phonons) by taking into account the quantum short-range correlations.

Finally, the condition for zero vibrations to be small may be violated only for separate forms of intracrystal motions. For instance, hydrogen dissolved in certain metals forms a quantum subsystem (sometimes a regular sublattice) in a classical crystal. The hydrogen atomic displacement should then be described by taking into account the quantum properties for such a type of motion, whereas the motion in the other degrees of freedom of a crystal may be regarded as quasi-classical.

8.2
The Ground State of Quantum Crystal

According to experimental data the "quantum feature" of solid helium (in particular, the specific role of zero vibrations) is not manifest when considering the dynamic crystal properties described by the harmonic approximation.

However, there are such physical properties of quantum crystals in which large zero atomic vibrations are dominant. There is, first of all the possibility of a *tunneling* atomic motion in a crystal lattice, which is completely determined by the purely quantum effect of particles tunneling through a potential barrier. The tunneling motion may cause a rearrangement of the ground state of a quantum crystal.

In describing the ground state of an ordinary crystal it is assumed that there is one atom at each lattice site. But when the zero-vibration amplitude levels are the same order of magnitude as the interatomic distance in a crystal, then the atom ceases to be localized at a definite site. A purely quantum situation can then be realized when in the ground state (at $T = 0$ K) the number of lattice sites N_0 is not the same as the number of atoms N.

For all existing "candidates" for quantum crystals, one should expect $N_0 \geq N$. When the value of N_0 exceeds N then the probability to find an atom at a given site is less than 1, although the crystal does not lose its periodicity. The density that determines the probability of various positions of a particle in space remains in this case a periodic function with a lattice period. Andreev and Lifshits (1969) were the first to predict the existence of such quantum crystals and to describe their basic properties.

To characterize the state of a quantum crystal, it is necessary to take into account additional degrees of freedom that are not present in a classical crystal – the quantum tunneling. The excess in the number of sites of a spatial lattice over the number of atoms caused by quantum tunneling is called *quantum dilatation*. The quantum dilatation is not associated with an external stretching action or thermal expansion. Let the function $\eta(\boldsymbol{r})$ describe this dilatation and we normalize this function by the condition

$$\frac{1}{V_0}\int \eta(\boldsymbol{r})\,dV = N_0 - N, \tag{8.2.1}$$

where the integral is calculated over the whole crystal volume. The probability of detecting an atom positioned in a definite unit cell with the volume V_0 is given by

$$\frac{1}{V_0}\int\limits_{V_0}(1-\eta)\,dV = \frac{1}{V}\int(1-\eta)\,dV = \frac{N}{N_0} < 1, \tag{8.2.2}$$

where the first integral is taken over the volume of one unit cell.

Thus, the ground state of a quantum crystal is a phonon vacuum that is characterized by certain quantum dilatation. In the long-range approximation the ground state can be regarded as homogeneous, i.e., it is admissible to set $\eta(\boldsymbol{r}) = \eta_0 = \text{const}$. The crystal density in the ground state is

$$\rho = \rho_0 (1 - \eta_0), \qquad \rho_0 = \frac{m}{V_0}, \tag{8.2.3}$$

where m is the mass of an atom (or the mass of atoms in the unit cell of a polyatomic lattice).

However, the presence of quantum dilatation in the ground state is not unique or the necessary property of a quantum crystal. The quantum properties are more transparent in the crystal dynamics.

8.3 Equations for Small Vibrations of a Quantum Crystal

If tunneling motion of atoms is possible, the ordinary equations of crystal motion are inapplicable. These equations cannot be derived by means of a classical approach and the problem requires a rigorous quantum-mechanical consideration. We will not derive here the equations of quantum crystal vibrations in the whole frequency range, but will confine ourselves to an analysis of long-wave (low-frequency) vibrations. For ordinary crystals this approximation results in the dynamic equations of elasticity theory (Section 2.9), i. e., in the equations of motion of a continuous medium.

The system of linear equations of the continuous medium dynamics includes two obvious equations

$$\frac{\partial j_i}{\partial t} = \nabla_k \sigma_{ki}, \tag{8.3.1}$$

$$\frac{\partial \rho}{\partial t} + \operatorname{div} \boldsymbol{j} = 0, \tag{8.3.2}$$

where $\boldsymbol{j}$ is a vector of the flow density of the substance mass that plays the role of the momentum of a unit volume. In a classical crystal

$$\sigma_{ik} = \lambda_{iklm} \nabla_l u_m; \qquad \boldsymbol{j} = \rho \boldsymbol{v}; \qquad \boldsymbol{v} = \frac{\partial \boldsymbol{u}}{\partial t}, \tag{8.3.3}$$

and (8.3.1), (8.3.2) take the standard form of the elasticity theory dynamic equation (2.9.18) and the continuum equation (2.9.6).

The physical characteristics for the medium state and the relations between them in (8.3.1), (8.3.2) must be redefined for the quantum crystal. In particular, these relations should describe two kinds of motion: a quasi-classical "solid state" and a purely quantum one. The latter motion is called a *superfluid*, emphasizing the similarity with a particular flow behavior of a quantum liquid.

To preserve notations and relations such as (8.3.3), we introduce the vector of site displacements in the quantum crystal lattice from their equilibrium positions $\boldsymbol{u} = \boldsymbol{u}(\boldsymbol{r})$, and associate it with the tensor of small distortions $u_{ik} = \nabla_i u_k$ and strain ε_{ik}. The vector $\boldsymbol{u}$ describes the normal form of classical crystal lattice motion.

In a nondeformed (ground) state, $u_{ik} = 0$, and the quantum dilatation is homogeneous: $\eta = \eta_0 =$ constant. When small vibrations arise, there appears a small lattice deformation ($|u_{ik}| \ll 1$) and a little deviation of η from equilibrium value: $\eta = \eta_0 + \eta'$, $|\eta'| \ll 1$. We shall further restrict ourselves to the approximation linear in small values of η' and ε_{ik}.

The internal stresses now characterized by the tensor σ_{ik} are generated both by the deformation of the lattice itself and by the inhomogeneous quantum dilatation distribution. For small gradients of the function η we may set

$$\sigma_{ik} = \lambda_{iklm} u_{lm} + p_{ik} \eta', \tag{8.3.4}$$

where the moduli λ_{iklm} are determined by the elastic properties of a quantum crystal and, thus, may be η_0 dependent. The second-rank tensor p_{ik} determines the crystal reaction to the inhomogeneity of the function $\eta(\boldsymbol{r}, t)$. As the quantum dilatation cannot be associated with the moment of forces that act on the whole crystal, the tensor p_{ik} should be regarded as symmetric: $p_{ik} = p_{ki}$. With the quantum dilatation the change in the crystal mass density ρ cannot be unambiguously determined by the geometric deformation of the lattice, i. e., by the tensor u_{ik}. In this situation a small deviation of the density $\rho(\boldsymbol{r}, t)$ from a homogeneous equilibrium value ρ_0 in the approximation linear in η' and u_{ik}, may be written as

$$\delta\rho \equiv \rho - \rho_0 = \frac{\partial \rho}{\partial u_{ik}} u_{ik} - \rho_0 \eta' = \rho_0 \left(\gamma_{ik} u_{ik} - \eta' \right),$$

where we denote $\rho_0 \gamma_{ik} = \partial\rho / \partial u_{ik}$. Since $u_{kk} = \operatorname{div} \boldsymbol{u}$ coincides with a relative increase in the lattice volume under deformation, then $\gamma_{ik} = -\delta_{ik}$ in an ordinary (classical) crystal. Assuming the quantum properties of a crystal to be weak we consider the expansion

$$\delta\rho = -\rho \left(\operatorname{div} \boldsymbol{u} + \eta' \right). \tag{8.3.5}$$

We now proceed to the second and third relations (8.3.3) that should be revised as the derivative $\partial \boldsymbol{u} \partial t$ does not coincide with the atomic motion velocity $\boldsymbol{v}$. The type of motion described by the vector $\partial \boldsymbol{u} \partial t$ is represented as spatial lattice vibrations in a certain medium created by quantum dilatation and capable of superfluid motion. A superfluid type of motion is characterized by the average velocity $\boldsymbol{v}^s$.

In the reference frame for which $\boldsymbol{v}^s = 0$ the momentum of the crystal unit volume coincides with the mass flow density of normal motion:

$$j_i^0 = (\rho \delta_{ik} + \rho_{ik}) \frac{\partial u_k}{\partial t},$$

where ρ_{ik} is the tensor of the adjoined mass of a crystal lattice in the above-mentioned quantum dilatation "medium" ($\rho_{ik} = \rho_{ki}$). By assumption, the quantum dilatation creates a negative mass density compared to a classical crystal, thus, the matrix ρ_{ik} should not be positively definite. In a laboratory reference frame, the momentum of a crystal unit volume is

$$j_i = \rho v_i^s + j_i^0 = (\rho\delta_{ik} + \rho_{ik}) \frac{\partial u_k}{\partial t} - \rho_{ik} v_k^s. \tag{8.3.6}$$

The formula for the kinetic energy of the crystal unit volume is obtained in a similar way:

$$\begin{aligned} K = & \frac{1}{2}\rho (v^s)^2 + v^s j^0 + \frac{1}{2} (\rho\delta_{ik} + \rho_{ik}) \left(\frac{\partial u_i}{\partial t} - v_i^s\right) \left(\frac{\partial u_k}{\partial t} - v_k^s\right) \\ = & \frac{1}{2} (\rho\delta_{ik} + \rho_{ik}) \frac{\partial u_i}{\partial t}\frac{\partial u_k}{\partial t} - \frac{1}{2}\rho_{ik} v_i^s v_k^s. \end{aligned} \tag{8.3.7}$$

Using (8.3.6), (8.3.7) the following effective mass densities can be assigned to superfluid and normal motions

$$\rho_{ik}^{(s)} = -\rho_{ik}, \qquad \rho_{ik}^{(n)} = \rho\delta_{ik} + \rho_{ik} = \rho\delta_{ik} - \rho_{ik}^{(s)}. \tag{8.3.8}$$

Then the crystal unit volume momentum and the kinetic energy density are

$$j_i = \rho_{ik}^{(n)} \frac{\partial u_k}{\partial t} + \rho_{ik}^{(s)} v_k^s; \tag{8.3.9}$$

$$K = \frac{1}{2}\rho_{ik}^{(n)} \frac{\partial u_i}{\partial t}\frac{\partial u_k}{\partial t} + \frac{1}{2}\rho_{ik}^{(s)} v_i^s v_k^s. \tag{8.3.10}$$

Since the energy is a positive quantity, the matrices $\rho_{ik}^{(s)}$ and $\rho_{ik}^{(n)}$ should be positively definite.

We substitute (8.3.9), (8.3.4) into (8.3.1)

$$\rho_{ik}^{(n)} \frac{\partial^2 u_i}{\partial t^2} - \lambda_{iklm} \nabla_k \nabla_l u_m = p_{ik} \nabla_k \eta + \rho_{ik} \frac{\partial v_k^s}{\partial t}, \tag{8.3.11}$$

and (8.3.5), (8.3.9) into (8.3.2)

$$\rho_0 \frac{\partial \eta}{\partial t} - \rho_{ik}^{(s)} \nabla_i v_k^s = p_{ik} \frac{\partial \epsilon_{ik}}{\partial t}. \tag{8.3.12}$$

It is clear that in the equations of quantum crystal motion there are two new dynamic quantities η and v^s. A relation between the superfluid motion v^s and the quantum dilatation density η cannot be established from general consideration and needs the solution of a corresponding quantum problem. However, if we assume a superfluid motion to be potential ($\mathrm{curl}\, v^s = 0$) and postulate that the quantum dilatation dynamics is described by one additional scalar function of coordinates and time, the necessary relation can be formulated phenomenologically.

We first write the density of the elastic energy of the system concerned

$$U = \frac{1}{2}\lambda_{iklm} u_{ik} u_{lm} + p_{ik}\epsilon_{ik}\eta' + \frac{1}{2} M (\eta')^2. \tag{8.3.13}$$

The second term in (8.3.13) describes the interaction of the normal deformation with quantum dilatation, while the third arises naturally in a theory linear in η' and describes the changes in the nondeformed crystal energy with η deviating from the equilibrium value η_0.

The quantum crystal total energy density should have the form of a sum

$$\mathcal{E} = K + U, \tag{8.3.14}$$

where K and U are given by (8.3.10), (8.3.13).

We require that (8.3.11), (8.3.12) arise from the mechanical action principle. To formulate this principle, we introduce the potential of a superfluid motion φ:

$$v^s = A \operatorname{grad} \varphi; \qquad \eta' = \frac{\partial \varphi}{\partial t} - \beta_{ik}\epsilon_{ik}, \tag{8.3.15}$$

determining the necessary ratios between the coefficients A, M, p_{ik}, β_{ik}, etc.

We now write the Lagrange function density assuming the displacement vector $\boldsymbol{u}$ and the potential φ to be generalized coordinates. We cannot use a standard mechanical definition of the Lagrange function as the difference between the kinetic (8.3.10) and potential energy (8.3.13)[2]. The Lagrange function density must be chosen in the form

$$\begin{aligned} \mathcal{L} = {} & \frac{1}{2}\rho_{ik}\left(\frac{\partial u_i}{\partial t} - v_i^s\right)\left(\frac{\partial u_k}{\partial t} - v_k^s\right) + \frac{1}{2}M\left(\frac{\delta\rho}{\rho}\right)^2 \\ & - \frac{1}{2}\lambda^0_{iklm}u_{ik}u_{lm} - \frac{1}{2}\rho\left(v^s\right)^2, \end{aligned} \tag{8.3.16}$$

where v^s and $\delta\rho/\rho$ are related to the derivatives of the generalized coordinates by (8.3.15), (8.3.5).

A correct expression for the crystal unit volume momentum in the reference frame moving with velocity v^s is given by (8.3.16)

$$p_i = \frac{\partial \mathcal{L}}{\partial \dot{u}_i} = \rho_{ik}^{(n)}\left(\frac{\partial u_k}{\partial t} - v_k^s\right).$$

In addition, if we take $p_{ik} = M\beta_{ik}$ and set

$$\lambda_{iklm} = \lambda^0_{iklm} + M\left(\delta_{ik}\beta_{lm} + \delta_{lm}\beta_{ik} - \delta_{ik}\delta_{lm}\right),$$

then the equation for the energy density (8.3.14) follows from (8.3.16).

Equations (8.3.11), (8.3.12) result from the Lagrange function density (8.3.16) when the coefficients are related by

$$M = A\rho_0, \qquad p_{ik} = A\rho_0\beta_{ik}.$$

2) One can see that with the given choice of generalized coordinates (8.3.15), (8.3.14) will not follow from such a density of the Lagrange function. In field theory the Lagrange function density is then, generally, a positively definite form in the time derivatives of the generalized coordinates (in our case in $\partial \boldsymbol{u}\partial t$ and $\partial\varphi\partial t$).

We rewrite (8.3.11), (8.3.12) as equations for the functions $\boldsymbol{u}(\boldsymbol{r},t)$ and $\varphi(\boldsymbol{r},t)$:

$$\rho_{ik}^{(n)}\frac{\partial^2 u_k}{\partial t^2} - \lambda_{iklm}\nabla_k\nabla_l u_m = A\mu_{ik}\nabla_k\frac{\partial\varphi}{\partial t}; \tag{8.3.17}$$

$$\rho_0\frac{\partial^2\varphi}{\partial t^2} - A\rho_{ik}^{(s)}\nabla_i\nabla_k\varphi = \mu_{ik}\frac{\partial\epsilon_{ik}}{\partial t}, \tag{8.3.18}$$

with $\mu_{ik} = \rho_0\beta_{ik} - \rho_{ik}^{(s)}$, and introduce a new elasticity modulus tensor

$$\tilde{\lambda}_{iklm} = \lambda_{iklm} - M\beta_{ik}\beta_{lm} = \lambda_{iklm}^0 - M\left(\delta_{ik} - \beta_{ik}\right)\left(\delta_{lm} - \beta_{lm}\right).$$

Equations (8.3.17), (8.3.18) are a complete set of equations for small mechanical quantum crystal vibrations. They involve, apart from the tensor parameter $\rho_{ik}^{(s)}$ specific for a quantum crystal, the quantity A having the dimension of the velocity squared. According to (8.3.15), this quantity is included in the linear differential relation

$$\frac{\partial v^s}{\partial t} = A\,\mathrm{grad}\left(\eta + \beta_{ik}\epsilon_{ik}\right), \tag{8.3.19}$$

derived by assuming the existence of the velocity potential for a superfluid motion velocity. The constant A is a macroscopic quantum-mechanical characteristic of the tunneling atomic motion in the crystal.

8.4
The Long-Wave Vibration Spectrum

Let us now discuss the dispersion laws for small vibrations of a quantum crystal. Assuming all the variables in (8.3.17), (8.3.18) are dependent on the coordinates and time through the multiplier $\exp(i\boldsymbol{kr} - i\omega t)$, we find

$$\begin{aligned}(\omega^2\rho_{ij}^{(n)} - \tilde{\lambda}_{imlj}k_l k_m)u_j + A\mu_{il}k_l\omega\varphi &= 0;\\ (\omega^2\rho_0 - A\rho_{il}^{(s)}k_i k_l)\varphi + \mu_{il}k_i\omega u_l &= 0.\end{aligned} \tag{8.4.1}$$

We have obtained a system of four homogeneous algebraic equations that allows us to find the four unknowns $(\boldsymbol{u},\varphi)$. The solvability condition for this system determines the dependence $\omega = \omega(\boldsymbol{k})$, i. e., the dispersion law. But the solvability condition for the system (8.4.1) is the zero determinant of a corresponding fourth-rank matrix. Hence, to each value of the wave vector $\boldsymbol{k}$ there correspond four eigenfrequencies ω, i. e., there are four branches of the quantum crystal eigenvibrations. Thus, a new branch of mechanical vibrations generated by additional degrees of freedom appears in a quantum crystal.

It follows from (8.4.1) that all the branches of the vibrations have the sound-type dispersion laws: $\omega = s_\alpha(\boldsymbol{\kappa})k$, $\boldsymbol{\kappa} = \dfrac{\boldsymbol{k}}{k}$, $\alpha = 1, 2, 3, 4$. If the quantum properties of the crystal are weakly manifest ($A \ll s_\alpha^2$ for $\alpha = 1, 2, 3$ and $|\mu_{ik}|, \left|\rho_{ik}^{(s)}\right| \ll \rho_0$), then in the main approximation the vibrations are divided into purely lattice ones and those of quantum dilatation. The equation for lattice vibrations

$$\left[\omega^2(\rho\delta_{ij} - \rho_{ij}^{(s)}) - \tilde{\lambda}_{ilmj}k_l k_m\right] u_j = 0, \tag{8.4.2}$$

formulated in the approximation (8.4.2), is equivalent to a set of equations

$$\left\{\omega^2\rho_0\delta_{ij} - \left(\tilde{\lambda}_{ilmj} + \frac{1}{\rho_0}\rho_{in}^{(s)}\lambda_{nlmj}\right) k_l k_m\right\} u_j = 0.$$

Thus, the dynamics of quantum dilatation in the approximation linear in $\rho^{(s)}/\rho_0$, can be taken into account in the renormalization of crystal elastic moduli. The effective elastic moduli $\lambda_{iklm}^* = \tilde{\lambda}_{iklm} + (1/\rho)\rho_{in}^{(s)}\lambda_{nklm}^0$. The sound velocities are renormalized to the same extent.

The fourth equation and the corresponding dispersion law can be obtained from the last equation of (8.4.1): $\omega^2 = (1/\rho)A\rho_{il}^{(s)}k_i k_l$.

This dispersion law corresponds to the crystal density vibrations at fixed lattice sites ($\boldsymbol{u} = 0$).

Part 4 Crystal Lattice Defects

The Crystal Lattice: Phonons, Solitons, Dislocations, Superlattices, Second Edition. Arnold M. Kosevich

ISBN: 3-527-40508-9

9
Point Defects

9.1
Point-Defect Models in the Crystal Lattice

In Chapter 1 it was mentioned that any distortion or violation of regularity in the crystal atomic arrangement can be considered as a crystal lattice defect. The presence of defects in a real crystal distinguishes it from an ideal crystal lattice and some properties of a real crystal are determined by its defect structure. The influence of defects on the physical properties of the crystal depends essentially on the defect dimensionality. This value (dimensionality) is the number of spatial dimensions along which the defect has macroscopic dimensions.

A *point* (or zero-dimensional) defect is a lattice distortion concentrated in a volume of the order of magnitude of the atomic volume. If a regular atomic arrangement is broken only in the small vicinity of a certain line, the corresponding defect will be called *linear* (or one-dimensional). Finally, when a regular atomic arrangement is violated along the part of some surface with a thickness of the order of interatomic distances a *surface* (or two-dimensional) defect exists in the crystal.

Any defect can have the following two functions affecting various crystal properties. First, a region of a "distorted" crystal arises near the defect and the defect looks like a *local inhomogenity* in the crystal. Then, the presence of a defect causes some stationary deformations in the crystal lattice at a distance from it, resulting in the displacement of atoms from their equilibrium positions in an ideal lattice. Thus, the defect is also a displacement field source in a crystal. The field of atomic displacements near the defect is dependent naturally on the character of the influence of the defect on the surrounding lattice (matrix).

The simplest types of point defects in a crystal are as follows: *interstitial atoms* are atoms occupying positions between the equilibrium positions of ideal lattice atoms (Fig. 9.1a); *vacancies* are lattice sites where atoms are absent (Fig. 9.1b); *interstitial impurities* are "strange" atoms incorporated in a crystal, i. e., those that occupy interstitial positions in a lattice (Fig. 9.2a); *substitutional impurities* are "strange" atoms or entire molecules that replace the host atoms in lattice sites (Fig. 9.2b).

The Crystal Lattice: Phonons, Solitons, Dislocations, Superlattices, Second Edition. Arnold M. Kosevich

ISBN: 3-527-40508-9

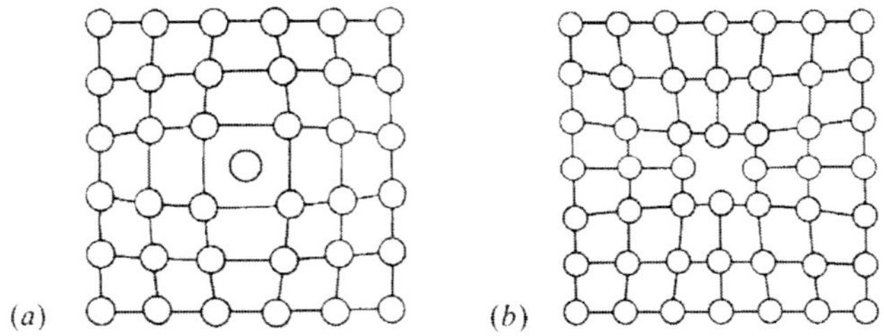

Fig. 9.1 "Proper" defects of a crystal lattice: (*a*) is an interstitial atom; (*b*) is a vacancy.

We consider the influence of a "proper" point defect on the surrounding matrix in a simple cubic lattice of a metal or a nonpolar dielectric (dielectric with a covalent bond). An interstitial atom incorporated in such a lattice locally breaks its ideality. The sites nearest to this atom are displaced due to the interstitial atom (Fig. 9.1a). In a simple cubic lattice this deformation has a cubic symmetry.

A vacancy in a simple crystal lattice leads to the displacement of the nearest atoms in the direction of the vacancy position (Fig. 9.1b).

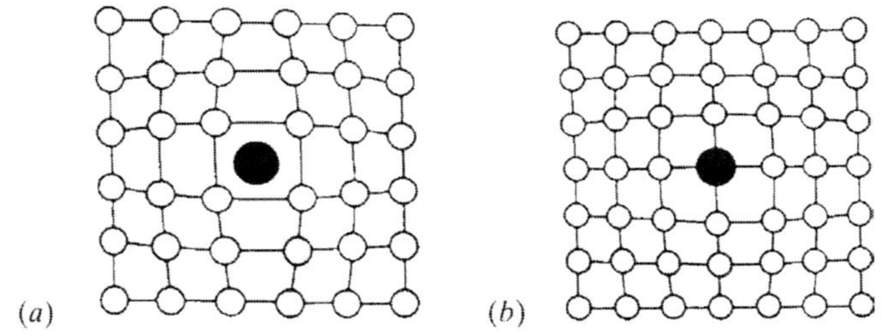

Fig. 9.2 Atoms of impurities: (*a*) – interstitial; (*b*) – substitutional.

It is seen from Fig. 9.1a,b that vacancies and interstitial atoms may be considered as defects of opposite "sign". In particular, a vacancy in a simple crystal lattice leads to the displacement of the nearest atoms towards the vacancy position (Fig. 9.1b). The annihilation of a vacancy and an interstitial atom may be effected, that is followed by the disappearance of the vacancy and interstitial atom.

Another process is possible when an interstitial atom leaves its position and goes over to an interstitial site, creating a pair of defects. This scheme is effected if a crystal is radiated by energetic particles when a particle passing through the crystal displaces an atom from its site, transferring it to an interstice.

A vacancy and an interstitial atom positioned close together are referred to as a *Frenkel pair* of defects (Ya. I. Frenkel was the first to describe these defects).

Let us note that the local environment of an interstitial atom (Fig. 9.1a) and the directions of the nearest-atom displacements for crystals with a primitive Bravais lattice (Fig. 9.3) are different from those observed in a body- or face-centered cubic lattice.

The displacements near the interstitial atom in such lattices may have no high degrees of symmetry, as shown in Fig. 9.3. In particular, in the vicinity of an "excess" atom the lattice configuration with two isolated atoms is possible (Fig. 9.4). Such a position of an interstitial atom generates a *dumb-bell-shaped configuration* of the defect.

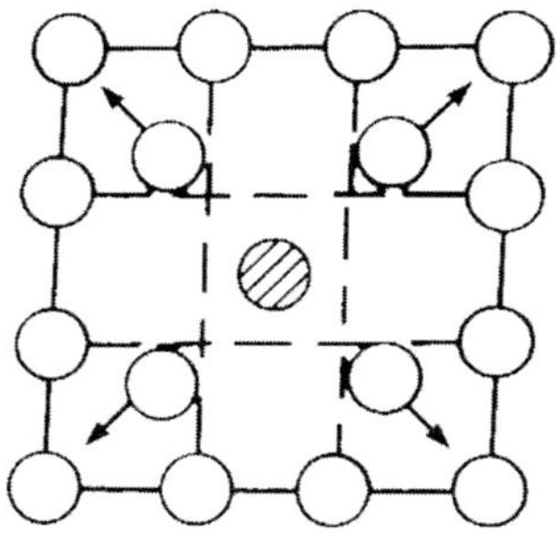

Fig. 9.3 Atomic displacement near an interstitial in a primitive cubic lattice.

Another possible position of an "extra" proper atom in a lattice is the crowdion configuration that is different in symmetry from the standard interstitial position. A crowdion in a 1D crystal is described in Section 5.2.

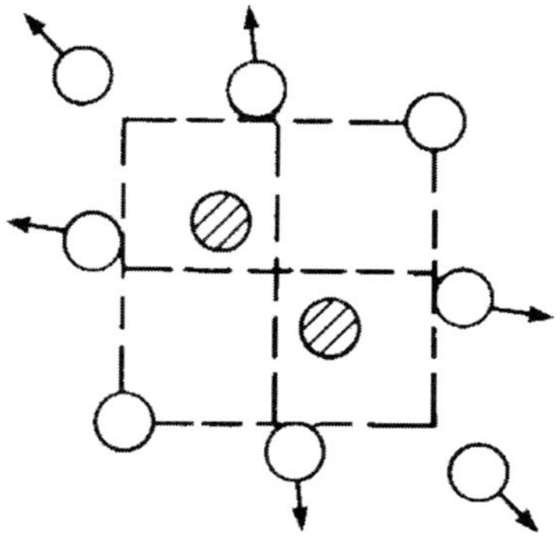

Fig. 9.4 Dumb-bell-shaped configuration of an interstitial atom.

Now we turn to a line of densely packed atoms in a 3D crystal (Fig. 9.5). Of the two possible types of defects with an extra atom of the same type (dumb-bell-shaped and crowdion configurations) the "dumb-bell" is an energetically more advantageous static point defect, while the crowdion is a dynamic defect capable of moving easily along a line of densely packed atoms. The comparatively easy motion of a crowdion is associated with the fact that with a sufficiently extended region of atomic condensation along a separate line, the displacement of the crowdion center of mass is achieved by an insignificant displacement of each atom in this line.

It is natural that in a low-symmetry lattice the vacancy may also be characterized by a local atomic rearrangement of a dipole or "anticrowdion" type.

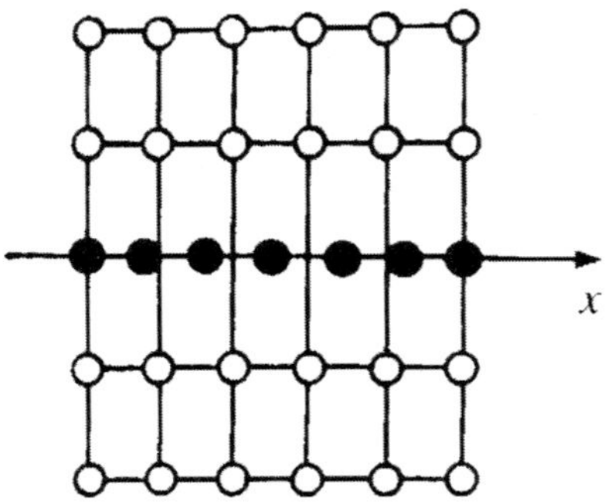

Fig. 9.5 A crowdion on the x-axis.

An anisotropic lattice deformation (indicated by arrows in Fig. 9.4) is also generated by an impurity molecule that has no spherical symmetry. The corresponding static point defect is sometimes called an *elastic dipole*. As a rule, the same dipole-type defect may be oriented in different ways with respect to the crystal axes and the processes of elastic dipole reorientation are possible where its dynamic properties are manifest.

Considering the localization of an interstitial impurity (or an interstitial atom) in a crystal, we note that the unit cell of each crystal lattice has one or several equivalent positions for interstitials that are determined by the geometrical structure of the lattice. For instance, in a primitive cubic lattice this position is the cube center, in a FCC lattice the positions for interstitials are localized either at the center of the cube or at the middle of the cube edges. In a BCC lattice these positions are at the centers of tetrahedrons constructed of two atoms at the cube vertices and two atoms at the centers of neighboring cubes or in the middle of the cube edges (in octahedron interstices).

If the local surroundings of interstitial vacancies have no cubic symmetry (as, e. g., in a BCC lattice) nonisotropic deformations arise around the interstitial atoms. First, a prerequisite for a dumb-bell or crowdion configuration of an interstitial atom appears and, second, the impurity in such a position behaves as an elastic dipole. A classical example of the latter is the iron crystal distortion around a carbon atom impurity. Carbon atoms get into octahedron interstices of a BCC iron lattice and behave as single-axis elastic dipoles oriented along the cube edges.

9.2 Defects in Quantum Crystals

In describing the point defects of a crystal lattice we proceeded from a seemingly obvious assumption of defect localization in a certain site or interstice. However, the existence of crystals with specific quantum properties (Chapter 8) makes such an assumption that is purely substantiated and even unreal in some cases. This refers first of all to the defects in quantum crystals and to the behavior of hydrogen atom impurities in the matrix of rather heavy elements.

Since defect localization is doubtful, it is necessary to make a more rigorous analysis of the physical situation arising in a crystal containing defects.

If we have a crystal with a single point defect, then its Hamiltonian function (or Hamiltonian) is a periodic function of the coordinate of the defect with the period of the crystal lattice. We shall study only those crystal degrees of freedom that are described by the defect coordinates, with the temperature assumed to be zero. We consider the localization of the defect at one of the equilibrium positions corresponding to the crystal-energy minimum. The dynamical properties of a defect are then manifest only in small oscillations near the fixed equilibrium position. Thus, the crystal state concept is unambiguously associated with the notion of a fixed coordinate of the defect.

Defect localization becomes impossible when quantum tunneling occurs. The defect coordinate as a characteristic of the crystal state ceases to be a well-defined quantity and various states of a crystal with a defect should be classified by the values of a quasi-wave vector $\boldsymbol{k}$. The crystal energy becomes a periodic function with respect to $\boldsymbol{k}$.

If we subtract from this energy the energy of a crystal without a defect, then we get the defect energy $\varepsilon_D(\boldsymbol{k})$. The different values of $\boldsymbol{k}$ inside the Brillouin zone determine different energies $\varepsilon_D(\boldsymbol{k})$. Thus, an energy band of some width $\Delta\varepsilon$ proportional to the probability for quantum tunneling of a defect from one equilibrium position to a neighboring one arises. This new part of a crystal energy spectrum (Fig. 9.6a) is due to the appearance of a movable quantum defect. Thus, the defect is associated with an additional branch of quantum single-particle excitations.

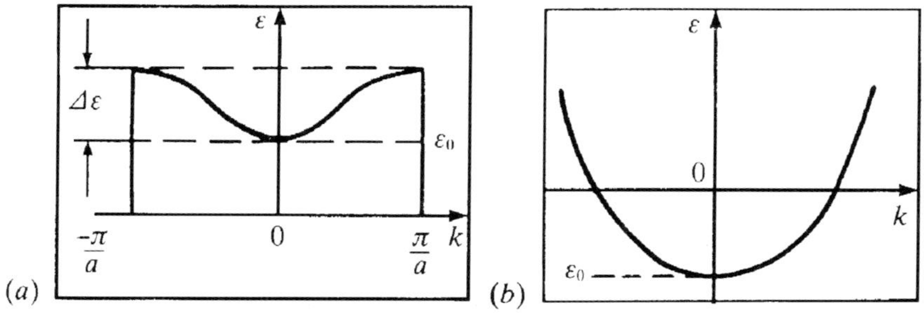

Fig. 9.6 The defecton dispersion law: (*a*) the appearance of a defecton needs energy expenditure ε_0; (*b*) the appearance of vacancies induces the rearrangement of the crystal ground state.

It is clear that a defect in a quantum crystal is delocalized and behaves as a free particle. It is called a *defecton* (Andreev and Lifshits, 1969) and the dependence of the energy $\varepsilon_D(\boldsymbol{k})$ on $\boldsymbol{k}$ is referred to as the defecton dispersion law. An example of a defecton is a ^{3}He isotope atom in solid ^{4}He.

The defectons can collide with one another at finite concentrations, and collisions with other crystal excitations (e. g., phonons) are also possible at finite temperatures.

An increasing number of collisions fundamentally affects the character of defecton motion. If the frequency of collisions is small enough, we practically have a freely moving defecton. With increasing frequency of collisions the defecton can approach equilibrium with the lattice during the time when it stays within a unit cell. We then speak about a practically localized defect.

In most cases the tunneling probability is relatively small. Therefore, to calculate the defecton dispersion law, we may use the strong coupling approximation known in electron theory. The function $\varepsilon_D(\boldsymbol{k})$ is found in this case explicitly at all values of $\boldsymbol{k}$. For instance, for a simple cubic lattice we have

$$\varepsilon_D(\boldsymbol{k}) = \varepsilon_1 - \varepsilon_2(\cos \boldsymbol{k a}_1 + \cos \boldsymbol{k a}_2 + \cos \boldsymbol{k a}_3), \tag{9.2.1}$$

where ε_1, ε_2 are constant values, $|\varepsilon_2|$ is proportional to the quantum-tunneling probability; $\boldsymbol{a}_\alpha$ are the fundamental lattice translation vectors, $\alpha = 1, 2, 3$. The width of the defecton energy band is $\Delta\varepsilon = 6\,|\varepsilon_2|$.

In an isotropic approximation, the expansion (9.2.1) near the minimum value of ε_0 (the energy band bottom) has the form

$$\varepsilon_D = \varepsilon_0 + \frac{\hbar^2 k^2}{2m^*}, \quad \varepsilon_0 = \varepsilon_1 - 3\varepsilon_2, \tag{9.2.2}$$

where m^* is the defecton effective mass (of order of magnitude $\hbar^2/m^* \sim a^2 \Delta\varepsilon$).

The presence of a defecton in a quantum crystal allows one to explain the physical nature of quantum dilatation (Chapter 8). We assume that in a crystal free of impurities the "defectiveness" arises only due to the excitation of vacancies. The possibility of tunneling transforms a vacancy into a defecton, or a *vacancy wave* with the dispersion relation (9.2.2).

Vacancy wave generation with $\boldsymbol{k} = 0$ does not break the ideal periodicity of a crystal. However, the number of crystal lattice sites becomes unequal to the number of atoms. The defecton energy with $\boldsymbol{k} = 0$, i. e., ε_0, is dependent on the state of a crystal, in particular on its volume V changing under the influence of an external pressure. It may turn out that at a certain volume V_k the parameter ε_0 vanishes. We assume that near this point

$$\varepsilon_0 = \lambda \frac{V_k - V}{V_k}. \tag{9.2.3}$$

We shall set, for definiteness, $\lambda > 0$ and consider the defectons obeying Bose statistics (such as vacancies in solid ^{4}He).

It then follows from (9.2.3) that for $V < V_k$, $\varepsilon_0 > 0$ and the energy spectrum of defectons is separated from the ground-state energy (without defectons) by a gap. A finite activation energy is needed to form a vacancy and thus at $T = 0$ the number of defectons equals zero.

For $V > V_k$, $\varepsilon_0 < 0$ (Fig. 9.6b) and defecton generation becomes advantageous even at $T = 0$. Vacancies tend to assemble into a state with the last energy ($\boldsymbol{k} = 0$) that is promoted by Bose statistics. The defects are accumulated (condensed) in this state until defect repulsion effects start to manifest themselves. It is clear that with any vacancy repulsion law the energy minimum at $T = 0$ corresponds to a nonzero equilibrium concentration of defectons with $\boldsymbol{k} = 0$. Since with a fixed number of atoms each vacancy increases the number of crystal sites by unity, a finite defect concentration actually generates a certain crystal dilatation. This is just the quantum dilatation.

A narrow energy band is typical for the defecton dispersion law (9.2.1). According to experiments, its width $\sim 10^{-5} - 10^{-4}$ K $\sim 10^{-9} - 10^{-8}$ eV $\sim 10^{-21} - 10^{-20}$ erg for the ^{3}He atom playing the role of an "impuriton" in solid ^{4}He. Thus, the energy band width of the quasi-particle motion of this well-studied defecton is very small compared to any energies on an atomic scale. This makes the defecton dynamics in external inhomogeneous fields that arise in a crystal, essentially different from the dynamics of ordinary free particles.

Let the defecton be in an external field providing the potential energy $U(\boldsymbol{x})$ but not influencing its kinetic energy (9.2.1). The total energy of the defecton $E(\boldsymbol{k}, \boldsymbol{x})$ can then be written as $E(\boldsymbol{k}, \boldsymbol{x}) = \varepsilon_D(\boldsymbol{k}) + U(\boldsymbol{x})$.

If, for the characteristic distances $l \gg a$, a the potential energy changes by $\delta U \gg \Delta\varepsilon$ the defecton energy will fill a narrow strongly distorted band (Fig. 9.7). The fixed energy of a defecton $E(\boldsymbol{k}, \boldsymbol{x}) = E =$ const corresponds to its motion localized in a space region with dimensions $\delta x \sim l\Delta\varepsilon/\delta U \ll l$. Such a localization is independent of the sign of grad U (Fig. 9.7).

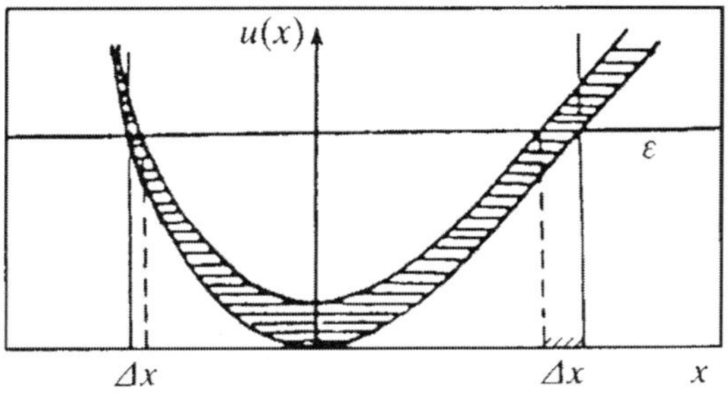

Fig. 9.7 Localization of defecton motion in an external potential field.

Let us assume that the external field is changed at a distance $l \sim 100a \sim 10^{-6}$ cm in a certain direction by δU eV. In this case, for $\Delta\varepsilon \sim 10^{-8}$ the defecton is localized along the direction of grad U at a distance $\Delta x \sim 10^{-8}$ cm and it moves in fact along the surface of constant potential energy $U(\boldsymbol{x}) =$ const (to be precise, in a thin layer with a thickness that is comparable with the interatomic distance).

We note that in a strong magnetic field the ^{3}He impuritons with different orientation of nuclear spin are in different energy bands. Indeed, the nuclear magnetic moment of

^{3}He in a magnetic field H has the energy (μ_0 is the Bohr magneton)

$$U_m = \pm\mu_{\text{nucl}} H \sim \pm 10^{-3} \mu_0 H.$$

For $H \sim 10^{-3}$ Oe we obtain $|U_m| \sim 10^{-19}$ erg $\gg \delta\varepsilon$, i. e., the energy bands of impuritons with magnetic moments directed along an external magnetic field or opposite to it do not overlap.

9.3 Mechanisms of Classical Diffusion and Quantum Diffusion of Defectons

Any point defects in a lattice are capable of migrating, i. e., moving in a crystal. For classical defects the only reason for migrating is a fluctuational thermal motion, and this arises through chaotic movement of a point defect in a lattice. If under the action of certain "driving forces" such a migration is effected directionally, then one can speak of the diffusion of point defects. However, sometimes the diffusion motion implies any thermal migration of the defects even if it is not characterized by a specific direction. In what follows we shall be interested not in the direction of diffusional migration of the defects but in the mechanisms of their migration.

Thus, the atomic process in which the defect performs more or less random walks by jumping from a certain position to the neighboring equivalent position underlies classical diffusion. What determines the probability of a separate jump?

A strong interaction between the classical defect and a crystal lattice makes it localized. As a result, the defect turns out to be in a deep potential well and performing here small oscillations with frequency ω_0.

Suppose that we transfer the defect quasi-statically from some point of localization to neighboring ones, defining the crystal energy E as a function of an instantaneous coordinate of the defect. If we denote by x a coordinate measured along this imaginary route of the defect migration, the crystal-energy plot in the simplest case will have the form shown in Fig. 9.8, where $x = x_0$ and $x = x_1$ are two neighboring positions of the defect localization. Varying the "routes" of the transition from x_0 to x_1, we can find a way through the "saddle point" characterized by the lowest barrier U_0 that divides the positions x_0 and x_1. The probability for a thermally activated fluctuation transition of the defect into the neighboring equilibrium position will then be proportional to $\exp(-U_0/T)$ and the diffusion coefficient

$$D = D_0 e^{-U_0/T}, \qquad D_0 \sim a^2\omega_0, \tag{9.3.1}$$

where a is the lattice constant determining the order of magnitude of the distance between the neighboring equivalent positions of the defect.

Comparing the defect migration activation energy U_0 with the parameter of the plot shown in Fig. 9.8, we should remember that this parameter is conventional. The existence of the function $E = E(x)$ assumes that in the process of defect migration, the crystal has time to get into an equilibrium state characterized by the defect coordinate x. Thus, the energy E can be regarded as a function of the coordinate only when the defect moves slowly in passing between the positions x_0 and x_1. If the "slowness" condition is not satisfied, the plot for the function of the variable x becomes meaningless. Equation (9.3.1) remains valid, but the activation energy U_0 becomes an independent diffusion parameter that determines the "saddle-point" energy in some multidimensional space of crystal lattice atomic configurations near the defect.

Finally, the migration of a complex point defect may be multistepped. The crowdion mechanism of "extra" atom migration is very often observed along a line of densely packed atoms. However, crowdion displacement under a strong external influence resembles mechanical motion rather than jump diffusion.

Quite a different mechanism is responsible for quantum defect (defecton) motion. We consider the behavior of an individual defecton in an almost ideal crystal with a very small concentration of both classical defects and defectons.

At absolute zero ($T = 0$) the defecton behaves as a quasi-particle with the dispersion law (9.2.1), moving freely in a crystal and a set of defectons has the properties of an ideal gas. In this case the defecton diffusion coefficient can be determined by a gas-kinematic expression $D \sim vl_0$ where v is the defecton velocity ($v \sim \Delta\varepsilon/k \sim \Delta\varepsilon/\hbar$); l_0 is the mean free path ($l_0 \gg a$) characterizing the collision of defectons with classical defects (e. g., impurities). The impurity free path (or defecton–defecton free path with a fixed number of defectons) is temperature independent in the low-temperature region. Therefore, one can always find an interval of extremely low temperatures in which the defecton diffusion coefficient is independent of T: $D = D(0) =$ constant.

The defecton energy band width $\Delta\varepsilon$ that is proportional to the defecton tunneling probability, is very small ($\Delta\varepsilon \ll \hbar\omega$) and thermal crystal oscillations can break it easily. The main reason for the defecton band breaking can be explained by means of the following rough scheme.

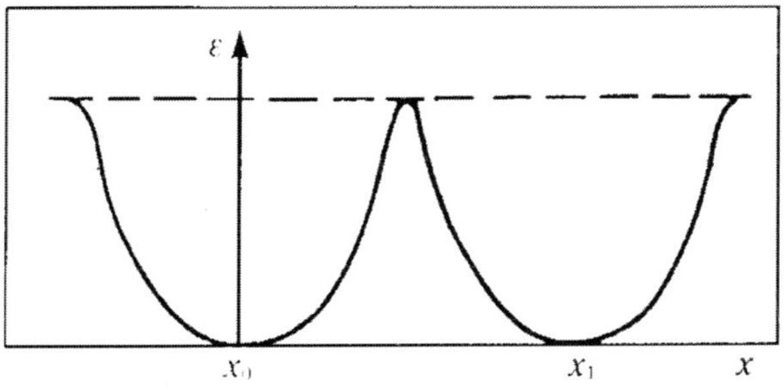

Fig. 9.8 The potential barrier separating equilibrium positions of the defect at point x_0 and x_1.

Let us make use of the concept of a site potential well as shown in Fig. 9.8. Neglecting weak quantum tunneling, we consider the defecton localized in one such well. As the well is rather deep, there exists a finite number of discrete energy levels of the defecton ε_n ($n = 1, 2, 3, \ldots$), where $\varepsilon_{n+1} - \varepsilon_n \sim \hbar\omega_0$. By virtue of the translational symmetry of the crystal, such energy levels are also observed in the other site wells (Fig. 9.9). The switching on of the tunneling effect generates in all systems of resonance levels, narrow energy bands inside which the energy of the defecton free motion depends on the quasi-wave vector $\boldsymbol{k}$ and is determined by an equation such as (9.2.1). The quantum phenomenon of the above-barrier reflection in a periodic structure also generates some energy bands of the defecton in the region of a classical continuous energy spectrum (S is one such band (Fig. 9.9). At $T = 0$, quantum tunneling occurs at the ground level, and due to the band character of defecton motion the tunneling time t is of the order of magnitude $t \sim a/v \sim \hbar/\Delta\varepsilon$.

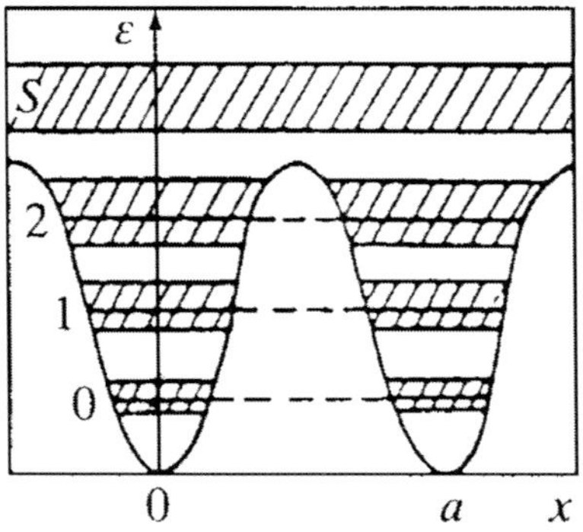

Fig. 9.9 Energy bands of free defecton motion in a crystal at $T = 0$.

Thermal crystal motion shifts separate wells relative to each other, and for $T \gg \Delta\varepsilon$ the energy levels in the neighboring site wells cease to be resonant. This does not destroy the defecton "pseudoband" motion. Indeed, the time of real tunneling $t \sim \hbar/\Delta\varepsilon$ is large compared to both $1/\omega_0$ and $\hbar/T$. Thus, tunneling occurs at the ground level that is averaged over the lattice oscillations (at $t \ll \hbar\omega_0$). Such a *coherent* motion of the defect at rather low temperatures ($\Delta\varepsilon \ll T \ll \Theta$) has a rather large free path $l \gg a$. The defecton scattering by phonons makes l temperature dependent: $l = l(T)$. With increasing temperature the function $l(T)$ decreases and at some temperature $T = T_1$ it becomes less than l_0; furthermore, at a certain temperature $T = T_2$ it is equal to the lattice period: $l(T_2) \sim a$. In the latter case the defecton band is broken dynamically, and the defect is localized at a site.

The diffusion coefficient of the localized defect is expressed through the probability of the transition w into the neighboring site by the relation $D \sim a^2 w$, taking into account random jumps of a distance a with frequency w. Under the condition $T \ll \hbar\omega_0$, the probability $w \approx w_0$, where w_0 is the transition probability at the ground level and is temperature independent. The diffusion coefficient is then also temperature independent and is of the order of magnitude $D \sim a^2 w_0 \sim a^2 \Delta\varepsilon/\hbar$.

Further increase in the temperature leads to the fact that the defect may, with high probability, be in an excited state with energy ε_n in the potential well. The probability of a transition into the neighboring site in this case is: $w = \sum_n w_n e^{-(\varepsilon_n - F)/T}$, $F = -T \log \sum_n e^{-\varepsilon_n/T}$ where w_n is the probability of a transition from the state n. With rising temperature, w increases from the value $w_0 \sim 1/t$ (corresponding to quantum tunneling at the ground level) up to a classical value to $w \sim \exp(-U_0/T)$ that is attained due to the above-barrier transitions.

Thus, by lowering the temperature the diffusion coefficient of the defects first falls exponentially (classical diffusion), then reaches a plateau (the quantum diffusion of localized defects) and then rises to the value $D(0)$ (Fig. 9.10).

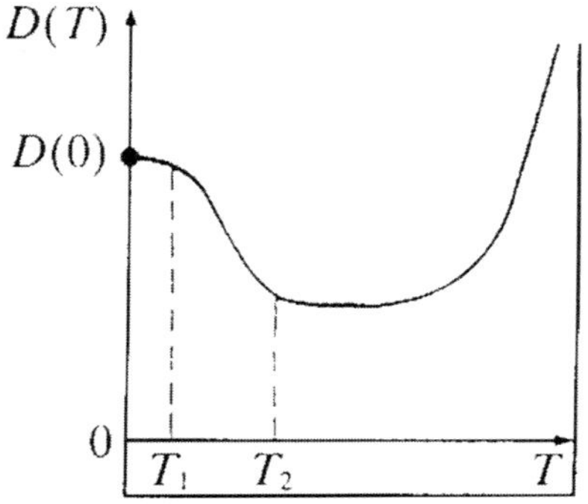

Fig. 9.10 The temperature dependence of the defecton diffusion coefficient; the defecton mobility is limited by the interaction with phonons in the interval (T_1, T_2).

9.4 Quantum Crowdion Motion

A crowdion moves mechanically, and the mechanical motion of atomic particles is quantized. Thus, quantum effects should manifest themselves in crowdion motion.

According to the classical equations of crowdion motion, the crowdion Hamiltonian function (1.7.6) is independent of the position of its center and this is a result of the continuum approximation in which this function is derived.

The simplest way to take into account the discreteness of the system concerned is the following. We use the solution (1.7.8) and calculate the static energy of an atomic chain with a crowdion at rest by using the formula

$$E = \sum_{n=-\infty}^{\infty} \left\{ \frac{1}{2}\alpha_0 \left(u_{n+1} - u_n \right)^2 + f(u_n) \right\}, \tag{9.4.1}$$

where $u_n = u(x_n) \equiv u(an)$.

It is easily seen that the static crowdion energy in the continuum approximation is equally divided between the interatomic interaction energy and the atomic energy in an external field. It can be assumed that the same equal-energy distribution remains in a discrete chain, and instead of (9.4.1) we then write

$$E = 2 \sum_{n=-\infty}^{\infty} F(u_n) = \tau^2 \sum_{n=-\infty}^{\infty} \sin^2\left(\frac{\pi u_n}{a}\right).$$

We substitute here (1.6.12), assuming the point $x = x_0$ to be a crowdion center:

$$E = \tau^2 \sum_{n=-\infty}^{\infty} \operatorname{sech}^2\left(\frac{an - x_0}{l_0}\right), \tag{9.4.2}$$

where $l_0 = as\sqrt{m}/(\pi\tau) = a^2\sqrt{\alpha_0}/(\pi\tau)$.

Using the Poisson summation formula,

$$\sum_{n=-\infty}^{\infty} f(n) = \sum_{m=-\infty}^{\infty} \int_{-\infty}^{+\infty} f(k) e^{2\pi i m k}\, dk\ , \tag{9.4.3}$$

we find

$$E = \frac{\tau^2}{a} \int_{-\infty}^{+\infty} \frac{dx}{\cosh^2\left(\dfrac{x}{l_0}\right)} + \frac{2\tau^2}{a} \sum_{m=1}^{\infty} e^{2\pi i m (x_0/a)} \int_{-\infty}^{+\infty} \frac{e^{2\pi i m (x/a)}}{\cosh^2\left(\dfrac{x}{l_0}\right)} dx. \tag{9.4.4}$$

The first term in (9.4.4) coincides with the energy E_0 of a crowdion at rest found in the continuum approximation (1.7.1), and the second term is a periodic function of the coordinate x_0 with period a: $E = E_0 + U(x_0)$, where

$$\begin{aligned} U(x) &= \frac{2 l_0 \tau^2}{a} \sum_{m=1}^{\infty} e^{2\pi i m (x/a)} \int_{-\infty}^{+\infty} \frac{\cos\left(\dfrac{2\pi m l_0 \xi}{a}\right)}{\cosh^2 \xi} d\xi \\ &= 4\alpha_0 a \sum_{m=1}^{\infty} \frac{m}{\sinh\left(\dfrac{\pi^2 m l_0}{a}\right)} \cos\left(2\pi m \frac{x}{a}\right). \end{aligned} \tag{9.4.5}$$

Since we have assumed that $l_0 \gg a$, it then suffices to keep one term with $m = 1$

$$U(x_0) = 8\alpha_0 a^2 e^{-\pi^2 l_0/a} \cos\left(2\pi \frac{x_0}{a}\right). \tag{9.4.6}$$

It is clear that the crowdion energy is periodically dependent on the coordinate of its center x_0 that may be regarded as a quasi-particle coordinate. We set in (1.6.15) $x_0 = Vt$ using the ordinary relation between the coordinate and the velocity (at $V =$ const $\neq 0$). The part of the energy (9.4.6) that is dependent on the coordinate plays the role of the crowdion potential energy. Thus, crowdion migration in a discrete atomic chain deals with overcoming of the potential relief (9.4.6). However, for $l_0 \gg a$, the potential energy curve (9.4.6) creates very weak potential barriers between the neighboring energy minima, and the crowdion may overcome them through quantum tunneling.

The Hamiltonian

$$\mathcal{H} = E_0 + \frac{p^2}{2m^*} + U_1 \cos\left(2\pi\frac{x}{a}\right), \tag{9.4.7}$$

is used for the quantum description of crowdion motion, where

$$U_1 = 8\alpha_0 a^2 e^{-\pi^2(l_0/a)}. \tag{9.4.8}$$

As both m^* and U_1 decrease with increasing parameter $l_0/a \sim a\sqrt{\alpha_0}/\tau$ the physical situations, where

$$\frac{\hbar^2}{m^* a^2} \gg U_1, \tag{9.4.9}$$

are quite reasonable. The inequality (9.4.9) means that the potential energy contribution that is dependent on the coordinate is a weak perturbation of the kinetic energy of free crowdion motion. In other words, the amplitude of zero crowdion vibrations in one of the potential minima (9.4.9) greatly exceeds the one-dimensional crystal period and the crowdion transforms into a *crowdion wave* (Pushkarov, 1973).

The energy spectrum of a crowdion wave with the Hamiltonian (9.4.7) is rather complicated and consists of many bands in each of which the energy is a periodic function of the quasi-wave number k with period $2\pi/a$. However, small crowdion wave energies for $k \ll 2\pi/a$ are not practically distinguished from the free particle energy with the Hamiltonian (1.7.6) under the condition (9.4.9). Indeed, if we calculate quantum-mechanical corrections to the free particle energy in the second order of perturbation theory in the potential (9.4.6), then

$$\varepsilon(k) = E_0 - U_1\frac{m^* a^2 U_1}{(2\pi\hbar)^2} + \frac{\hbar^2 k^2}{2M},$$

$$M = m^*\left\{1 + \frac{1}{2}\left[\frac{m^* a^2 U_1}{(\pi\hbar)^2}\right]^2\right\}.$$

Thus, in spite of the presence of a potential energy curve (9.4.6), the crowdion wave moves through a crystal as a free particle with a mass close to the crowdion effective mass.

9.5 Point Defect in Elasticity Theory

When we consider the problems concerned with macroscopic mechanical properties of solids, as a rule, it is necessary to take the static lattice distortions caused by the point defect into account. In calculations of such displacements the point defect is the source of an elastic field.

In the atomic model of an interstitial atom (Fig. 9.1a, or 9.3), the atoms of the nearest surroundings of an interstitial atom are repelled by the forces distributed very symmetrically in each coordination sphere. The system of forces has a zero resultant and a zero total moment. From a macroscopic point of view, their action is equivalent to the action of three pairs of equal forces, applied to the point where an interstitial atom is located and directed along the coordinate axes. In elasticity theory such a system is described by the spatial distribution of forces such as (the defect is at the coordinate origin)

$$\boldsymbol{f}(\boldsymbol{r}) = -K\Omega_0 \, \mathrm{grad} \, \delta(\boldsymbol{r}), \tag{9.5.1}$$

where k is the total compression modulus; Ω_0 is a constant multiplier with the dimensions of volume, whose physical meaning will now be elucidated.

We note that if the distribution of bulk and surface forces in a cubic crystal is known an elastic change in the crystal volume is calculated directly. We can determine a total change in the deformed crystal volume without solving the problem of its deformed state.

The equation of the elastic medium equilibrium that follows from (2.9.17) has the form $\nabla_i \sigma_{ik} + f_k = 0$.

We multiply this relation by x_i and integrate over the whole crystal volume

$$\int x_i \nabla_k \sigma_{ki} \, dV = -\int \boldsymbol{r f} \, dV. \tag{9.5.2}$$

The l.h.s. of (9.5.2) can be transformed simply to give

$$\begin{aligned} \int x_i \nabla_k \sigma_{ki} \, dV &= \int \nabla_k (x_i \sigma_{ki}) \, dV - \int \sigma_{ik} \nabla_k x_i \, dV \\ &= \oint_S x_i \sigma_{ki} ds_k - \int \sigma_{ik} \delta_{ik} \, dV \\ &= -\int \sigma_{kk} \, dV + \oint \boldsymbol{rp} \, dS, \end{aligned} \tag{9.5.3}$$

where $\boldsymbol{p}$ is the vector of the surface force density acting on the external surface of a solid body S.

We substitute (9.5.3) into (9.5.2) and use the relation (2.9.14) in a cubic crystal and also (1.9.5)

$$\nabla V = \frac{1}{3K} \int \boldsymbol{rf} \, dV + \frac{1}{3K} \int \boldsymbol{rp} \, dS. \tag{9.5.4}$$

If the external surface of a solid is free ($\boldsymbol{p} = 0$), it follows from (9.5.4) that

$$\Delta V = \frac{1}{3K} \int \boldsymbol{rf} \, dV. \tag{9.5.5}$$

We use (9.5.5) and calculate the change in the crystal volume induced by the density of forces (9.5.1), and that experiences no external influence

$$\Delta V = -\frac{1}{3}\Omega_0 \int \boldsymbol{r}\,\mathrm{grad}\,\delta(\boldsymbol{r})\,dV = \frac{\Omega_0}{3}\int \delta(\boldsymbol{r})\,\mathrm{div}\,\boldsymbol{r}\,dV = \Omega_0 . \qquad (9.5.6)$$

Thus, in a cubic crystal (or in an isotropic medium) the quantity Ω_0 has a simple physical meaning and its value equals the increase in the crystal volume caused by one interstitial atom. The extra atom can only enlarge the crystal volume and therefore $\Omega_0 > 0$. Usually the volume increase caused by an interstitial atom has the order of magnitude of the atomic volume, and hence $\Omega_0 \sim V_0 = a^3$.

According to the classification of the elastic fields in the isotropic medium the defect described by (9.5.1) is called a *dilatation center*. We have thus used the dilatation model for the interstitial atom.

A vacancy is different from an interstitial atom in that the deformation of the crystal lattice is connected with the displacement of the nearest atoms towards the defect (Fig. 9.1a). This displacement is induced by forces whose symmetry in a simple cubic lattice is the same as in the case of an interstitial atom. In other words, the formula (9.5.1) is useful for describing vacancies, but the dilatation intensity should be regarded as negative ($\Omega_0 < 0$).

The deformation near a dipole-type point defect is described in a somewhat different way. Such a deformation is generated by a system of forces whose macroscopic equivalent is represented by three pairs of forces applied at the point where the defect is localized, with zero moment for each pair but with different values of the forces. In elasticity theory such a system can be described by the density of forces

$$f_i(\boldsymbol{r}) = -K\Omega_{ik}\nabla_k\delta(\boldsymbol{r}), \qquad (9.5.7)$$

and the absence of a total moment of these forces is contained in the symmetry of the tensor Ω_{ik} ($\Omega_{ik} = \Omega_{ki}$).

It is clear that the latter property is inherent only to a static point defect that is in equilibrium with the surrounding lattice. If the elastic dipole orientation is a dynamic characteristic, then the tensor Ω_{ik} is not necessarily symmetric.

Performing calculations analogous to (9.5.6), it is easy to verify that the density of forces (9.5.7) causes changes in the volume $\Delta V = (1/3)\,\Omega_{ll}$. Thus, for an interstitial impurity it is natural to take $\Omega_{ll} > 0$, assuming $\Omega_{ll} \sim a^3$, and for the vacancy to take $\Omega_{ll} < 0$.

If the elastic dipole has axial symmetry it is characterized by a tensor Ω_{ik} of the form

$$\Omega_{ik} = \Omega_0\delta_{ik} + \Omega_1\left(l_i l_k - \frac{1}{3}\delta_{ik}\right),$$

where $\boldsymbol{l}$ is the unit vector of the dipole axis. In a cubic crystal the first term describes the pure dilatation properties of an elastic dipole and the second term contains a new parameter Ω_1 that determines the deviator part of the tensor Ω_{ik}.

For the defect described by the density of forces (9.5.7) the convolution Ω_{ll} only has an obvious interpretation in a cubic crystal. If a crystal has no cubic symmetry, a simple physical meaning even of this quantity is lost and the tensor Ω_{ik} should be regarded as a certain effective characteristic of the point defect.

Let us find the elastic field generated by a point defect and see that it really is extended over macroscopic distances (otherwise, a proposed treatment would make no sense). We denote by G_{ik} the static Green tensor of elasticity theory, i. e., a solution to the equation vanishing at infinity

$$\lambda^*_{iklm}\nabla_k\nabla_l G_{mj}(\boldsymbol{r}) + \delta_{ij}\delta(\boldsymbol{r}) = 0.$$

The displacement vector induced in an unbounded crystal by the density of forces $f(\boldsymbol{r})$ is

$$u_i(\boldsymbol{r}) = \int G_{ik}(\boldsymbol{r}-\boldsymbol{r}')f_k(\boldsymbol{r}')\,dV'. \tag{9.5.8}$$

We substitute here (9.5.7) and integrate by parts:

$$u_i(\boldsymbol{r}) = -K\Omega_{kl}\nabla_l G_{ik}(\boldsymbol{r}). \tag{9.5.9}$$

In the isotropic approximation, the Green tensor for an unbounded space can be written explicitly

$$G_{ik}(\boldsymbol{r}) = \frac{1}{8\pi G}\left(\delta_{ik}\Delta - \frac{1}{2(1-\nu)}\nabla_i\nabla_k\right) r, \tag{9.5.10}$$

where Δ is the Laplace operator ($\Delta \equiv \nabla_k^2$); ν is the Poisson coefficient (Poisson's ratio).

We substitute (9.5.10) into (9.5.9) and restrict ourselves to the case of a dilatation center ($\Omega_{ik} = \delta_{ik}\Omega_0$):

$$\boldsymbol{u}(\boldsymbol{r}) = -\frac{\Omega_0}{12\pi}\frac{1+\nu}{1-\nu}\,\mathrm{grad}\,\frac{1}{r}. \tag{9.5.11}$$

As it follows from (9.5.11) that $\mathrm{div}\,\boldsymbol{u} = \varepsilon_{kk} = 0$ at all the points off the region occupied by an interstitial atom (off the point $\boldsymbol{r} = 0$), the dilatation center in an unbounded isotropic medium causes a pure shear deformation. It is natural that the latter conclusion is valid only when the following conditions are satisfied simultaneously: first, the medium is unbounded, second, the medium is purely isotropic, third, the point defect is equivalent to the dilatation center. If even one of these conditions is not satisfied, the elastic point defect field is not strictly a shear field.

We now discuss the first of the above-mentioned conditions. It is connected with the fact that static elastic fields similar to a Coulomb field decrease very slowly with distance. That is why it is necessary to take into account the finite dimensions of the crystal specimen even in considering the field of an isolated defect.

The displacement field that is generated by the dilatation center is described by (9.5.11) only in an infinite-dimension specimen. In any arbitrarily large, yet finite specimen, this formula needs to be improved. Indeed, we note that the dilatation is concentrated at the defect:

$$\operatorname{div} \boldsymbol{u} = \pi\gamma\Omega_0\delta(\boldsymbol{r}), \quad \gamma = -\frac{1}{3\pi}\frac{1+\nu}{1-\nu}.$$

Therefore, the increase in the volume of the whole specimen calculated from (9.5.11) is given by

$$\Delta V = \int \operatorname{div} \boldsymbol{u}\, dV = \pi\gamma\Omega_0. \tag{9.5.12}$$

It can be verified that $\pi\gamma = (1 + 4G/3K)^{-1} < 1$. Thus, the increase in the crystal volume obtained by a direct calculation appears to be less than the initial Ω_0. This is due to the fact that the solution (9.5.11) does not actually satisfy the requirement $\sigma_{nn} = 0$ on the external surface of a solid. It is clear that if $\boldsymbol{n}$ is the unit vector of the normal to the spherical surface, then it follows from (9.5.11) (at $r = R$) that

$$\sigma_{ik}n_k = 2G_{ik}n_k = -G\Omega_0\gamma\frac{n_i}{R^3}. \tag{9.5.13}$$

In order for the external surface of a solid to be free, the quantity (9.5.13) should be compensated by a surface force density

$$\boldsymbol{p} = G\Omega_0\gamma\frac{\boldsymbol{n}}{R^3},$$

generating in a crystal volume a homogeneous stressed state with the stress tensor

$$\sigma^1_{ik} = G\Omega_0\gamma\frac{\delta_{ik}}{R^3}. \tag{9.5.14}$$

Sometimes a field such as (9.5.14) is called an image field or a *field of imaginary* sources, by analogy with the electrostatic charge field arising near a conducting surface and equivalent to the field of the charge mirror image.

If we set $R = \infty$, then σ^1_{ik} will vanish, but at any finite R (9.5.14) results in a homogeneous expansion of a solid body

$$\varepsilon^1_{ll} = \frac{1}{3K}\sigma^1_{ll} = \Omega_0\gamma\frac{G}{K}\frac{1}{R^3}. \tag{9.5.15}$$

It is natural that (9.5.15) determines such an increase in the volume of the whole sample ΔV^1 that, being added to (9.5.15), will give Ω_0:

$$\Delta V + \Delta V^1 = \pi\gamma\Omega_0 + \frac{4\pi}{3}\gamma R^3\varepsilon^1_{ll} = \Omega_0. \tag{9.5.16}$$

The equality (9.5.16) proves the consistency of all our calculations. We, however, point out the other aspect of the problem discussed.

Let us consider a specimen with a uniform distribution of identical dilatation centers with small concentration and introduce the averaged characteristics of elastic fields varying at distances that greatly exceed the average distance between the defects. Then we calculate the mean field of all images of the defects multiplying (9.5.15) by the number of defects in a specimen and disregarding surface effects

$$\langle \varepsilon_{ll}^{\text{im}} \rangle = \frac{4\pi}{3} R^3 n \varepsilon_{ll}^1 = \pi\gamma \frac{4G}{3K} \Omega_0 n, \qquad (9.5.17)$$

where n is the number of dilatation centers in a unit volume.

We see that although in an unbounded crystal the dilatation centers do not generate relative volume change in the elastic medium between defects, in a finite-dimension specimen there always exists a finite dilatation ($\Omega_0 > 0$) or compression ($\Omega_0 < 0$) proportional to the defect concentration. However, since the result (9.5.17) is independent of the specimen dimension R, it should also remain valid in the limit $R \to \infty$ for $n =$ constant.

9.5.1
Problem

1. Find the hydrostatic compression of an isotropic medium around an elastic dipole with a unit vector $\boldsymbol{l}$ of its axis.

Solution.

$$\varepsilon_{kk} \equiv \operatorname{div} \boldsymbol{u}(\boldsymbol{r}) = -\frac{\Omega_0}{4\pi} l_i l_k \nabla_i \nabla_k \frac{1}{r}, \qquad r \neq 0.$$

10
Linear Crystal Defects

10.1
Dislocations

Dislocations are linear defects in a crystal near which the regular atomic arrangement is broken. In a theoretical treatment, dislocations in real crystals perform as important a role as electrons do in metals.

There are many microscopic models of dislocations. In the simplest model the dislocation is taken to be the edge of an "extra" half-plane present in the crystal lattice. In the conventional atomic scheme of this model where the trace of the half-plane coincides with the upper semiaxis Oy (Fig. 10.1), the edge of the extra half-plane on the z-axis, is called an *edge* dislocation. The regular crystal structure is then greatly distorted only in the near vicinity of the isolated line (the dislocation axis) and the region of irregular atomic arrangement has transverse dimensions of the order of a lattice constant. If we surround the dislocation with a tube of radius of the order of several interatomic distances, the crystal outside this tube may be regarded as ideal and subject only to elastic deformations (the crystal planes are connected to one another almost regularly) and inside the tube the atoms are considerably displaced relative to their equilibrium positions and form the *dislocation core*. In Fig. 10.1 the atoms of the dislocation core are distributed over the contour of the shaded pentagon.

Nevertheless, deformation even occurs far from the dislocation. The deformation at a distance from the dislocation axis may be seen by tracing a path in the plane xOy (Fig. 10.1) through the lattice sites along the closed contour around the dislocation core. We consider the displacement vector of each site from its position in an ideal lattice and find the total increment of this vector in the path. We go around the dislocation axis along the external contour starting from the upper left angle (atom 1) and see that the atomic displacement at the end of the path is nonzero and equal to one lattice period along the x-axis. This singularity of the dislocation deformation can be considered as the initial one when we define a dislocation in a crystal.

The Crystal Lattice: Phonons, Solitons, Dislocations, Superlattices, Second Edition. Arnold M. Kosevich

ISBN: 3-527-40508-9

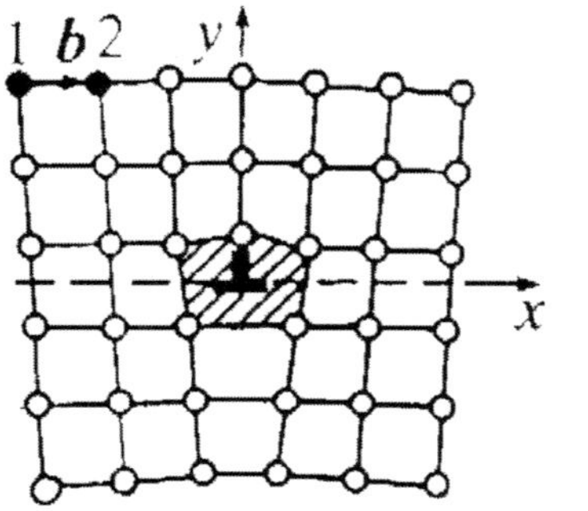

Fig. 10.1 A scheme of atomic arrangement in the vicinity of an edge dislocation.

We denote the vector connecting atoms 1 and 2 by $\boldsymbol{b}$. This vector is called the *Burgers vector* of a dislocation. The possible values of the Burgers vectors in an anisotropic solid are determined by its crystallographic structure and correspond, as a rule, to a small number of certain directions in a crystal. The dislocation lines are arranged arbitrarily, although their arrangement is limited by a set of definite crystallographic planes.

Let $\boldsymbol{\tau}$ be the unit vector of a tangent to the dislocation line. For an edge dislocation $\boldsymbol{\tau} \perp \boldsymbol{b}$. Edge dislocations with opposite directions of $\boldsymbol{b}$ differ in that the "extra" crystal half-plane lies above and below the xOz plane (Fig. 10.2a). If dislocations such as 1 and 2 are observed in a crystal simultaneously, they are called opposite-sign dislocations (for instance, the first one may be called a positive edge dislocation). When opposite-sign dislocations merge, annihilation takes place, resulting in the elimination of two defects and in a "reunification" of the regular atomic plane.

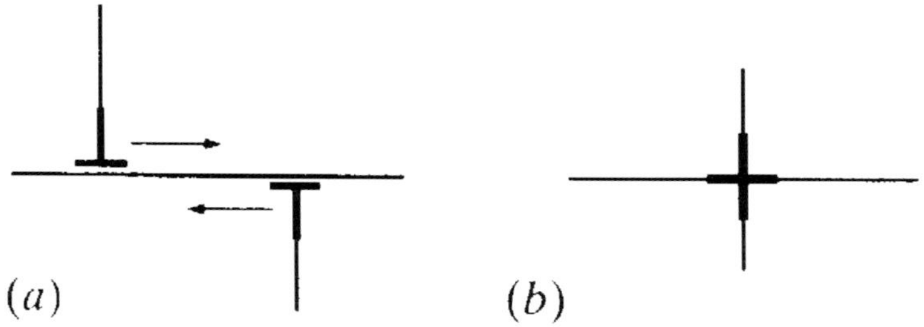

Fig. 10.2 Annihilation of edge dislocations: (*a*) two dislocation of opposite signs; (*b*) reproduction of an atomic plane after the dislocations merge.

When $\boldsymbol{\tau} \parallel \boldsymbol{b}$ the corresponding dislocation is called a *screw* dislocation. The presence of a linear screw dislocation in a crystal converts the lattice planes into a helicoidal surface similar to a spiral staircase. Figure 10.3 shows a scheme of atomic-plane arrangement in the presence of a screw dislocation coinciding with the OO' line.

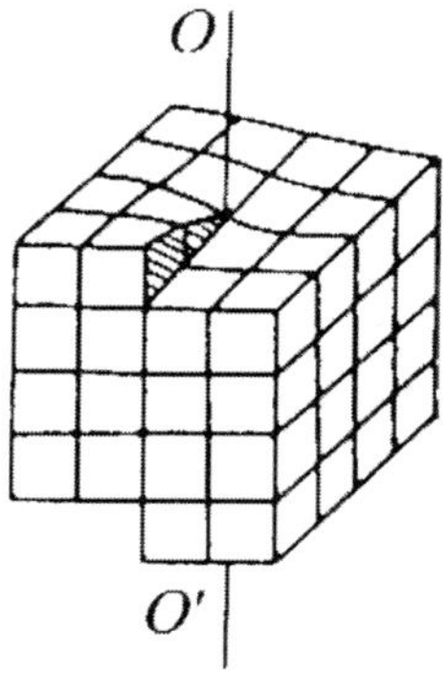

Fig. 10.3 Screw dislocation in a crystal.

If the dislocation line is not perpendicular and not parallel to the Burgers vector, it is called a segment of mixed type. Dislocation segments of an edge, screw and mixed type can arrange themselves continuously along a line forming a dislocation line. The dislocation line cannot end inside a crystal. It must either leave the crystal with each end at the crystal surface or (as is generally observed in real cases) form a closed *dislocation loop*. It is clear that the Burgers vector is constant along the dislocation line.

A crystal lattice with dislocations will sometimes be called a *dislocated* lattice (or a *dislocated* crystal).

10.2 Dislocations in Elasticity Theory

The main property of a dislocation implies that when a circuit is made around the dislocation line the total increment of the elastic displacement vector is nonzero and equal to the Burgers vector. Thus, a dislocation in a crystal will be said to be a specific line D having the following general property: after a circuit around the closed contour L enclosing the line D (Fig. 10.4) the elastic displacement vector $\boldsymbol{u}$ changes by a certain finite increment $\boldsymbol{b}$ equal (in value and direction) to one of the lattice periods. This property is written as

$$\oint_L du_i = \oint_L \frac{\partial u_i}{\partial x_k}\, dl_k = -b_i, \qquad (10.2.1)$$

assuming that the direction of the circuit is related by the corkscrew rule with a chosen direction of the tangent vector $\boldsymbol{\tau}$ to the dislocation line. The dislocation line is in this case a line of singular points of the fields of strains and stresses[1].

1) We do not consider the dislocation line as a local inhomogeneity of a crystal.

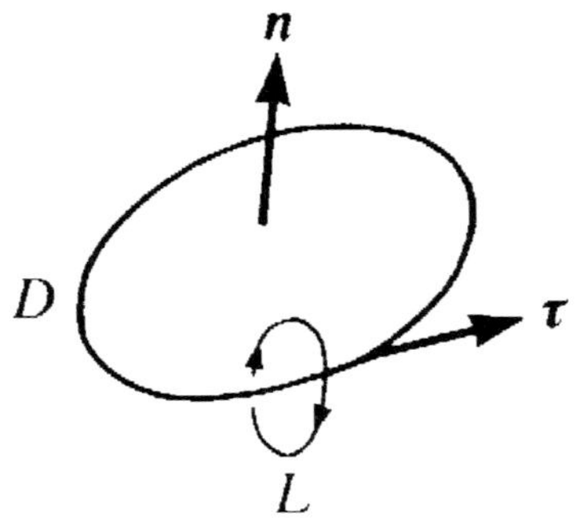

Fig. 10.4 Mutual orientations of the vector $\boldsymbol{n}$ and $\boldsymbol{\tau}$.

The majority of the essential physical properties of dislocations is not connected with microscopic models and can be described phenomenologically in the framework of elasticity theory using a similar definition.

From a mathematical point view, the condition (10.2.1) means that in the presence of a dislocation the displacement vector is a many-valued function of the coordinates that receives an increment in a passage around the dislocation line. In this case there is no physical ambiguity: the increment $\boldsymbol{b}$ denotes an additional displacement of crystal atoms by one lattice period that, due to the translational invariance, does not change its state. In particular, the stress tensor σ_{ik} characterizing the elastic crystal state is a single-valued and continuous function of the coordinates.

In a crystal with a separate dislocation, instead of many-valued function $\boldsymbol{u}(\boldsymbol{r})$, we can regard the displacement vector $\boldsymbol{u}$ as a single-valued function that undergoes a fixed jump $\boldsymbol{b}$ on some arbitrary chosen surface S_D spanning the dislocation loop D:

$$\delta \boldsymbol{u} \equiv \boldsymbol{u}^{+} - \boldsymbol{u}^{-} = \boldsymbol{b}, \qquad (10.2.2)$$

where $\boldsymbol{u}^{+}$ and $\boldsymbol{u}^{-}$ are the $\boldsymbol{u}(\boldsymbol{r})$ values from the upper and lower sides of the surface S_D, respectively. The "upward" direction (positive) is determined by the direction of a normal $\boldsymbol{n}$ to the surface S_D (this direction is connected by the corkscrew rule with the $\boldsymbol{\tau}$-vector direction, Fig. 10.4). If the jump $\delta\boldsymbol{u}$ is the same at all points of the S_D surface the distortion tensor u_{ik} is a continuous and differentiable function on this surface.

Using (10.2.2) we can formally give another definition of a dislocation, namely, by defining it as the line D on which the surface S_D with given jump (10.2.2) of the vector $\boldsymbol{u}(\boldsymbol{r})$ is spanned. In some cases this definition is more convenient than the initial one, e. g., using it we can easily find the field around the dislocation. If we calculate the strain tensor for a crystal with a dislocation, i. e., in the presence of the jump (10.2.2) on the S_D surface, it will have on this surface a δ-like singularity

$$\epsilon_{ik}^{(S)} = \frac{1}{2}(n_i b_k + n_k b_i)\delta(\zeta), \qquad (10.2.3)$$

where ζ is the coordinate along the normal $\boldsymbol{n}$. The value $\zeta = 0$ corresponds to the S_D surface.

As there is no physical singularity in the space near a dislocation the stress tensor σ_{ik} as noted above should be a continuous function everywhere. Meanwhile, the stress tensor $\sigma_{ik}^{(S)} = \lambda_{iklm}\epsilon_{lm}^{(S)}$ having a singularity on the surface S_D is formally associated with the strain tensor (10.2.3). To eliminate this stress tensor, it is necessary to define the function's body forces distributed over the surface S_D with a specially chosen density $f^{(S)}$:

$$f_i^{(S)} = -\nabla_k \sigma_{ki}^{(S)} = -\lambda_{iklm}\nabla_k \epsilon_{lm}^{(S)}. \tag{10.2.4}$$

Thus, the problem of finding a many-valued function $\boldsymbol{u}(\boldsymbol{r})$ is equivalent to that of finding a single-valued but discontinuous function in the presence of body forces (10.2.3), (10.2.4). Substituting (10.2.4) into (9.5.1) and performing the integration we obtain

$$u_i(\boldsymbol{r}) = -\lambda_{iklm} b_m \int_{S_D} n_l \nabla_k G_{ij}(\boldsymbol{r} - \boldsymbol{r}')\, dS'. \tag{10.2.5}$$

In principle, (10.2.5) allows one to obtain elastic displacements in a crystal when the form of the dislocation loop is arbitrary. The general formula (10.2.5), however, is complicated and the calculation of a displacement field even with simple dislocation line shapes is quite cumbersome. In the case of a straight-line dislocation, when we deal with the plane problem of elasticity theory, it is simpler to solve an equilibrium equation under the condition (10.2.1).

Studying the lattice dynamics, we used a scalar model to simplify the calculations. To the same aim we clarify to what the dislocation-type linear defect corresponds in a scalar elastic field model.

Let b characterize the linear defect intensity of a scalar field u and the defect itself is defined by

$$\oint_L du = \oint_L \frac{\partial u}{\partial x_k}\, dx_k = -b, \tag{10.2.6}$$

which plays the role of a boundary condition for the field equation (2.9.21). In a static case, (2.9.21) reduces to the Laplace equation

$$\Delta u = 0. \tag{10.2.7}$$

If we introduce the vector $\boldsymbol{h} = \operatorname{grad} u$ as a characteristic of the field state then the fixed circulation of this vector along any closed contour enclosing the defect line will be determined by (10.2.6). A similar defect is a *vortex* of the field $\boldsymbol{h}$. Thus, a dislocation in an elastic field is an analog of a vortex of some scalar field.

It follows from (10.2.6), (10.2.7) that the dislocation field in a scalar model coincides with the vortex field in a liquid up to the notations.

In particular, a rectangular vortex perpendicular to the plane xOy and a dislocation in a scalar model have potential field

$$u = \frac{b}{2\pi}\theta, \qquad \tan\theta = \frac{y}{x}. \tag{10.2.8}$$

The field (10.2.8) generates the vector field $\boldsymbol{h}$

$$h_x = \frac{b}{2\pi}\frac{y}{r}, \qquad h_y = -\frac{b}{2\pi}\frac{x}{r}, \qquad r^2 = x^2 + y^2. \tag{10.2.9}$$

Since a screw dislocation with Burgers vector $\boldsymbol{b}$ parallel to the Oz-axis in an isotropic medium is a singularity of the scalar field, u_z and its displacement field are $u_x = u_y = 0, u_z = u$, where the function u is given by (10.2.8).

10.3
Glide and Climb of a Dislocation

The definition of a dislocation (10.2.2) is a formal tool allowing us to solve some static elasticity problems in a medium with dislocations. If we associate the vector $\boldsymbol{u}(\boldsymbol{r})$ having a discontinuity (10.2.2) with real atomic displacements in a crystal and try to reproduce the real process of dislocation generation (via relative displacements of atomic layers on both sides of the surface S_D by the value $\boldsymbol{b}$), we run into certain difficulties of a physical character. Indeed, when the condition (10.2.2) was formulated we supposed that crystal continuity is conserved along the surface S_D. In particular, the interatomic distances remain unchanged (up to elastic deformations). However, when (10.2.2) is understood formally it is clear that crystal continuity is violated. In fact, when the cut boundaries are displaced by $\boldsymbol{b}$ the crystal volume changes inelastically

$$\delta V = \boldsymbol{n}\boldsymbol{b}\delta S, \tag{10.3.1}$$

per each element δS of a discontinuity surface. Therefore, the condition (10.2.2) implies that we "eject" the material where the atomic layers overlap under displacement and fill in the remaining "gaps" with additional material. However, a crystal has no mechanisms of automatic removal or supply of material in a solid. Thus, a purely mechanical way of dislocation generation through displacement of the atomic layers along an arbitrary surface S_D without discontinuities appearing in physical quantities is impossible.

However, it follows from (10.3.1) that in a crystal there exists a specified surface S_{sl} at each point of which $\boldsymbol{n}\boldsymbol{b} = 0$ and the displacement described is shear-like with no effect on the crystal continuity. It is clear that it is a cylindrical surface whose elements are parallel to the vector $\boldsymbol{b}$ and its directrix is a dislocation loop. It is called a slip surface of the dislocation concerned and is an envelope of the family of slip planes of all dislocation line elements. By the *slip plane* of a dislocation element we understand a surface tangent to the corresponding element of the dislocation line and this plane is determined by the vectors $\boldsymbol{\tau}$ and $\boldsymbol{b}$. Possible systems of slip planes in an anisotropic medium are determined by the structure of a corresponding crystal lattice.

A slip plane is singled out physically because a dislocation-induced shift is possible along it (the interatomic distances in the vicinity of a slip plane surface remain unchanged after a shift). A comparatively easy mechanical displacement of the dislocation is possible in this plane. The latter follows directly from a microscopic picture of the dislocation defect and is demonstrated by means of a scheme with an extra half-plane (Fig. 10.5). Let an edge dislocation be generated by a shift $\boldsymbol{b}$ along a slip plane whose trace in Fig. 10.5 coincides with a crystallographic direction AB. We consider two atomic configurations near the dislocation core when an extra atomic half-plane is in the position MM' (atoms are marked with black circles) and also when the role of an extra crystal half-plane is played by the atomic layer occupying the position NN' (atoms are shown with light circles).

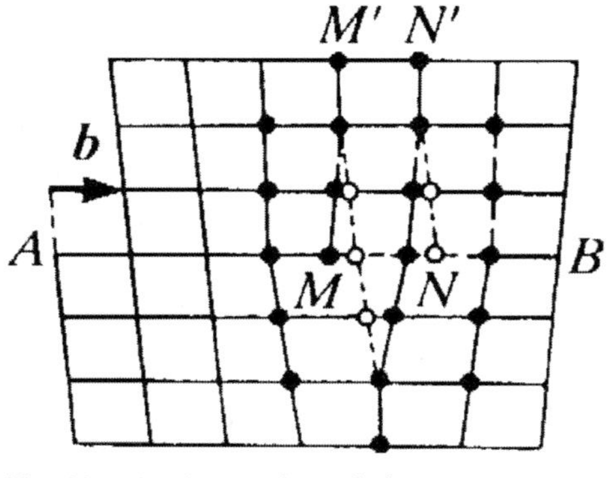

Fig. 10.5 A scheme of atomic-layer rearrangement with the edge dislocation gliding.

Although the transition from the first atomic configuration to the second one is connected with the dislocation migrating one interatomic distance to the right in the slip plane, the displacements of individual atoms are small compared to the value of $\boldsymbol{b}$. This means that such collective atomic rearrangements that provide a dislocation migration may be realized under the action of comparatively small forces. If we compare such forces with macroscopic loads, it turns out that the corresponding shear stresses σ_s necessary to initiate dislocation motion are less by a factor of $10^2 - 10^4$ than the shear modulus of a monocrystal G. The smallness of the parameter σ_s/G is a crucial physical factor that allows us to use linear elasticity theory to describe the mechanical processes accompanied by the motion of dislocations. Thus, a dislocation may move comparatively easily in its own slip plane. This motion of a dislocation is generally called *sliding* or *gliding*, or just *glide*. Finally, it is often called a *conservative motion*.

A simple mathematical model allows us to understand certain features of the dislocation dynamics and to explain qualitatively the high mobility of the dislocations.

Let us imagine that a chain of atoms (Fig. 1.11) is an edge series of one half of a plane crystal ($y > 0$, Fig. 10.6) displaced in a certain way with respect to another half of a crystal ($y < 0$, Fig. 10.6). Then the influence of the nondisplaced half of a crystal on the atoms distributed along the x-axis can be qualitatively described using the energy (1.6.4).

Crystallographic atomic rows with an extra half of the atomic row (Fig. 10.6) above the x-axis represent a model of the dislocation in a crystal. Thus, the scheme described may be an analog of the problem of dislocations in a two-dimensional and three-dimensional crystal. Such an interpretation of this model was performed by its authors (Frenkel and Kontorova, 1938).

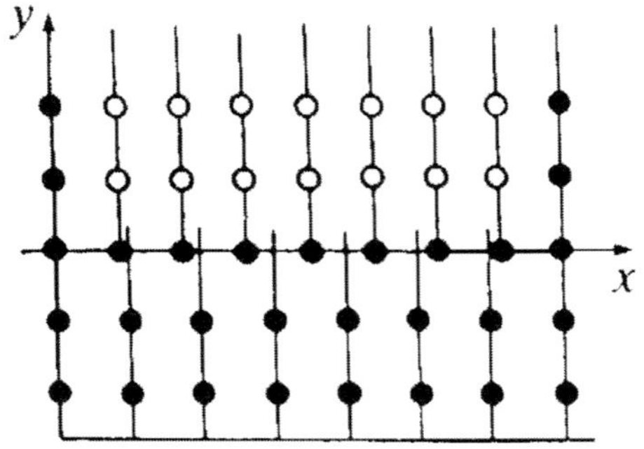

Fig. 10.6 Edge dislocation in the Frenkel–Kontorova model.

We remind ourselves of an important result obtained in analyzing the crowdion dynamics (Section 9.4). In the continuum approximation (a crystal is considered as a continuous elastic medium) the crowdion energy is independent of the coordinate of its center, and thus the crowdion may move freely along a 1D crystal. A dislocation has the same property.

However, if we take into account the discreteness of the crystal, a resistance force arises that must be overcome to start the motion of either a crowdion or of a dislocation, i. e., a *starting stress* σ_s appears.

We estimate it from (9.4.7) determining the crystal energy dependent on a crowdion coordinate. The derivative of this energy with respect to the crowdion center coordinate x_0 yields the force

$$f = -16\pi\alpha_0 a \exp\left(-\frac{\pi^2 l_0}{a}\right) \sin\left(2\pi\frac{x_0}{a}\right),$$

where l_0 is the crowdion width; a is the interatomic distance ($l_0 \gg a$). This force exists by virtue of the lattice discreteness ($a \neq 0$) and is exponentially small with respect to the parameter l_0/a determining the degree of macroscopicity of a crowdion core.

Extending the results obtained in a 1D model to the case of a linear dislocation in a 3D crystal, it is possible to expect that they are valid as qualitative statements. The latter makes it possible to explain the smallness of a starting stress of the dislocation by the macroscopicity of its core and understand why glide is comparatively easy.

Quite a different physical nature is inherent to the real migration of a dislocation in a direction perpendicular to the slip plane. We consider an arbitrarily small displacement δX of an element of the dislocation loop length dl, assuming δX to have a component normal to the slip surface. Such a migration of a dislocation element results in the surface area S_D increasing by the crystal volume, a value that may be characterized by an axial vector $\delta \boldsymbol{S} = [\delta \boldsymbol{X} \boldsymbol{\tau}] dl$.

As a result, the crystal volume exhibits an inelastic local increase that, from (10.3.1), is equal to

$$\delta V = \boldsymbol{b} \delta \boldsymbol{S} = \boldsymbol{b} [\delta \boldsymbol{X} \boldsymbol{\tau}]\, dl = -[\boldsymbol{b} \boldsymbol{\tau}] \delta \boldsymbol{X}\, dl. \tag{10.3.2}$$

If $\delta V \neq 0$, then the deficiency ($\delta V > 0$) or excess ($\delta V < 0$) of material cannot be balanced in the ideal crystal volume in a mechanical way (if continuity is conserved). However, in a real crystal a slow-acting mechanism of condensation or rarefaction of a substance exists, requiring no macroscopic breaks of continuity. We are referring to the processes of formation and diffusion migration of point defects: atoms in the interstices (the material condensed) and vacant sites (the material rarefied). Thus, an inelastic increase in the volume (10.3.2) on the dislocation axis should be compensated by an equal decrease in the crystal volume due to the formation of a corresponding number of point defects in the vicinity of a dislocation core. As each crystal site is associated with unit cell volume V_0, the quantity (10.3.2) should be associated with the number $|\delta V| \,/\, V_0$ arising from vacancies or vanishing interstitial atoms. However, since point defects of both types may vanish or be generated, the change in their number is related to the displacement of the dislocation line element by

$$\delta N = \frac{[\boldsymbol{b} \boldsymbol{\tau}]}{V_0} \delta \boldsymbol{X}\, dl, \tag{10.3.3}$$

where δN is the difference in numbers of interstitial atoms and vacancies produced.

The point defects for which (10.3.3) is written appear or vanish just near the dislocation core. Therefore, in macroscopically describing the dislocation motion the total variation in crystal volume can be concentrated at the dislocation line. Thus, the migration of a dislocation in a direction perpendicular to the slip plane is accompanied by a local increase in crystal volume with a relative value given by

$$\delta \epsilon_{kk}^{0} = \delta \boldsymbol{X} [\boldsymbol{b} \boldsymbol{\tau}] \delta(\boldsymbol{\xi}), \tag{10.3.4}$$

where $\delta(\boldsymbol{\xi})$ is a 2D δ-function; $\boldsymbol{\xi}$ is a 2D radius vector in the plane perpendicular to the vector $\boldsymbol{\tau}$ at a given point of the dislocation loop with its origin at the dislocation axis.

Dislocation migration with a velocity limited by the diffusion processes that provide changes in the volume (10.3.4) is called *climb* or *nonconservative motion* of a dislocation.

10.4 Disclinations

The definition of a dislocation based on (10.2.2) reflects an important property of deformation in a continuous medium. If the function $\boldsymbol{u}(\boldsymbol{r})$ exhibits a jump (10.2.2) on the surface S, then where $\boldsymbol{b}$ is a fixed vector identical at all points on the surface S, the distortion tensor u_{ik} remains a continuous and differentiable function of the coordinate everywhere except for a closed line on which the surface S is spanned.

However, the requirement of continuity for the distortion tensor is, to some extent, excessive as the physical state of the elastic body depends only on stresses and elastic strains proportional to them. Thus, for studying the elastic fields generated by discontinuities $\delta \boldsymbol{u}$ on surfaces not distinguished by their physical properties in the body volume we restrict ourselves to the requirement of unambiguity and continuity of the strain tensor ϵ_{ik}. It then turns out that (10.2.2) does not present a general form of the change of the vector $\boldsymbol{u}$ on the surface S at which the strain tensor ε_{ik} conserves continuity and differentiability as a function of the coordinates. The most general form of the discontinuity at the surface S is

$$\delta \boldsymbol{u} = \boldsymbol{u}^{+} - \boldsymbol{u}^{-} = \boldsymbol{b} + [\boldsymbol{\Omega r}], \tag{10.4.1}$$

where $\boldsymbol{b}$ and $\boldsymbol{\Omega}$ are constant vectors ($\boldsymbol{b}$ is a translation vector and $\boldsymbol{\Omega}$ is a rigid rotation vector); $\boldsymbol{r}$ is the point radius vector on the surface S. The vector $\boldsymbol{\Omega}$ in a crystal should coincide with one of the elements of crystal rotational symmetry.

Under the condition (10.4.1) an antisymmetric part of the distortion tensor is broken on the surface. On reminding ourselves of the definition of the rotation vector under the deformation (1.9.4) we see that the rotation vector ω undergoes a step at the surface:

$$\delta \boldsymbol{\omega} \equiv \boldsymbol{\omega}^{+} - \boldsymbol{\omega}^{-} = \boldsymbol{\Omega}. \tag{10.4.2}$$

The jumps (10.4.1), (10.4.2) generate an elastic field singularity in the crystal concentrated along the line D on which the surface S is spanned.

If $\boldsymbol{\Omega} = 0$, then (10.4.1) is transformed into (10.2.2), and the translation vector $\boldsymbol{b}$ coincides with the Burgers dislocation vector.

If $\boldsymbol{b} = 0$, but $\boldsymbol{\Omega} \neq 0$, the singularity arising in a solid is called a *disclination*. The disclination vector Ω is sometimes called the *disclination power*, and sometimes the *Frank vector*. We shall use the latter name.

It follows from (10.4.2) that in the presence of a disclination the Frank vector describes a relative rigid rotation of two parts of a solid positioned on both sides of the surface S. It is clear that for an unambiguous definition of $\delta \boldsymbol{u}$ in (10.2.1), the space position of the vector $\boldsymbol{\Omega}$, i. e., the position of the disclination axis should be fixed. The displacement of the disclination rotation axis by the vector $\boldsymbol{R}$ amounts to adding an ordinary dislocation with Burgers vector $\boldsymbol{b} = [\boldsymbol{\Omega R}]$. Therefore, the definition of a disclination in (10.4.1) becomes unambiguous if we rewrite it as

$$\delta u_i(r) = e_{ilm}\Omega_l(x_m - x_m^0), \tag{10.4.3}$$

where $\boldsymbol{r}^0$ is the radius vector of the point through which the rotation axis given by the vector $\boldsymbol{\Omega}$ runs. It follows from a direct calculation that (10.4.2) is a consequence of (10.4.3).

As the disclination is a linear singularity of the elastic deformation field, it may be defined in a form that does not use the notion of an arbitrary surface S, i. e., in a form analogous to the definition of a dislocation (10.2.1). Indeed, we introduce a continuous and differentiable function $\omega(\boldsymbol{r})$ (the medium element rotation at point $\boldsymbol{r}$ as a result of an elastic deformation of a solid). The disclination will then be said to be a specific line D with the following property: in passing around any closed contour L enclosing the line D, the elastic rotation vector $\boldsymbol{\omega}$ gets a certain finite increment $\boldsymbol{\Omega}$. This property is written as

$$\oint_L d\omega_i = \oint \frac{\partial \omega_i}{\partial x_k} \, dx_k = -\Omega_i. \qquad (10.4.4)$$

The Frank vector $\boldsymbol{\Omega}$ unambiguously determines the properties of the disclination D only when the point through which the disclination rotation axes runs is specified.

Let $\boldsymbol{\tau}$ be a unit vector tangent to the line of the disclination. If $\boldsymbol{\tau} \parallel \Omega$, the disclination is called a *wedging* or a *slope disclination*. If $\boldsymbol{\tau} \perp \boldsymbol{\Omega}$ we have a *twisting disclination*.

A disclination in a crystal is most vividly exemplified by a 60° wedging disclination in a hexagonal crystal when this defect is parallel to the six-fold symmetry axis. Analyzing the atomic arrangement in a plane perpendicular to the axis of this disclination in a nondefective crystal (Fig. 10.7a) and also their distortion in the presence of a *positive* (Fig. 10.7b) and a *negative* (Fig. 10.7c) wedging disclination, we note the following peculiarities. Choosing the sign of a wedging disclination, unlike choosing it for an edge dislocation, has an absolute character: the atomic displacements in the vicinity of a positive disclination is inverse to the atomic displacements in the vicinity of a negative disclination. In the first case, crystal stretching is observed along the contour that encloses the disclination, and in the second case, we have crystal compression.

Another important peculiarity of the wedge disclination observed is the change in the crystal lattice symmetry in the vicinity of a disclination. Indeed, for a 60° positive wedging disclination there arises a five-fold symmetry axis coincident with the vector $\boldsymbol{\Omega}$ (Fig. 10.7b), and a 60° negative disclination generates pseudosymmetry with a seven-fold symmetry axis (Fig. 10.7c). In a perfect crystal such rotational symmetry is impossible.

The two types of linear defects considered here (dislocations and disclinations) are in fact two independent forms of the same family of peculiarities inherent to the deformation of continuous media that are called *Volterra dislocations*. The dislocations in a crystal are translational Volterra dislocations, and disclinations are rotational Volterra dislocations. Generally, the Volterra dislocation may have a mixed character, i. e., simultaneously represent a translational dislocation and a disclination.

We now turn to finding an elastic field around a separate disclination. Note that a simple tool for calculating this field can be obtained on the basis of (10.4.3). If we consider the disclination as a line limiting the surface S with given rotation vector

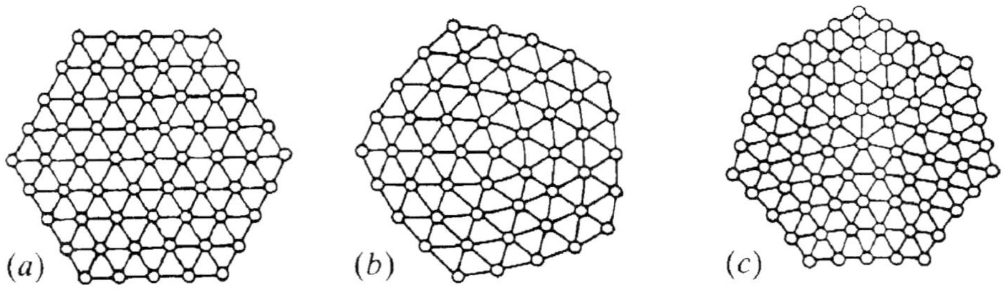

Fig. 10.7 A 60° wedge dislocation parallel to the six-fold symmetry axis: (*a*) ideal structure in a basal plane; (*b*) structure with a positive disclination; (*c*) structure with a negative disclination.

jump (10.4.2), then it corresponds formally to a dislocation with "Burgers vector" distributed over the surface S

$$b_i^{(S)} = e_{ilm}\Omega_l(x_m - x_m^0). \tag{10.4.5}$$

It is true that in contrast to the Burgers vector of an ordinary dislocation distribution (10.4.5) is dependent on the coordinate on the surface S. However, this does not prevent us from using (10.2.5), without taking the Burgers vector outside the integral sign. Hence, the problem of finding the displacement vector $\boldsymbol{u}(\boldsymbol{r})$ around a separate disclination loop is reduced to calculation of the integral

$$u_i(\boldsymbol{r}) = -\lambda_{jklm}e_{mpq}\Omega_p \int_S (x'_q - x_q^0)\nabla_k G_{ij}(\boldsymbol{r} - \boldsymbol{r}')\, dS'_l. \tag{10.4.6}$$

The expression (10.4.6) allows us to determine the elastic deformations of a crystal with an arbitrary disclination loop.

10.5 Disclinations and Dislocations

In some cases a simple isolated disclination can easily be represented by a planar "pile-up" of continuously distributed dislocations. Equation (10.4.5) allows us to clarify how the "Burgers vector" of dislocations is distributed along the surface S, which is necessary for these dislocations to be equivalent to a disclination with Frank vector $\boldsymbol{\Omega}$.

The above point is illustrated by the relationship between a wedge disclination and a so-called dislocation wall. By "wall" we mean a large number of identical parallel linear edge dislocations distributed in the same plane perpendicular to their Burgers vectors (Fig. 10.8a). The dislocations are in parallel slip planes, the distances between which in the simplest case are the same, i. e., equal to h.

The geometrical meaning of a crystal deformation generated by this system of dislocations is easily understood. The presence of a wall leads to a misorientation of the two parts of a crystal that are divided by the system of dislocations concerned (Fig. 10.8a). If h is the distance between the dislocations (in a macroscopic theory it is necessary that $h \gg b$), the misorientation angle between the two parts of a crystal is $\psi = b/h$. Thus, the dislocation wall is a model of the boundary between two blocks or subgrains with small misorientation. If the boundary consists of edge dislocations, then the axis around which the neighboring subgrains are inclined is in the plane dividing them. Such a low-angle grain boundary is called an *inclination boundary.*

Let a half-infinite inclination boundary end at a straight line A (Fig. 10.8b), the dislocations being distributed in it continuously with the Burgers vector density b/h. If we now imagine a closed contour enclosing the line A and intersecting the inclination boundary at the point y, then such a contour encloses dislocations with the total Burgers vector

$$B_x = \frac{by}{h}, \tag{10.5.1}$$

where the coordinate y is measured from the line A.

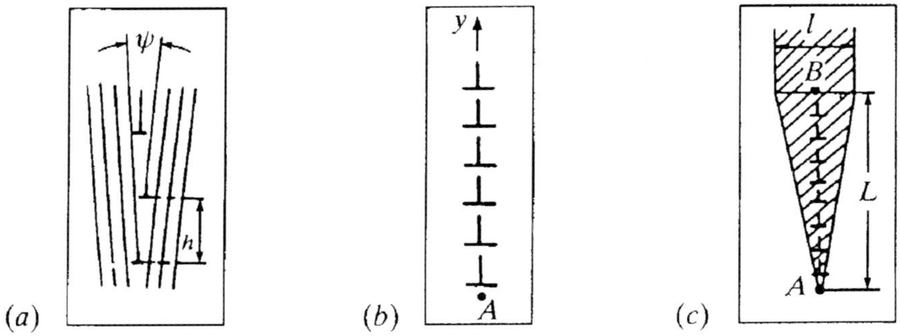

Fig. 10.8 Dislocation wall and a wedging disclination: (*a*) low-angle grain boundary, (*b*) dislocation wall ends at the line A, (*c*) representation of the dislocation wall by a disclination dipole.

Comparing (10.5.1) and (10.4.6), we can conclude that when the dislocation wall (Fig. 10.8b) is treated macroscopically then it is equivalent to a negative wedge dislocation with Frank vector $\Omega_z = -b/h$ and rotation axis coincident with the line A. Naturally, the Frank vector in this case is equal to the misorientation angle between the parts of a crystal.

In this situation the rotation by an angle Ω should not be an element of the group symmetry of the ideal crystal, but the boundary between two misoriented parts of a crystal is then taken as a plane *stacking fault.*

If the inclination boundary ends at a certain line B (Fig. 10.8c), the latter should be coincident with the rotation axis of the second disclination. Its Frank vector $\boldsymbol{\Omega}'$ then satisfies the requirement that the rotation through the angle $\Omega + \Omega'$ be an element of the *group symmetry* of the crystal lattice. In particular, the most interesting case is $\Omega' = -\Omega$.

A pair of disclinations of the same type but with opposite sign and parallel rotation axes forms a *disclination dipole*. In the example concerned (Fig. 10.8c) when $\Omega_B = -\Omega_A$, the wedging disclinations positioned along the lines A and B form a dipole. If the distance between the lines A and B is equal to L, the deformation caused by the dipole described can be thought of as a peculiar wedging-out of a crystal: i. e., a semi-infinite planar plate of thickness $l = \Omega L$ whose edge has a triangular cross section with angle Ω at the vertex is set into the crystal cut.

At distances greatly exceeding L, the dipole of wedging disclinations is regarded as an edge dislocation with Burgers vector $b_x = l = \Omega L$ and the "core" enclosing the whole region of an inclination boundary of length L.

10.5.1
Problems

1. Obtain an expression for the displacement vector around the dislocation in an isotropic medium, isolating explicitly the contour integration over the dislocation loop.

Hint. Use (10.2.5) and (9.5.10) for the Green tensor in an isotropic medium and also the Stokes theorem.

Solution.

$$u = \boldsymbol{b}\frac{O}{4\pi} + \oint \frac{[\boldsymbol{b}\boldsymbol{\tau}]\,dl}{R} + \operatorname{grad}\psi; \qquad \psi = \frac{\lambda+\mu}{4\pi(\lambda+2\mu)}\oint [\boldsymbol{b}\boldsymbol{R}]\boldsymbol{\tau}\frac{dl}{R},$$

where O is the solid angle subtended at the observer by the dislocation loop.

11
Localization of Vibrations

11.1
Localization of Vibrations near an Isolated Isotope Defect

We begin studying the influence of defects on lattice vibrations by using a scalar model, treating independently one branch of crystal vibrations only. An equation for stationary vibrations of frequency ω in this model is

$$\omega^2 u(\boldsymbol{n}) - \frac{1}{m}\sum_{\boldsymbol{n}'}{}' \alpha(\boldsymbol{n}-\boldsymbol{n}')u(\boldsymbol{n}') = 0. \tag{11.1.1}$$

The dispersion law for an ideal lattice follows from (11.1.1)

$$\omega^2 = \omega_0^2(\boldsymbol{k}) \equiv \frac{1}{m}\sum \alpha(\boldsymbol{n})\cos \boldsymbol{k}\boldsymbol{r}(\boldsymbol{n}). \tag{11.1.2}$$

Equation (11.1.1) describes acoustic crystal vibrations for which

$$\omega_0^2 = \frac{1}{m}\sum \alpha(\boldsymbol{n}) = 0, \tag{11.1.3}$$

and where the possible vibration frequencies are in a finite range $(0, \omega_m)$.

If we remove the requirement of (11.1.3), then (11.1.1) can be used to analyze qualitatively the optical lattice vibrations (Chapter 3) upon determination of the extremely long-wave frequency of vibrations by

$$\omega_0^2(0) = \frac{1}{m}\sum \alpha(\boldsymbol{n}) \neq 0. \tag{11.1.4}$$

The eigenfrequencies of (11.1.1) will then be within a certain interval (ω_1, ω_2), where $\omega_1 > 0$.

The simplest point defect arises in a monatomic species when one of the lattice sites is occupied by an isotope of the atom making up the crystal. Since the isotope atom differs from the host atom in mass only, it is natural to assume that the crystal

The Crystal Lattice: Phonons, Solitons, Dislocations, Superlattices, Second Edition. Arnold M. Kosevich

ISBN: 3-527-40508-9

perturbation does not change the elastic bond parameters. Let the isotope be situated at the origin ($\boldsymbol{n} = 0$) and have a mass M different from the mass of the host atom m. With such a defect we get, instead of (11.1.1),

$$\begin{aligned} &\omega^2 M u(0) - \sum_{n'} \alpha(\boldsymbol{n}') u(\boldsymbol{n}') = 0, \qquad \boldsymbol{n} = 0; \\ &\omega^2 m u(\boldsymbol{n}) - \sum_{n'} \alpha(\boldsymbol{n} - \boldsymbol{n}') u(\boldsymbol{n}') = 0, \quad \boldsymbol{n} \neq 0. \end{aligned} \tag{11.1.5}$$

Equations (11.1.5) can be written more compactly as

$$m\omega^2 u(\boldsymbol{n}) - \sum_{n'} \alpha(\boldsymbol{n} - \boldsymbol{n}') u(\boldsymbol{n}') = (m - M)\omega^2 \delta_{\boldsymbol{n}0}, \tag{11.1.6}$$

by introducing the 3D Kronecker delta $\delta_{\boldsymbol{n}\boldsymbol{n}'}$.

We denote

$$\Delta m = M - m, \qquad U_0 = -\frac{\Delta m}{m}\omega^2, \tag{11.1.7}$$

and rewrite (11.1.6) in a form typical for such problems

$$\omega^2 u(\boldsymbol{n}) - \frac{1}{m} \sum_{n'} \alpha(\boldsymbol{n} - \boldsymbol{n}') u(\boldsymbol{n}') = U_0 \omega^2 \delta_{\boldsymbol{n}0}. \tag{11.1.8}$$

We write a formal solution to (11.1.8) as

$$u(\boldsymbol{n}) = U_0 G^0_\varepsilon(\boldsymbol{n}) u(0), \tag{11.1.9}$$

where G^0_ε is the Green function for ideal lattice vibrations, $\varepsilon = \omega^2$; $u(0)$ is a constant multiplier still to be defined.

If we reject the scalar model and proceed to the general case of a simple lattice, it is easy to write a formula analogous to (11.1.9). With an isotope defect at the site $\boldsymbol{n} = 0$, the displacement vector of any atom in the crystal is

$$u^i(\boldsymbol{n}) = U_0 G^{ik}_\varepsilon(\boldsymbol{n}) u^k(0). \tag{11.1.10}$$

It is also easy to generalize (11.1.9), (11.1.10) for the case of a polyatomic lattice. However, in order not to complicate the formula, we return to the scalar model.

Setting $\boldsymbol{n} = 0$ in (11.1.9), we find that (11.1.9) is consistent only when

$$1 - U_0 G^0_\varepsilon = 0. \tag{11.1.11}$$

The expression (4.5.12) for the Green function is substituted into (11.1.11):

$$1 - \frac{U_0}{N} \sum_{\boldsymbol{k}} \frac{1}{\varepsilon - \omega^2(\boldsymbol{k})} = 0. \tag{11.1.12}$$

Finally, after a transition from summation to integration over quasi-wave vectors and then changing to integration over frequencies:

$$1 - U_0 \int \frac{g_0(z)\,dz}{\varepsilon - z} = 0. \tag{11.1.13}$$

Here, $g_0(\varepsilon)$ is the ideal lattice vibration density.

Before we proceed to analyze (11.1.13), we derive a corresponding general equation, taking into account the different polarizations of vibrations when it is necessary to use (11.1.10).

We set $\boldsymbol{n} = 0$ in (11.1.10) and note that the resulting homogeneous system of three algebraic equations for three independent $u_i(0)$ is solvable if

$$\text{Det}\left\|\delta_{ik} - U_0 G_\varepsilon^{ik}(0)\right\| = 0. \tag{11.1.14}$$

An explicit expression for the Green tensor (4.5.14) does not allow us to reduce (11.1.14) to the relation containing only the ideal lattice vibration density $g_0(\varepsilon)$ and that is independent of the polarization vectors. In a cubic crystal or in the isotropic approximation, however, we have

$$G_\varepsilon^{ik}(0) = \frac{1}{3}\delta_{ik} G_\varepsilon^{ll}(0),$$

and (11.1.14) is reduced to three identical equations of the type (11.1.13).

As (11.1.13) is a condition for solvability of the corresponding equation of motion, it is an equation to determine the squares of frequencies ε at which the atomic displacement around an isotope has the form of (11.1.9). In a theory of crystal vibrations with a point defect, equations such as (11.1.11)–(11.1.13) were first obtained by Lifshits (1947).

We start to analyze (11.1.13) for the case of acoustic vibrations when the unperturbed crystal frequencies are in the interval $(0, \omega_m)$. It is clear that in this case (11.1.13) is meaningful only for $\varepsilon > \varepsilon_m = \omega_m^2$, but the denominator in the integral (11.1.13) is then always positive and (11.1.13) can only have a solution for $U_0 < 0$, i. e., for a light isotope ($M < m$). In a 3D crystal, however, the availability of the necessary sign of the perturbation does not guarantee the existence of a solution to (11.1.13). This is easily seen when we graphically analyze (11.1.13). Introducing the notation

$$F(\varepsilon) = \varepsilon \int \frac{g_0(z)\,dz}{\varepsilon - z},$$

and taking (11.1.7) into account, we rewrite (11.1.13) as

$$F(\varepsilon) = -\frac{m}{\Delta m}. \tag{11.1.15}$$

Since $g_0(z) \sim \sqrt{\varepsilon_m - z}$ for $z \to \varepsilon_m$ the function $F(\varepsilon)$ is finite and positive at the point $\varepsilon = \varepsilon_m$ ($F(\varepsilon_m) = F_m > 0$) and, in the small range $\varepsilon - \varepsilon_m \ll \varepsilon_m$, has a negative derivative

$$F'(\varepsilon) \simeq -\varepsilon_m \int \frac{g_0(z)\,dz}{(\varepsilon - z)^2}.$$

For $\varepsilon \gg \varepsilon_m$ due to the chosen normalization of the vibration density $F(\varepsilon) \approx 1$, so that a plot of the function $F(\varepsilon)$ has the shape of a curve (Fig. 11.1) and a plot of the r.h.s. of (11.1.15) is a horizontal line. Thus, the solution $\varepsilon > \varepsilon_m$ exists if $|\delta m/m|\, F_m > 1$. If this condition is satisfied, the solution to (11.1.13) gives a discrete frequency ω_d lying outside the continuous spectrum band (Fig. 11.1, $\varepsilon_d = \omega_d^2$).

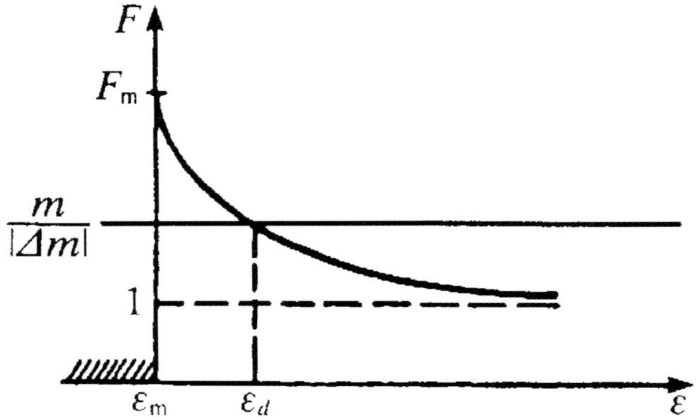

Fig. 11.1 A graphical method of finding the frequency of local vibrations.

Thus, in order for a vibration to appear at a discrete frequency lying near the acoustic band, an isotope should have the mass M satisfying the condition $M < m(1 - 1/F_m) < m$.

Consider the case when the solutions to (11.1.13) determine the discrete frequencies near the optical band of an ideal lattice. Let the continuous spectrum of an ideal crystal occupy the interval $(\varepsilon_1, \varepsilon_2)$, where $\varepsilon_1 = \omega_1^2$ and $\varepsilon_2 = \omega_2^2$ and the defect intensity is characterized by the parameter $U = \varepsilon W_0$. Then, instead of (11.1.15), we write

$$F(\varepsilon) = -\frac{1}{W_0}, \tag{11.1.16}$$

assuming that the function $F(\varepsilon)$ is defined for $\varepsilon < \varepsilon_1$ and $\varepsilon > \varepsilon_2$ (Fig. 11.2, $F_1 = F(\varepsilon_1) < 0$; $F_2 = F(\varepsilon_2) > 0$). Simple analysis leads to the conclusion that the existence of discrete solutions to (11.1.16) is provided only by the defects whose parameter W_0 satisfies certain conditions. If $W_0 > 0$, it suffices to require that $W_0 > 1/|F|$. If $W_0 < 0$, the absolute value of the parameter W_0 is within $1/F_2 < |W_0| < 1$. In the first case the solution ε_d' is to the left of the point ε_1 (Fig. 11.2), and in the second case to the right of the point ε_2.

Thus, depending on the sign of the parameter W_0, discrete frequencies can arise either to the left of the continuous spectrum band ($\omega_d' < \omega_1$ at $W_0 > 0$) or to the

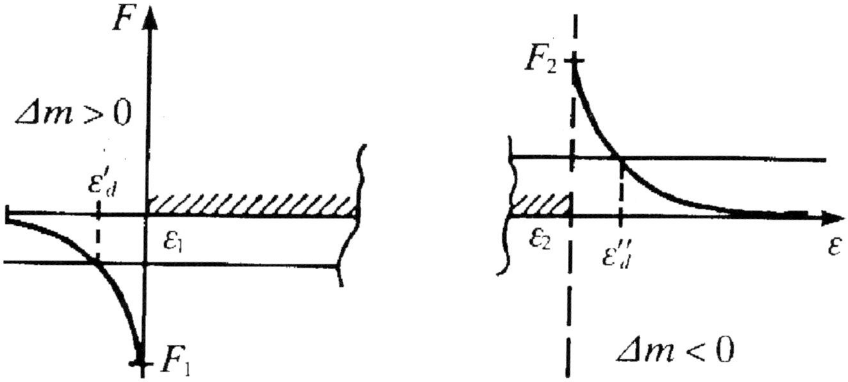

Fig. 11.2 Determination of the position of local frequencies for different signs of Δm.

right of it ($\omega_d'' > \omega_2$ at $W_0 < 0$). Crystal vibrations with the frequencies described are called *local vibrations*, and the frequencies ω_d – *local frequencies*. This name is attributed to the fact that the amplitude of the corresponding vibration is only nonzero in a certain vicinity near the point defect. Let us analyze the letter of the cases considered, assuming that $W_0 > 0$ and the local frequency ω_d is to the left of the continuous spectrum band ($\omega_d < \omega_1$). The local vibration amplitude is given by (11.1.9), implying its coordinate dependence is completely determined by the behavior of the ideal crystal Green function. We rewrite the Green function as

$$G_\varepsilon^0(\boldsymbol{n}) = \frac{1}{N}\sum_{\boldsymbol{k}} \frac{e^{i\boldsymbol{k}\boldsymbol{r}(\boldsymbol{n})}}{\varepsilon - \omega^2(\boldsymbol{k})} = \frac{V_0}{(2\pi)^3}\int \frac{e^{i\boldsymbol{k}\boldsymbol{r}(\boldsymbol{n})}d^3\boldsymbol{k}}{\varepsilon - \omega^2(\boldsymbol{k})}. \tag{11.1.17}$$

It is known that the behavior of such integrals at large distances ($r \gg a$) is mainly determined by the form of the integrand at small values of k ($ak \ll l$). In other words, the major contribution to the integral (11.1.17) at large r comes from the integration in the region of small wave vectors where the dispersion law of an ideal crystal is quadratic in k. To avoid possible complications we write down the quadratic dispersion law for the isotropic model in the form

$$\omega^2(\boldsymbol{k}) = \omega_1^2 + \gamma^2 k^2, \qquad \gamma^2 \sim (\omega_2^2 - \omega_1^2)a^2.$$

We take into account this dispersion law by changing $\varepsilon_1 = \omega_1^2$ and performing the integration required in (11.1.17)

$$\begin{aligned}\int \frac{e^{i\boldsymbol{k}\boldsymbol{r}(\boldsymbol{n})}d^3\boldsymbol{k}}{\varepsilon_1 - \varepsilon + \gamma^2 k^2} &= 2\pi \int_0^\infty \frac{k^2 d^k}{\varepsilon_1 - \varepsilon + \gamma^2 k^2} \int_0^\pi e^{ikr\cos\theta}\, d\theta \\ &= \frac{2\pi^2}{\gamma^2 r} \exp\left(-\frac{r}{\gamma}\sqrt{\varepsilon_1 - \varepsilon}\right).\end{aligned}$$

We substitute the result obtained in (11.1.17) and denote $\boldsymbol{r} = \boldsymbol{r}(\boldsymbol{n})$

$$G^0_\varepsilon(\boldsymbol{n}) \sim \frac{1}{r} \exp\left(-\frac{r}{\gamma}\sqrt{\omega_1^2 - \varepsilon}\right). \tag{11.1.18}$$

Substituting (11.1.18) into (11.1.9) to obtain

$$u(\boldsymbol{n}) = \frac{U_0 V_0}{4\pi\gamma^2} \frac{u(0)}{r} \exp\left(-\frac{r}{l}\right), \tag{11.1.19}$$

where the characteristic length l determining the localization region of vibrations with a discrete frequency is introduced:

$$l = \frac{\gamma}{\sqrt{\varepsilon_1 - \varepsilon_d}} = \frac{\gamma}{\sqrt{\omega_1^2 - \omega_d^2}}. \tag{11.1.20}$$

Thus, the local vibration amplitude decreases very quickly if the distance r increases and the decay length has the following order of magnitude

$$l \sim a\sqrt{\frac{\omega_2^2 - \omega_1^2}{\omega_1^2 - \omega_d^2}} \sim a\sqrt{\frac{\omega_2^2 - \omega_1^2}{\delta\omega}},$$

where $\omega_2 - \omega_1$ is the continuous spectrum frequency band width; $\delta\omega = \omega_1 - \omega_d$.

Thus, the localization of vibrations is determined by the ratio of the gap size $\delta\omega$ separating a discrete frequency from the edge of the continuous spectrum to the width of the continuous spectrum band.

Note that local vibrations are concerned with many physical effects observed in crystals and indicate two aspects of this relation.

On the one hand, a discrete frequency separated by a finite range of frequencies from the continuous spectrum band results in singularities in the frequency dependencies of different characteristics of the scattering and absorption processes. Corresponding resonance singularities arise in the scattering amplitudes of different particles (e. g., neutrons) on local defects. Infrared absorption in ionic crystals also shows peaks corresponding to local vibrations of various centers specific to these crystals.

On the other hand, a local vibration has a finite weight in the expansion of the displacement vector of an impurity atom into normal modes of a defect lattice, unlike each mode of the continuous spectrum vibrations that has an infinitely small weight (we remind ourselves that the contribution of a separate vibration of the quasi-continuous spectrum is proportional to $1/\sqrt{N}$, where N is the number of atoms in a crystal). Therefore, the effects connected with the displacement of impurity atoms and their nearest environment (e. g., optical transitions in impurity atoms, Mössbauer effect connected with impurity atoms, etc.) are very sensitive to local vibrations.

11.2 Elastic Wave Scattering by Point Defects

In general, the scattering problem can be subdivided into two parts: a calculation of the effective cross section for scattering and a study of the shape of the wave surface after scattering (at large distances from the scattering center). The first part of the problem requires knowledge of the local inhomogeneity structure (details of the point defect model). The second part concerns itself with the wave-surface shape at large distances and can be solved by very general assumptions concerning the character of the point defects.

A knowledge of the isofrequency surface shape is sufficient for studies of the asymptotic behavior of scattering waves. As we are interested in the scattering wave asymptotics, we restrict ourselves to the simplest model of a point defect, focusing on singularities of the isofrequency surface of a vibrating crystal.

The stationary vibrations of an ideal lattice are described by an equation represented symbolically by (2.2.2). The equation of crystal vibrations with an isolated defect is given in symbolic form by

$$\varepsilon u - \frac{1}{m} A u = \boldsymbol{U} u, \qquad \varepsilon = \omega^2, \tag{11.2.1}$$

where A is a linear Hermitian operator corresponding to the dynamic matrix of an ideal crystal

$$A u(\boldsymbol{n}) = \sum_{\boldsymbol{n}'} A(\boldsymbol{n} - \boldsymbol{n}')\, u(\boldsymbol{n}'),$$

$\boldsymbol{U}$ is the perturbation matrix

$$\boldsymbol{U} u(\boldsymbol{n}) = \sum_{\boldsymbol{n}'} \boldsymbol{U}(\boldsymbol{n}, \boldsymbol{n}') u(\boldsymbol{n}'). \tag{11.2.2}$$

For an isotope defect positioned at the site $\boldsymbol{n}_1$ of a monatomic lattice, we have

$$U^{ik}(\boldsymbol{n}, \boldsymbol{n}') = U_0 \delta_{\boldsymbol{n}\boldsymbol{n}_1} \delta_{\boldsymbol{n}'\boldsymbol{n}_1} \delta_{ik} \qquad U_0 = -\varepsilon \frac{\Delta m}{m}. \tag{11.2.3}$$

We consider the scattering of the vibration mode of an ideal crystal

$$\boldsymbol{u}(\boldsymbol{n}, t) = \boldsymbol{u}_0(\boldsymbol{n}) e^{-i\omega t}, \qquad \boldsymbol{u}_0(\boldsymbol{n}) = \frac{\boldsymbol{e}(\boldsymbol{k}_0)}{\sqrt{N}} e^{i\boldsymbol{k}_0 \boldsymbol{r}(\boldsymbol{n})}, \tag{11.2.4}$$

by the point defect with a perturbation potential of the type (11.2.3) where we set $\boldsymbol{n}_1 = 0$.

Since the defect has no internal degrees of freedom, scattering will occur without change of frequency, and the coordinate part of the scattering wave is a solution to (11.2.1). We write the perturbed solution to this equation (in usual vector notations) as

$$\boldsymbol{u}(\boldsymbol{n}, t) = \boldsymbol{u}_0(\boldsymbol{n}) + \boldsymbol{\chi}(\boldsymbol{n}), \tag{11.2.5}$$

where $\chi(\boldsymbol{n})$ describes the scattered wave and is a solution to the matrix equation (in the column notations)

$$\left(\varepsilon - \frac{1}{m}\boldsymbol{A}\right)\chi = \boldsymbol{U}(u_0 + \chi). \tag{11.2.6}$$

A formal solution to this equation is found [similar to (11.1.9)] by using the Green tensor of an ideal crystal (the index ε is omitted here)

$$\chi = \boldsymbol{G}^0\boldsymbol{U}(u_0 + \chi). \tag{11.2.7}$$

Taking the noncommutativity of the operators in (11.2.7) into account we find that

$$\chi = (1 - \boldsymbol{G}^0\boldsymbol{U})^{-1}\boldsymbol{G}^0\boldsymbol{U}u_0. \tag{11.2.8}$$

In the isotropic approximation or in a cubic crystal with only one isotope defect producing the perturbation potential (11.2.3), the matrix $\boldsymbol{G}^0\boldsymbol{U}$ is diagonal, hence, $1 - \boldsymbol{G}^0\boldsymbol{U} = 1 - \boldsymbol{G}^0_\varepsilon(0)\boldsymbol{U}$.

Changing from the operator form of (11.2.8) to the coordinate form we find for the scattering wave (in the vector component notations)

$$\chi^i(\boldsymbol{n}) = \frac{U_0}{\sqrt{N}D(\varepsilon)}G_0^{ij}(\boldsymbol{n})e^j(\boldsymbol{k}_0), \tag{11.2.9}$$

where

$$D(\varepsilon) = 1 - U_0 G^0_\varepsilon(0) = 1 + \varepsilon\left(\frac{\Delta m}{m}\right)G^0_\varepsilon(0), \tag{11.2.10}$$

and for a scalar model the r.h.s. of (11.2.10) should be understood literally, while for a cubic crystal $G^0(0) = (1/3)G_0^{ll}(0)$.

It follows from (11.2.9) that the form of a scattering wave at large distance is determined unambiguously by the asymptotic behavior of the Green tensor of an ideal lattice.

There is an expression (4.5.12) for the Green function in a scalar model. However, one can really use this formula for calculations only when (4.5.12) is rewritten in the form of an integral (4.5.13), but since the frequency ω is assumed to belong to the eigenfrequency band of an ideal crystal, (4.5.13) should be regularized

$$G^0_\varepsilon(\boldsymbol{n}) = \frac{V_0}{(2\pi)^3}\int \frac{e^{i\boldsymbol{k}\boldsymbol{r}(\boldsymbol{n})}d^3\boldsymbol{k}}{\varepsilon - \omega^2(\boldsymbol{k}) - i\gamma}. \tag{11.2.11}$$

Regularization should result in a diverging wave at infinity, i. e., it should be the same as that used to determine the retarded Green function (Section 4.6).

Denote

$$I(\boldsymbol{r}) = \int \frac{e^{i\boldsymbol{k}\boldsymbol{r}(\boldsymbol{n})}d^3\boldsymbol{k}}{\varepsilon - \omega^2(\boldsymbol{k}) - i\gamma}, \tag{11.2.12}$$

and transform from an integration with respect to $\boldsymbol{k}$ to an integration with respect to $z = \omega^2(\boldsymbol{k})$ and over the isofrequency surface $z = \text{const}$

$$I(\boldsymbol{r}) = \int \frac{J(\boldsymbol{r}, z)\,dz}{\varepsilon - z - i\gamma}, \tag{11.2.13}$$

where

$$J(\boldsymbol{r},z) = \oint_{\omega^2(\boldsymbol{k})=z} \frac{e^{i\boldsymbol{kr}(\boldsymbol{n})}\,dS_k}{\left|\nabla_k\omega^2(\boldsymbol{k})\right|}. \qquad (11.2.14)$$

The main contribution to the asymptotic (at $r \to \infty$) values of the integral (11.2.14) comes from the points of the stationary phase of the integrand. To elucidate the geometrical meaning of these points we write $\boldsymbol{kr} = r\boldsymbol{kn} = rh$ ($\boldsymbol{n}$ is a unit vector in the direction $\boldsymbol{r}$, and $h = \boldsymbol{kn}$) and introduce the orthogonal curvilinear coordinate ξ_i ($i = 1, 2$) on the surface $\omega^2(\boldsymbol{k}) = z$. The stationary phase points are then determined by the conditions

$$\frac{\partial h}{\partial \xi_i} = 0, \qquad i = 1, 2. \qquad (11.2.15)$$

The conditions (11.2.15) are satisfied at the points where the supporting plane $h \equiv \boldsymbol{kn} =$ constant touches the isofrequency surface (Fig. 11.3a).

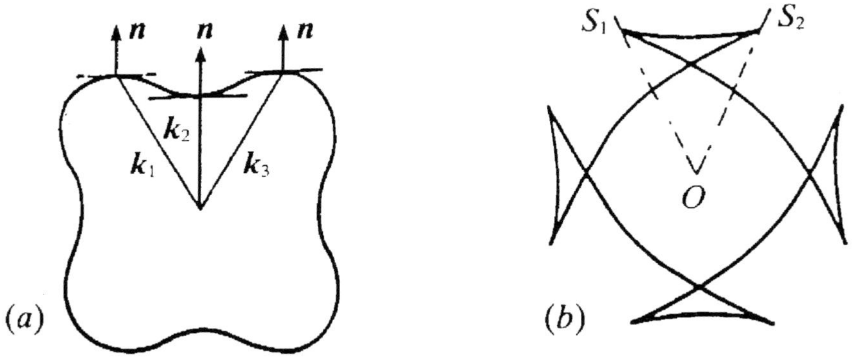

Fig. 11.3 Cross sections: (*a*) of an isofrequency surface on which the support points corresponding to the direction $\boldsymbol{n}$ are indicated; (*b*) of the wave surface on which the rays OS_1 and OS_2 limit the "folds".

We denote the contact points by $\boldsymbol{k}_\nu$ (sometimes they are called support points). On an isofrequency surface with an inversion center such points are positioned in pairs $\pm\boldsymbol{k}_\nu$.

We consider one of the supporting points measuring the coordinates ξ_i from it. In the close vicinity of this point the following expansion holds:

$$h(\xi) = \boldsymbol{k}_\nu\boldsymbol{n} + \frac{1}{2}\alpha^\nu_{ik}\xi_i\xi_k, \qquad \alpha^\nu_{ik} = \left(\frac{\partial^2 h}{\partial\xi_i\partial\xi_k}\right)_\nu.$$

We choose a network of curvilinear coordinates, so that the matrix α^ν_{ik} is diagonal, i. e., we choose the local coordinate axes ξ_1 and ξ_2 along the main directions of the surface curvature $\omega^2(\boldsymbol{k}) = z$ at the point $\boldsymbol{k} = \boldsymbol{k}_\nu$. Then

$$h(\xi) = \boldsymbol{k}_\nu\boldsymbol{n} + \frac{1}{2}\left(\alpha_1\xi_1^2 + \alpha_2\xi_2^2\right), \qquad \alpha_1 = \alpha_{11}, \qquad \alpha_2 = \alpha_{22}. \qquad (11.2.16)$$

With such a choice of 2D coordinates in $\boldsymbol{k}$-space, the product $\alpha_1\alpha_2$ determines the total (Gaussian) curvature of the surface at the tangent point; $K_\nu = \alpha_1\alpha_2$. If $K_\nu > 0$, the corresponding point is called *elliptical*, and if $K_\nu < 0$ it is called *hyperbolic*.

All the points of a convex surface are elliptical. If the surface is not convex (such as shown in Fig. 4.3 or Fig. 11.3), there exist parts of the surface with points of different types (either elliptical or hyperbolic), but the parts of an isofrequency surface of the first and second types are divided by the lines along which one of the coefficients (α_1 or α_2) vanishes. These are the lines of *parabolic* points. Finally, at the intersection of the lines of parabolic points are the flat regions where $\alpha_1 = \alpha_2 = 0$.

We start by analyzing the asymptotic form of the integral (11.2.14) when the supporting points are elliptical or hyperbolic. Reducing the integral (11.2.14) to the sum of integrals in the vicinities of the tangent points (the points of the stationary phase), we get

$$J(\boldsymbol{r}, z) = \sum_\nu \frac{e^{i\boldsymbol{k}_\nu \boldsymbol{r}}}{|\nabla\omega^2(\boldsymbol{k}_\nu)|} \int \exp\left[\frac{i}{2} r \left(\xi_1^2 + \xi_2^2\right)\right] d\xi_1\, d\xi_2. \qquad (11.2.17)$$

The integrals in (11.2.17) are calculated simply to be

$$\int \exp\left(\frac{i}{2} r\alpha\xi^2\right) d\xi = \sqrt{\frac{2}{r\,|\alpha|}} \int_0^\infty \frac{e^{\pm ix}\, dx}{\sqrt{x}} = \sqrt{\frac{2\pi}{r\,|\alpha|}} \exp\left(\pm\frac{i\pi}{4}\right), \qquad (11.2.18)$$

where the sign in the exponent index coincides with the sign of the main curvature α.

Thus we obtain an asymptotic expression proportional to $1/r$ (Lifshits, 1950)

$$J(\boldsymbol{r}, z) = \frac{2\pi}{r} \sum_\nu \frac{\exp\left(i\boldsymbol{k}_\nu \boldsymbol{r} \pm \dfrac{i\pi}{4}\right)}{|\nabla\omega^2(\boldsymbol{k}_\nu|\, \sqrt{|K_{nu}|}}. \qquad (11.2.19)$$

Substituting (11.2.19) into (11.2.13), we note that as $r \to \infty$ the integral with respect to z is reduced to an integral over the close proximity of the point $z = \varepsilon$, where the integrand has a singularity. The main part of the integral (11.2.13) is obtained by integrating near the pole $z = \varepsilon$, so that the final asymptotic expression for the Green function (11.2.11) is

$$G_\varepsilon^0(\boldsymbol{n}) = \frac{iV_0}{4\pi r} \sum_\nu \frac{\exp\left(i\boldsymbol{k}_\nu \boldsymbol{r} \pm \dfrac{i\pi}{4}\right)}{|\nabla\omega^2(\boldsymbol{k}_\nu|\, \sqrt{|K_{nu}|}}, \qquad (11.2.20)$$

where the summation is over the support points on the isofrequency surface $\omega^2(\boldsymbol{k}) = 0$ with $\boldsymbol{n}\boldsymbol{v}_\nu \equiv \boldsymbol{n}\nabla\omega(\boldsymbol{k}_\nu) > 0$.

The asymptote (11.2.20) shows the necessary amount of decrease with distance ($\sim 1/r$), providing a finite value of the scattering wave energy flow in any finite interval of a solid angle dO. Indeed, the flow of the energy density of the elastic field is quadratic in space and time derivatives of $\kappa(\boldsymbol{r}, t)$, i. e., at large distances it is proportional to $1/r^2$. Thus, the flow per area element $r^2 dO$ is finite.

Thus, the scattered wave represents a superposition of several diverging waves whose number equals the number of possible solutions to (11.2.15) at $z = \omega^2$. Each of these waves has its own shape and its own propagation velocity. The imaging of the spatial distribution of a scattered wave can be obtained by studying the so-called *wave surface*. The wave surface in coordinate space is, in a certain sense, polar with respect to the isofrequency surface and is constructed as follows. From the defect position (point O in Fig. 11.3b) a ray is drawn in the direction $\boldsymbol{n}$ and along it the length $r = 1/(\boldsymbol{n}\boldsymbol{k}_\nu)$ is plotted, where $\boldsymbol{k}_\nu = \boldsymbol{k}_\nu(\boldsymbol{n})$ are the support points.

It the isofrequency surface is convex, there is one supporting point with $\boldsymbol{n}\boldsymbol{v}_\nu > 0$. If it is not convex there can be several points of this kind. In the last case "folds" and recursion points arise on the wave surface. A tangent plane in the vicinity of each support point generates its own region of a wave surface. On the boundary of neighboring folds there is a transition from the region of elliptic points to that of hyperbolic points on the isofrequency surface. The boundaries are the parabolic point lines ($K_\nu = 0$). There always exists a continuous multitude of directions (a conical surface) corresponding to $K_\nu = 0$. These directions are shown as straight lines OS_1 and OS_2 in Fig. 11.3b; at the points S_1 and S_2 a pair of wave surface parts merges and breaks. Such singularities are classified in a theory of catastrophes and it is shown for elastic wave scattering in crystals that only *catastrophes* of the fold and the reversion point types are possible. The *catastrophe* implies that the energy flow density calculated formally by (11.2.20) in the directions considered goes to infinity ($K_\nu = 0$). Actually, at these points (more exactly on corresponding conical surfaces) the asymptotic behavior of the scattered wave changes, i. e., dependence of the scattered wave amplitude on the inverse distance $1/r$ becomes another: a power of the distance r decreases in the denominator of the function $I(\boldsymbol{r}, z)$ or the function $I(\boldsymbol{r})$.

As an illustration we consider a simple parabolic point $\boldsymbol{k}_0$ in the vicinity of which the function $h = \boldsymbol{k}\boldsymbol{n}$ has the expansion

$$h = \boldsymbol{k}_0\boldsymbol{n}_0 + \frac{1}{2}\alpha\xi_1^2 + \beta\xi_2^2, \tag{11.2.21}$$

where $\boldsymbol{n}_0$ is a unit vector of the direction whose supporting planes are tangential to the isofrequency surface at the parabolic point $\boldsymbol{k}_0$ (we have chosen the coordinate axes ξ_1 and ξ_2 along the main directions of the isofrequency surface curvature). Then, in calculating (11.2.17) apart from the integral (11.2.18) for the direction ξ_1 we have another integral for the direction ξ_2, namely

$$\int \exp\left(ir\beta\xi^3\right) d\xi = \frac{2}{3\sqrt[3]{r\,|\beta|}} \int_0^\infty \frac{\cos x}{\sqrt[3]{x^2}}\, dx = \sqrt[3]{\frac{3}{r\,|\beta|}}\,\Gamma\left(\frac{4}{3}\right), \tag{11.2.22}$$

where $\Gamma(m)$ is the gamma-function. Thus, instead of (11.2.19), we get an asymptotic expansion (Lifshits and Peresada, 1955):

$$J(\boldsymbol{r}, z) = \frac{\sqrt{6\pi}\Gamma\left(\frac{4}{3}\right)}{r^{\frac{5}{6}}} \frac{\exp\left(i\boldsymbol{k}_0\boldsymbol{r} \pm \frac{i\pi}{4}\right)}{\left|\omega^2(\boldsymbol{k}_0)\right| |\alpha|^{\frac{1}{2}} |\beta|^{\frac{1}{3}}}. \tag{11.2.23}$$

Calculating by means of (11.2.23) the asymptotes of the Green function and then the energy flow density, we find that the latter is proportional to $r^{-5/3}$. If there are no additional singularities on the line of parabolic points, then $|\alpha|\sqrt{|\beta|} \sim K_*$ where K_* is the Gaussian curve at an arbitrary point of the isofrequency surface. Thus, the energy flow density along the direction $\boldsymbol{n}_0$ "catastrophically" exceeds at large r the energy flow density from the other points in the ratio $r^{1/3}\,|\beta|^{1/6}$.

However, a solid angle inside which this energy flow density exists decreases with increasing r. Indeed, consider a scattered wave in the direction $\boldsymbol{n}$ inclined by an angle $\delta\theta_2$ (along ξ_2) to $\boldsymbol{n}_0$ The supporting point will then be displaced by the value $\delta\xi_2$ determined by $\theta_2 = 3\beta(\delta\xi_2)^2$. In the new supporting point, the Gaussian curvature equals $K = 6\alpha\beta\delta\xi_2$. Comparing with (11.2.19), where this Gaussian curvature is used with (11.2.23) for a parabolic supporting point, we see that they match by the order of magnitude $\delta\xi_2 \sim (|\beta|\,r)^{-1/3}$. Thus, the angle inside which an increased energy intensity is observed can be evaluated as $\delta\theta_2 \sim |\beta|^{1/3}\,r^{-2/3}$. It is clear that the angle $\delta\theta_2$ decreases with increasing r faster than the energy density increases. Thus, the total energy flow per angle $\delta\theta_2$ decreases with distance proportional to $r^{-1/3}$.

We ultimately calculate the contribution to the integral (11.2.14) generated by the flattening point in the vicinity of which an expansion of the function h is

$$h = \boldsymbol{k}_0\boldsymbol{n}_0 + \beta_1\xi_1^3 + \beta_2\xi_2^3.$$

Without repeating the calculations, we write down the corresponding part of the integral (11.2.14) as

$$J(\boldsymbol{r}, z) = \frac{3\Gamma^2\left(\frac{4}{3}\right)}{r^{\frac{2}{3}}} \frac{e^{i\boldsymbol{k}_0\boldsymbol{r}}}{\left|\nabla\omega^2(\boldsymbol{k}_0)\right| |\beta_1\beta_2|^{\frac{1}{3}}}.$$

In the given case the energy flow density in the scattered wave exceeds that in the ordinary conditions in the ratio $r^{2/3}\,|\beta_1\beta_2|^{-1/6}$ Accordingly, the solution of a solid angle, where the flow of such density is concentrated, decreases with the distance as $|\beta_1\beta_2|^{1/3}\,r^{-4/3}$.

Having discussed the singularities of the scattered wave shape and its asymptotic behavior at infinity, we shall analyze in brief the frequency dependence of the scattered wave amplitude. The frequency dependence is primarily given by the multiplier $D(\omega^2)$ in the denominator (11.2.9). The required regularization of this expression can be performed using the relation (4.7.4)

$$D(\varepsilon) = 1 - U_0 G^0_{\varepsilon - i\gamma} = 1 - U_0\,\text{P.V.}\int \frac{g_0(z)\,dz}{\varepsilon - z} - i\pi U_0 g_0(\varepsilon), \tag{11.2.24}$$

where P.V. means the principal value of the integral; $g_0(\varepsilon)$ is the vibration density of a crystal without defects.

We denote

$$R(\varepsilon) = \mathrm{Re}\{D(\varepsilon)\} = 1 - U_0\,\mathrm{P.V.}\int \frac{g_0(z)\,dz}{\varepsilon - z},$$

and represent the complex function $D(\varepsilon)$ as

$$\begin{aligned} D(\varepsilon) = R(\varepsilon) - i\pi U_0 g_0(\varepsilon) &= \sqrt{R^2(\varepsilon) + [\pi U_0 g_0(\varepsilon)]^2}e^{i\varphi(\varepsilon)}, \\ \tan\varphi(\varepsilon) &= \frac{\pi U_0 g_0(\varepsilon)}{R(\varepsilon)}. \end{aligned} \tag{11.2.25}$$

Strictly speaking, the function (11.2.25), has no specific properties for real ε, but it may turn out that at a certain frequency $R(\varepsilon) = 0$. If the vibration density $g_0(\varepsilon)$ at the corresponding frequency is small, $|D(\varepsilon)|^{-1}$ will be very large. The resonance amplitude of the scattered wave increases. It is known that the appearance of resonance denominators in the scattering amplitude at certain frequencies ω_q is evidence of the presence of quasi-stationary eigenvibrations in a medium through which the wave travels. Such a quasi-stationary state is analogous to a *resonance* state in quantum mechanics and its lifetime increases if $g_0(\varepsilon)$ becomes smaller. Since in a 3D case $g_0(\varepsilon) \sim \sqrt{\varepsilon}$, the resonance will be pronounced at a small quasi-stationary state frequency. Thus, a point defect that provides the equality $R(\varepsilon) = 0$ at low frequency may generate the resonance vibrational states in a crystal (Kagan and Iosilevskii, 1962). A more consistent theory of resonance states near a point defect is presented in Section 11.5, but we shall now try to obtain general conditions for the existence of a quasi-stationary state.

At small ε ($\varepsilon \ll \omega_D^2$), we set in (11.2.24)

$$\begin{aligned} \int \frac{g_0(z)\,dz}{\varepsilon - z} &\approx -\int \frac{g_0(z)}{z}\,dz = -\frac{1}{\varepsilon_*}, \quad \varepsilon_* \sim \omega_D^2, \\ g_0(\varepsilon) &= \frac{V_0}{(2\pi)^3}\frac{\sqrt{\varepsilon}}{s_0^3} \sim \frac{\sqrt{\varepsilon}}{\omega_D^3}. \end{aligned} \tag{11.2.26}$$

Therefore, the function $R(\varepsilon)$ is represented by

$$R(\varepsilon) = 1 + \frac{U_0}{\varepsilon_*} = 1 - \left(\frac{\Delta m}{m}\right)\frac{\varepsilon}{\varepsilon_*} = \frac{\Delta m}{m\varepsilon_*}\left(\frac{m\varepsilon_*}{\Delta m} - \varepsilon\right).$$

Thus, the resonance frequency $\varepsilon_k = (m/\Delta m)\varepsilon_*$ satisfies the condition $\varepsilon \ll \omega_D^2$ for a large enough parameter $\Delta m/m \gg 1$. The last inequality provides results in a resonance state in a crystal with heavy isotopic defects.

11.3
Green Function for a Crystal with Point Defects

The equations of crystal vibrations with a single defect (11.2.8) involve a perturbation in the form of (11.1.7). The Green tensor of a crystal with a defect is a resolvent for

(11.1.6), i. e., it is found as a solution to the following equation vanishing at infinity

$$\varepsilon G_\varepsilon - (A_0 + U)G_\varepsilon = I. \tag{11.3.1}$$

We rewrite (11.3.1) in the form $(\varepsilon I - A_0)G_\varepsilon = I + UG_\varepsilon$ and represent a formal solution to this equation by means of the Green tensor of an ideal crystal $G = G^0 + G^0UG$.

Performing calculations analogous to those used in deriving (11.2.8) we get

$$G = G^0 + G^0TG^0, \tag{11.3.2}$$

where the matrix T is determined by (11.3.2)

$$T = \left(I - UG^0\right)^{-1} U. \tag{11.3.3}$$

In the present case of an isolated defect-isotope, $T = [D(\varepsilon)]^{-1} U$, where the definition (11.2.10) is used. Substituting (11.3.3) into (11.3.2) we obtain

$$G^{ik}(n, n') = G_0^{ik}(n - n') + \frac{U_0}{D(\varepsilon)} G_0^{il}(n - n_1) G_0^{lk}(n_1 - n'). \tag{11.3.4}$$

For a scalar model the formula (11.3.4) is not simplified much (only indices i, k, l are not used). The poles of the Green function as functions of the variable ε determine the squares of the eigenfrequencies of vibrations of the corresponding system (Section 4.5). It follows from (11.3.4) that the Green function for a crystal with a point defect may have an additional pole at the point where the function $D(\varepsilon)$ is zero. Thus, the additional poles of the Green function (with respect to the ideal crystal Green function) determine the frequencies of local vibrations of a crystal with a defect considered in previous sections.

We shall now point out a specific structure of the Green function in a crystal with a single defect (11.3.4). By definition, the Green function concerned gives us a displacement of the n-th site in a lattice under the action of a unit force, which is periodic in time $\omega^2 = \varepsilon$ and applied to the site n'. In an ideal lattice, the stationary interaction is transmitted directly from the site n' to the site n (the first term in (11.3.4)). In the presence of an isolated defect at the site n_1 an additional excitation transfer channel through a defect appears (Fig. 11.4a) (this is the second term in (11.3.4)). In the second term the vector going to a defect or from it is associated with a multiplier G^0, and the presence of the defect results in an additional multiplier $U_0/D(\varepsilon)$.

A remarkable feature of the excitation transfer through a defect possessing a local frequency is the resonance character of the transfer. By deriving (11.1.18) we have shown that the Green function $G_\varepsilon^0(n)$ decays exponentially with distance for the values of the parameter ε lying outside the band of the squares of the eigenfrequencies of an ideal crystal. Thus, at frequencies close to a discrete local frequency, the first term in (11.3.4) at large distances ($r(n - n') \gg l$) becomes negligibly small. The

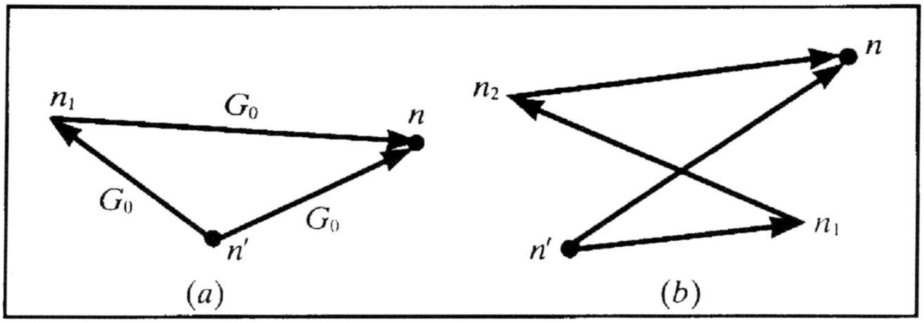

Fig. 11.4 Diagram for calculating the Green function in a crystal: (*a*) with one impurity at the point n_1; (*b*) with two impurities at the points n_1 and n_2.

second term has a resonance denominator $D(\varepsilon)$ and for ε can essentially exceed the first term. Thus, if the crystal has a defect with a discrete frequency in the forbidden band of frequencies of an ideal lattice, it promotes a resonance transfer of excitations at large distances. However, the second term in (11.3.4) also vanishes in the limit $|\boldsymbol{n}-\boldsymbol{n}'| \to \infty$ at any finite $D(\varepsilon)$. The situation will be different if we go over from a crystal with one defect to a crystal with small, but finite concentration of defects.

Let us now see whether it is possible to use the method presented to find the Green function in a crystal with a system of point defects.

Unfortunately, the relation (11.2.22) between the matrix $\boldsymbol{T}$ and the matrix $\boldsymbol{U}$ is nonlinear, and the contributions of separate defects cannot be summarized in an equation such as (11.3.4). However, a solution generalizing (11.3.4) can also be found in this case.

We restrict ourselves to a scalar model and choose the perturbation matrix in the form

$$U(\boldsymbol{n},\boldsymbol{n}') = u_0 \sum_{\alpha} \delta_{\boldsymbol{n}\boldsymbol{n}_\alpha} \delta_{\boldsymbol{n}'\boldsymbol{n}_\alpha'} \tag{11.3.5}$$

where $\boldsymbol{n}_\alpha$ are the number vectors of the sites where the defects are positioned.

For the matrix $\boldsymbol{T}$ one should expect a representation such as

$$T(\boldsymbol{n},\boldsymbol{n}') = \sum_{\alpha\beta} T_{\alpha\beta} \delta_{\boldsymbol{n}\boldsymbol{n}_\alpha} \delta_{\boldsymbol{n}'\boldsymbol{n}_\beta}. \tag{11.3.6}$$

Indeed, rewriting (11.3.3) in the form

$$(\boldsymbol{I} - \boldsymbol{U}\boldsymbol{G}_\varepsilon^0)\boldsymbol{T} = \boldsymbol{U},$$

and substituting here (11.3.5), (11.3.6) we obtain a condition unambiguously determining a set of parameters $T_{\alpha\beta}$:

$$D(\varepsilon) T_{\alpha\beta} - U_0 \sum_{\gamma \neq \alpha} G_\varepsilon^0(\boldsymbol{n}_\alpha - \boldsymbol{n}_{gamma}) T_{\gamma\beta} = u_0 \delta\alpha\beta. \tag{11.3.7}$$

If all $T_{\alpha\beta}$ are determined from this set of linear algebraic equations, the Green function will be derived as

$$G(\boldsymbol{n},\boldsymbol{n}') = G^0(\boldsymbol{n}-\boldsymbol{n}') + \sum_{\alpha\beta} G^0(\boldsymbol{n}-\boldsymbol{n}_\alpha)T_{\alpha\beta}G^0(\boldsymbol{n}_\beta-\boldsymbol{n}'). \tag{11.3.8}$$

Equations (11.3.7) are simple, but writing their solution for an arbitrary number of defects is rather cumbersome, and their quantitative analysis is complicated. Therefore we restrict ourselves to a qualitative characterization of the Green function (11.3.8).

If the distance between all neighboring defects $r_{\alpha\beta}$ greatly exceeds the dimension l of the localization region of vibrations near an isolated defect, there are values of the parameter ε such that $\left|U_0 G^0_\varepsilon(\boldsymbol{n}_\alpha-\boldsymbol{n}_\beta)\right| \ll D(\varepsilon)$ for $\alpha \neq \beta$. In the relation (11.3.7), we can then omit the terms with $\gamma \neq \alpha$. Subsequently, it is simple to solve (11.3.7):

$$T_{\alpha\beta} = \frac{U_0}{D(\varepsilon)}\delta_{\alpha\beta}. \tag{11.3.9}$$

The solution obtained does not, in fact, differ from (11.3.2) and the Green function is written analogously (11.3.4)

$$G(\boldsymbol{n},\boldsymbol{n}') = G^0(\boldsymbol{n}-\boldsymbol{n}') + \frac{U_0}{D(\varepsilon)}\sum_{\alpha} G^0(\boldsymbol{n}-\boldsymbol{n}_\alpha)G^0(\boldsymbol{n}_\alpha-\boldsymbol{n}'). \tag{11.3.10}$$

If the defects are distributed randomly but uniformly, (11.3.10) yields an approximate expression for the Green function, which is valid under the assumption that the average distance between the defects greatly exceed the length l. The expression (11.3.10) is inapplicable near the boundaries of the continuous spectrum of an ideal crystal and in the vicinity of a local frequency.

Coming back to an analysis of Fig. 11.4a we see that (11.3.10) takes into account the excitation transfer from site $\boldsymbol{n}'$ to site $\boldsymbol{n}$ through a simple use of all the defects. This situation resembles a kinematic X-ray scattering theory taking into account only simple X-ray scattering by each of the crystal atoms, but even with two defects available (at sites $\boldsymbol{n}_1$ and $\boldsymbol{n}_2$) there exists an excitation transfer channel ($\boldsymbol{n}' \to \boldsymbol{n}_1 \to \boldsymbol{n}_2 \to \boldsymbol{n}$) disregarded in (11.3.10) and shown in Fig. 11.4b.

Assuming the contribution of two defect processes to be small, it is easy to describe this. We find a correction to the solution (11.3.9) by setting

$$T_{\alpha\beta} = \frac{U_0}{D(\varepsilon)}\delta_{\alpha\beta} + t_{\alpha\beta}, \qquad \left|t_{\alpha\beta}\right| \ll \left|\frac{U_0}{D(\varepsilon)}\right|. \tag{11.3.11}$$

A small correction $t_{\alpha\beta}$ is easily found by substituting (11.3.11) into (11.3.7)

$$t_{\alpha\beta} = \left[\frac{U_0}{D(\varepsilon)}\right]^2 G^0_\varepsilon(\boldsymbol{n}_\alpha-\boldsymbol{n}_\beta). \tag{11.3.12}$$

Thus, a specified expression (11.3.10) that takes into account simple and double "scatterings" by the defects takes the form of the following expansion

$$\begin{aligned} G(\boldsymbol{n},\boldsymbol{n}') = \; & G^0(\boldsymbol{n}-\boldsymbol{n}') + \frac{U_0}{D(\varepsilon)} \sum_{\alpha} G^0(\boldsymbol{n}-\boldsymbol{n}_\alpha) G^0(\boldsymbol{n}_\alpha - \boldsymbol{n}') \\ + \; & \left[\frac{U_0}{D(\varepsilon)}\right]^2 \sum_{\alpha \neq \beta} G^0(\boldsymbol{n}-\boldsymbol{n}_\alpha) G^0(\boldsymbol{n}_\alpha - \boldsymbol{n}_\beta) G^0(\boldsymbol{n}_\beta - \boldsymbol{n}'). \end{aligned} \tag{11.3.13}$$

We shall consider (11.3.13) as the beginning of an infinite series of successive approximations. The structure of terms of different approximations in (11.3.13) is such that the possibility to construct the Green functions $G(\boldsymbol{n},\boldsymbol{n}')$ using a simple *diagram technique* is quite obvious. The rule for constructing the whole series is as follows. The points $\boldsymbol{n}$ and $\boldsymbol{n}'$ are linked by all possible straight rays broken down by the defects (the simplest of them are shown in Fig. 11.4). Each ray starting from the point $\boldsymbol{n}'$ does not return to it and coming to the point $\boldsymbol{n}$ never leaves it again. The ray may come back to the defect-occupation points many times. The number of breaks of a ray determines the order of the term in an expansion such as (11.2.13). Each segment of the ray $\boldsymbol{n}_{\alpha\beta}$ in the corresponding term is associated with a factor $G^0(\boldsymbol{n}_{\alpha\beta})$ and each defect with a factor $U_0/D(\varepsilon)$. After the summation over all rays, we get the desired Green function.

The calculation scheme described above can really only be used for an approximate calculation of the Green function in a crystal with small concentration of impurities, since the higher-order terms in (11.3.13) should be proportional to higher degrees of concentration of the defects.

We come back to (11.3.10) for the Green function of a crystal with randomly distributed impurities and consider it in the linear approximation with respect to concentration. Being interested first of all in extremely long-wave vibrations of a defect crystal, we change the summation over the numbers of impurities in (11.3.10) to an integration over the impurity coordinates, taking into account their homogeneous space distribution

$$\sum_{\alpha} f(\boldsymbol{r}) = \frac{c}{V_0} \int f(\boldsymbol{r})\, d^3\boldsymbol{r}, \tag{11.3.14}$$

where $\boldsymbol{r}_\alpha = \boldsymbol{r}(\boldsymbol{n}_\alpha)$ is the radius vector of an impurity atom; V_0 is the unit cell (atomic) volume; c is the concentration of defects, equal to the ratio of the number of the defects to N ($c \ll 1$). Note that the procedure (11.3.14) is equivalent to an averaging over the random positions of impurities.

We use (11.3.14) to carry out the summation using the $\boldsymbol{k}$-representation of the crystal Green function

$$\begin{aligned} & \sum_{\alpha} G^0(\boldsymbol{n}-\boldsymbol{n}_\alpha) G^0(\boldsymbol{n}_\alpha - \boldsymbol{n}') \\ = \; & \frac{c}{N^2} \sum_{\boldsymbol{k}\boldsymbol{k}'}{}' \frac{\exp\left[i\boldsymbol{k}\boldsymbol{r}(\boldsymbol{n}) - i\boldsymbol{k}'\boldsymbol{r}(\boldsymbol{n}')\right]}{\left[\varepsilon - \omega^2(\boldsymbol{k})\right]\left[\varepsilon - \omega^2(\boldsymbol{k}')\right]} \frac{1}{V_0} \int \exp\left[-i\left(\boldsymbol{k}-\boldsymbol{k}'\right)\boldsymbol{x}\right] d^3\boldsymbol{x}. \end{aligned}$$

We then apply the formula $\int \exp(i\boldsymbol{k}\boldsymbol{r})\, d^3r = V\delta_{k0}$ and write

$$\sum_\alpha G^0(\boldsymbol{n}-\boldsymbol{n}_\alpha)G^0(\boldsymbol{n}_\alpha-\boldsymbol{n}')$$
$$= \frac{c}{N}\sum_k \frac{\exp\{i\boldsymbol{k}\,[\boldsymbol{r}(\boldsymbol{n})-\boldsymbol{r}(\boldsymbol{n}')]\}}{\left[\varepsilon-\omega^2(\boldsymbol{k})\right]^2} = -c\frac{\partial}{\partial\varepsilon}G^0_\varepsilon(\boldsymbol{n}-\boldsymbol{n}'). \tag{11.3.15}$$

Substituting (11.3.15) in (11.3.10) we obtain an expansion of the Green function (we are interested only in dependence on the variable ε)

$$G(\varepsilon) = G^0(\varepsilon) - c\Pi(\varepsilon)\frac{\partial G^0(\varepsilon)}{\varepsilon}, \tag{11.3.16}$$

where the notation

$$\Pi(\varepsilon) = \frac{U_0}{D(\varepsilon)} = \frac{(\Delta m/m)\varepsilon}{1+(\Delta m/m)\varepsilon G^0_\varepsilon(0)} \tag{11.3.17}$$

is introduced.

If we restrict ourselves to a linear approximation with respect to concentration, (11.3.16) can be represented as

$$G(\varepsilon) = G^0\left[\varepsilon - c\Pi(\varepsilon)\right]. \tag{11.3.18}$$

We now remind ourselves that the diagram technique described above allows us to obtain the term of an arbitrary order in $\Pi(\varepsilon)$ in the expansion (11.3.13). The structure of these terms is such that after changing from the summation to the integration by the rule (11.3.14), they are expressed through a derivative of a corresponding order with respect to ε of the ideal crystal Green function and contain the necessary power of concentration. Thus, the validity of (11.3.18) is connected only with the possibility of replacing the summation over the numbers of impurities by an integration (11.3.15). It can be derived without assuming an approximation linear in c.

However, there is a requirement that must be satisfied in order to use the Green function (11.3.18). Using it we can obtain only those crystal characteristics that are determined by quantities averaged over the defect positions. Since the averaging over random configurations of a system of impurities results in a considerable loss of information contained in the initial form of the Green function, this requirement is extremely important.

11.4 Influence of Defects on the Density of Vibrational States in a Crystal

Having the Green function (11.3.4), we can use the recipe presented in Chapter 4 to determine the density of lattice vibrations in the presence of point defects. According to this method it is necessary to use (4.8.9) in the given case. For a monatomic lattice the density of vibrational states $g(\varepsilon)$ normalized to unity equals (4.8.9),

$$g(\varepsilon) = \frac{1}{\pi} \lim_{\gamma \to +0} \operatorname{Im} \frac{1}{3N} \operatorname{Tr} \left(1 - \frac{dU}{d\varepsilon}\right) G_{\varepsilon - i\gamma}. \tag{11.4.1}$$

For a scalar model the factor $3N$ in the denominator must be replaced by N.

Of much physical interest is generally the density of vibrational states in a crystal containing a large number of equivalent point defects. Just this quantity can characterize the macroscopic properties of the crystal. In this case, when the average distance between the defects is large (the concentration is small) the Green function of a scalar model of crystal vibrations is calculated from (11.3.10). Substituting (11.3.5), (11.3.10) into (11.4.1) we first calculate the trace:

$$\begin{aligned} S &= \frac{1}{N} \operatorname{Tr} \left(1 - \frac{dU}{d\varepsilon}\right) \boldsymbol{G} = \frac{1}{N} G(\boldsymbol{n}, \boldsymbol{n}) + \left(\frac{\Delta m}{m}\right) \frac{1}{N} \sum_{\alpha} G(\boldsymbol{n}_\alpha, \boldsymbol{n}_\alpha) \\ &= G^0(0) + \frac{U_0}{N D(\varepsilon)} \sum_{n\,\alpha} G^0(\boldsymbol{n} - \boldsymbol{n}_\alpha) G^0(\boldsymbol{n}_\alpha - \boldsymbol{n}) \\ &+ c \left(\frac{\Delta m}{m}\right) G^0(0) + \left(\frac{\Delta m}{m}\right) \frac{U_0}{N D(\varepsilon)} \sum_{\alpha\beta} G^0(\boldsymbol{n}_\alpha - \boldsymbol{n}_\beta) G^0(\boldsymbol{n}_\beta - \boldsymbol{n}_\alpha). \end{aligned} \tag{11.4.2}$$

Preserving the accuracy with which (11.3.10) for the Green function was derived we omit the term with $\alpha \neq \beta$ in the last sum of (11.4.2)

$$\frac{1}{N} \sum_{\alpha\beta} G^0(\boldsymbol{n}_\alpha - \boldsymbol{n}_\beta) G^0(\boldsymbol{n}_\beta - \boldsymbol{n}_\alpha) = c \left[G^0(0)\right]^2. \tag{11.4.3}$$

In addition, it is easy to verify, using the $\boldsymbol{k}$-representation for the Green function of an ideal lattice, that

$$\sum_{n} G^0(\boldsymbol{n} - \boldsymbol{n}_\alpha) G^0(\boldsymbol{n}_\alpha - \boldsymbol{n}) = \frac{1}{N} \sum_{k} \frac{1}{\left[\varepsilon - \omega_0^2(\boldsymbol{k})\right]^2} = -\frac{d}{d\varepsilon} G^0_\varepsilon(0).$$

Therefore,

$$\frac{1}{N} \sum_{n\,\alpha} G^0(\boldsymbol{n} - \boldsymbol{n}_\alpha) G^0(\boldsymbol{n}_\alpha - \boldsymbol{n}) = -c \frac{d}{d\varepsilon} G^0_\varepsilon(0). \tag{11.4.4}$$

By substituting (11.4.3), (11.4.4) into (11.4.2), and remembering the definition (11.2.10) for $D(\varepsilon)$, we have

$$\begin{aligned} S &= G^0_\varepsilon(0) + \frac{c(\Delta m/m)}{N D(\varepsilon)} \left[\varepsilon \frac{d}{d\varepsilon} G^0_\varepsilon(0) + G^0_\varepsilon(0)\right] \\ &= G^0_\varepsilon(0) + \frac{c(\Delta m/m)}{N D(\varepsilon)} \frac{d}{d\varepsilon} \left[\varepsilon G^0_\varepsilon(0)\right] = G^0_\varepsilon(0) + c \frac{d}{d\varepsilon} \log D(\varepsilon). \end{aligned} \tag{11.4.5}$$

We now return to the initial formula (11.4.1) as applied to a scalar model

$$g(\varepsilon) = g_0(\varepsilon) + \frac{c}{\pi} \lim_{\gamma \to +0} \operatorname{Im} \frac{d}{d\varepsilon} \log D(\varepsilon). \tag{11.4.6}$$

In a linear approximation with respect to the concentration c, the expression (11.4.6) determines the variation of the density of vibrational states due to point defects. The applicability of the final formula (11.4.6) is not limited to a scalar model. The presence of three vibrational polarizations in a cubic crystal does not affect (11.4.6). It is sufficient to take account of a simple replacement $G^0(0) \rightarrow (1/3)G_0^{ll}(0)$ in the definition of $D(\varepsilon)$. For a monatomic lattice having no cubic symmetry $D(\varepsilon)$ is expressed through the determinant appearing in (11.1.14):

$$\log D(\varepsilon) = \frac{1}{3} \log \operatorname{Det} \left\| \delta_{ik} - U_0 G_0^{ik}(0) \right\| .$$

If the value of ε lies outside the continuous spectral band for an ideal crystal, then $g_0(\varepsilon) = 0$, and $G_\varepsilon^0(0)$ is a real function of the real variable ε:

$$\lim_{\gamma \to 0} G^0_{\varepsilon - i\gamma}(0) = G^0_\varepsilon(0) \equiv \operatorname{Re}\{G^0_\varepsilon(0)\}.$$

However, in this case the density of vibrations for a crystal with point defects can be nonzero outside the continuous spectrum only at the points where the argument of the logarithm in (11.4.6) vanishes: $D(\varepsilon) = 0$, i. e., for the values of ε coincident with the squares of local frequencies. We represent in close proximity of the point $\varepsilon = \varepsilon_d$, the function $D(\varepsilon - i\gamma)$ as

$$D(\varepsilon - i\gamma) = (\varepsilon - \varepsilon_{\mathrm{d}} - i\gamma) \frac{dD(\varepsilon_{\mathrm{d}})}{d\varepsilon_{\mathrm{d}}},$$

remembering that ε_D is the solution to the equation $D(\varepsilon) = 0$.

We then have from (11.4.6)

$$\delta g(\varepsilon) = \frac{c}{\pi} \lim_{\gamma \to +0} \operatorname{Im} \frac{1}{\varepsilon - \varepsilon_{\mathrm{d}} - i\gamma}. \tag{11.4.7}$$

However, this limit is equivalent to a δ-function. Therefore, in a corresponding frequency range the density of vibrational states has a δ-like character

$$\delta g(\varepsilon) = c\delta(\varepsilon - \varepsilon_{\mathrm{d}}), \tag{11.4.8}$$

showing that the local frequency is discrete.

The result contained in (11.4.8) has already been discussed, so that we shall now pass to analyzing the density of "defect" crystal vibrations in the continuous spectrum range of an ideal lattice ($\varepsilon_1 < \varepsilon < \varepsilon_2$).

In writing (11.2.24), we have taken into account that in the continuous spectrum region

$$\lim_{\gamma\to 0} G^0_{\varepsilon - i\gamma}(0) = \text{P.V.} \int \frac{g_0(z)\,dz}{\varepsilon - z} + i\pi g_0(\varepsilon), \tag{11.4.9}$$

where P.V. means the principal value of the integral.

In order to separate real and imaginary parts of the derivative $(d/d\varepsilon)G^0_\varepsilon(0)$ it is sufficient to differentiate (11.4.9) with respect to ε for all ε except for those getting into near the edges of the continuous spectrum interval and the van Hove specific point in which the function $g_0(\varepsilon)$ loses its analyticity. Thus

$$\begin{aligned} &\lim_{\gamma\to +0} D(\varepsilon - i\gamma) = R(\varepsilon) - i\pi U_0 g_0(\varepsilon), \\ &\lim_{\gamma\to +0} \frac{d}{d\varepsilon} D(\varepsilon - i\gamma) \frac{d}{dR(\varepsilon)\varepsilon} - i\pi \frac{d}{d\varepsilon}\left[U_0 g_0(\varepsilon)\right]. \end{aligned} \tag{11.4.10}$$

The first of these relations is consistent with (11.2.25).

Performing the operations indicated in (11.4.6), it is easy to obtain

$$\delta g(\varepsilon) = \frac{c}{|D(\varepsilon)|^2}\left\{U_0 g_0(\varepsilon)\frac{dR(\varepsilon)}{d\varepsilon} - R(\varepsilon)\frac{d}{d\varepsilon}\left[U_0 g_0(\varepsilon)\right]\right\}. \tag{11.4.11}$$

The expression (11.4.11) describes the deformation of the continuous spectrum of frequency squares under the influence of point defects. It has nonphysical singularities at the points where the function $g_0(\varepsilon)$ loses its analyticity and has root singularities. Thus, (11.4.11) is applicable only away from the vicinity of these points. The appearance of singularities in (11.4.11) results from the fact that an approximation that is linear in concentration is not sufficient for examining the spectral deformation near its singular points.

11.5 Quasi-Local Vibrations

We have mentioned that (11.4.11) has nonphysical singularities, because the linear approximation with respect to the defect concentration is inapplicable when describing $\delta g(\varepsilon)$ in the vicinity of certain values of ε. However, the function $\delta g(\varepsilon)$ may have another singularity having a physical meaning. Thus, with a defect-isotope present in the crystal, the real part $D(\varepsilon)$ can vanish for a certain value of $\varepsilon = \varepsilon_q$ lying inside the continuous spectrum:

$$R(\varepsilon_q) \equiv 1 + \left(\frac{\Delta m}{m}\right)\varepsilon_q \int \frac{g_0(z)\,dz}{\varepsilon_q - z} = 0. \tag{11.5.1}$$

Generally, at $\varepsilon = \varepsilon_q$ the function $\delta g(\varepsilon)$ has no singularities in a formal mathematical sense, but if the point $\varepsilon = \varepsilon_q$ is near the edges of the continuous spectrum band where the density of states $\delta g_0(\varepsilon)$ is very small, then the denominator in (11.4.11) becomes anomalously small in the vicinity of ε_q and the function $\delta g(\varepsilon)$ increases.

This situation corresponds to the appearance of a quasi-stationary vibration with $\varepsilon = \varepsilon_q$.

To determine the form of the resonance peak of the function $\delta g(\varepsilon)$, we use an expansion of the type (11.4.7): $R(\varepsilon) = (\varepsilon - \varepsilon_q) dR(\varepsilon_q)/d\varepsilon_q)$ and restricting ourselves to a small region near the point considered we set $\varepsilon = \varepsilon_q$ in the numerator of the r.h.s. of (11.4.11)

$$\delta g(\varepsilon) = \frac{c}{\pi} \frac{\Gamma_q}{(\varepsilon - \varepsilon_q)^2 + \Gamma_q^2},$$
$$\Gamma_q = -\pi \left(\frac{\Delta m}{m}\right) \frac{\varepsilon_q g_0(\varepsilon_q)}{R'(\varepsilon_q)} \quad \delta g_0(\varepsilon) = -\pi \left(\frac{\Delta m}{m}\right) \frac{\varepsilon_q g_0(-\varepsilon_q)}{R'(\varepsilon_q)}. \tag{11.5.2}$$

The formula (11.5.2) has the characteristic form of an expression describing the frequency distribution of a quasi-stationary state of the system. The parameter Γ_q gives the half-width of a corresponding resonance curve and may be associated with the imaginary part of ε in the conventional notation of (11.2.7). If Γ_q is small, the expression (11.5.2) results in a sharp peak in the vibration density. The frequency $\omega_q = \sqrt{\varepsilon_q}$ is then called a *quasi-local frequency* and the vibration corresponding to it a *quasi-local vibration*.

Analysis of (11.5.2) is of most interest in the case when the quasi-local vibration frequencies are near the long-wave edge of an acoustic spectrum (such vibrations will be shown to be generated by heavy isotopes with $\Delta m \gg m$). If $\varepsilon \ll \omega_D^2$, where ω_D is the Debye frequency, the main part of the expansion of $R(\varepsilon)$ in powers of ε is obtained from (11.2.26)

$$R(\varepsilon) = 1 - \left(\frac{\Delta m}{m} \frac{\varepsilon}{\varepsilon_*}\right), \qquad \varepsilon_* \sim \omega_D^2. \tag{11.5.3}$$

We note that negativeness of the derivative $R'(\varepsilon)$ and positiveness of the parameter Γ_q for $\Delta m > 0$ follow from (11.5.3).

Substituting (11.5.3) into (11.5.1), we see that for a large enough value of $\Delta m/m$ there is necessarily a solution to (11.5.1) in the region of small ε

$$\varepsilon_q = (\Delta m/m)\varepsilon_* \sim (\Delta m/m)\omega_D^2, \quad \Delta m \gg m. \tag{11.5.4}$$

It follows from (11.5.4) that with increasing $\Delta m/m$, the quasi-local frequency position approaches the long-wave edge of the vibration of the continuous spectrum ($\varepsilon_q \to 0$). Simultaneously, the peak in the spectral density narrows. Indeed, since at small ε we may use the estimate (11.2.26) for $g_0(\varepsilon)$, the resonance peak half-width is

$$\Gamma_q = \pi \varepsilon_* \varepsilon_q g_0(\varepsilon_q) \sim (m/\Delta m)^{3/2} \omega_D^2. \tag{11.5.5}$$

Thus, the relative half-width of the Lorentz curve (11.5.2) for heavy isotopes is determined by

$$\frac{\Gamma_q}{\varepsilon_q} = \pi\varepsilon_* g_0(\varepsilon_q) \sim \sqrt{\frac{m}{\Delta m}} \ll 1,$$

confirming the resonance peak narrowing with increasing $\Delta m/m$. Hence, with increasing $\Delta m/m$ the resonance peak acquires δ-function character. The possible appearance of a sharp peak in the function $\delta g(\varepsilon)$ in the region of low crystal vibrational frequencies with heavy impurity isotopes was first noted by Kogan and Iosilevskii (1962), and Brout and Visher (1962).

It is natural that the Lorentz-type peak (11.5.2) also arises in a plot of the frequency spectrum

$$\frac{\delta\nu(\omega)}{N} = c\frac{2\omega_q}{\pi}\frac{\Gamma_q}{(\omega^2-\omega_q^2)^2+\Gamma_q^2}, \qquad (11.5.6)$$

where $\omega_q^2 = \varepsilon_q$. A typical plot of the function $\nu(\omega) = \nu_0(\omega) + \delta\nu(\omega)$ for $\omega \ll \omega_D$ is shown in Fig. 11.5.

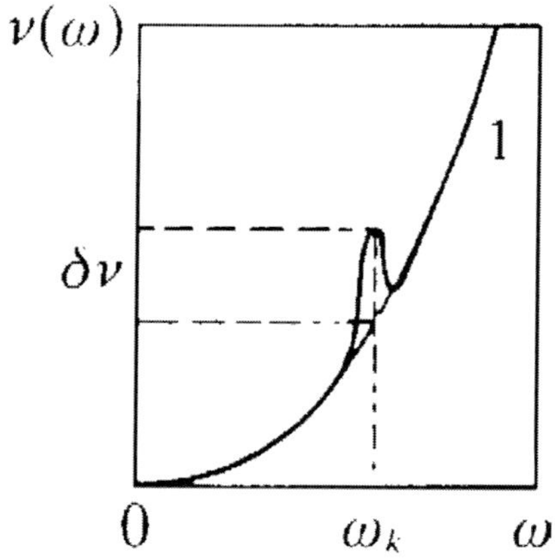

Fig. 11.5 The frequency spectrum near a quasi-local frequency (the curve 1: $\nu_0(\omega) = \text{const}\,\omega^2$).

It is easy to estimate the impurity concentration c_* at which the peak height (Fig. 11.5) calculated by (11.5.6) is comparable with the frequency distribution function $\nu_0(\omega)$ of an ideal crystal at a quasi-local frequency $\omega = \omega_q$. From the relation

$$\frac{\delta\nu(\omega_q)}{\nu_0(\omega_q)} \sim \delta\nu(\omega_q)\frac{\omega_D^3}{\omega_q\Gamma_q} \sim 1,$$

we get for the concentration to be estimated:

$$c_* \sim \Gamma_q\omega_q/\omega_D^3. \qquad (11.5.7)$$

We substitute in (11.5.7) the estimates (11.5.4), (11.5.5) to obtain

$$c_* \sim (m/\Delta m)^2 \ll 1. \tag{11.5.8}$$

Since, by assumption, the parameter $m/\Delta m$ is very small, the quasi-local vibration density peak may already be equal in order of magnitude to the vibration density of an ideal crystal at small concentrations of heavy impurities ($c \sim c_* \ll 1$). However, the derivation of (11.5.2), (11.5.6) and the calculation scheme were based on using the linear approximation ($c \ll 1$) with respect to concentrations for which additions (proportional to c) to the quantities studied are assumed to be small. Since for $c \sim c_* \ll 1$ the last assumption is not justified, it is necessary to analyze more consistently the influence of heavy isotopic impurities on the spectrum of crystal vibrations even at their smallest concentrations.

Quasi-local vibrations affect the thermodynamic and kinetic properties of a crystal.

The singularities in the amplitudes of elastic wave scattering near quasi-local frequencies ω_q lead to resonance anomalies in the ultrasound absorption. Coherent neutron scattering (with emission or absorption of a simple phonon) in the presence of the corresponding impurity in a crystal has certain specific features at frequencies of emitted (or absorption) phonons close to ω_q. The differential cross section of this neutron scattering has an additional characteristic factor of the type (11.5.6) increasing anomalously near a quasi-local frequency. It is natural that similar peculiarities should be observed in the infrared absorption spectrum of crystals with impurities that produce quasi-local vibration.

Finally, much interest has been generated in quasi-local vibrations, related to the Mössbauer effect for the nuclei of impurity atoms. The Mössbauer phenomenon in impurities is connected with the specific relation between the momentum and the energy are transferred to an impurity nucleus. This relation is determined by those possible motions in which an impurity atom is capable of participating, i. e., by the expansion of the impurity displacement vector with respect to normal modes of the defect crystal. Among the vibration modes there is a large group of vibrations with very close frequencies (quasi-stationary wave packets of these oscillations cause the quasi-local vibrations). This leads to the fact that in expanding the displacement vector of an impurity atom with respect to normal vibrations the relative contribution of a quasi-local vibration greatly exceeds the relative contribution of ordinary crystal vibration modes with frequencies of a continuous spectrum. Thus, quasi-local crystal vibrations should manifest themselves in the spectrum of phonon transitions of the Mössbauer effect.

It should be noted that in phenomena such as the Mössbauer effect or neutron scattering the contribution of true local vibrations is quite pronounced and larger than the contribution of the quasi-local vibrations (reduced to the same frequency). However, in the range of low frequencies (considered, for instance, while studying the influence of relatively heavy impurities on the crystal properties) there exist quasi-local vibrations only: local vibrations cannot appear near the low-frequency boundary of an acoustic spectrum.

11.6
Collective Excitations in a Crystal with Heavy Impurities

The Green tensor $\boldsymbol{G}^0$ of an ideal crystal in the site representation is a function of the difference in the numbers of sites $\boldsymbol{n}$ and $\boldsymbol{n}'$. It follows from (11.3.18) that the averaged matrix $\boldsymbol{G}$ for a crystal with randomly distributed point defects is also a function of this difference. This reflects the macroscopic homogeneity of a crystal with defects that appear as a result of averaging over random distributions of impurity atoms.

For the Green function of a homogeneous system the long-wave $\boldsymbol{k}$-representation is introduced in the usual way. Thus, we just write the averaged Green function (11.3.18) of a crystal with point defects in $(\varepsilon, \boldsymbol{k})$-representation.

$$G(\varepsilon, \boldsymbol{k}) = \frac{1}{\varepsilon - \omega_0^2(\boldsymbol{k}) - c\prod(\varepsilon)}. \tag{11.6.1}$$

We first show that the validity of (11.6.1) is not really restricted by the linear approximation with respect to c. We focus on a scalar model and assume that the defect-isotope introduces a point perturbation (11.3.5). Discussing the Green function of the defect crystal, it is necessary to find the matrix (11.3.6) as a solution to (11.3.7). In studying the long-wave vibrations when the distances $|\boldsymbol{r}_\alpha - \boldsymbol{r}_\beta| \gg a/c^{1/3}$ give the main contribution, the quantity

$$U_0 \sum_{\gamma \neq \alpha} G^0(\boldsymbol{n}_\alpha - \boldsymbol{n}_\gamma) T_{\gamma\beta},$$

on the l.h.s. of (11.3.7) at fixed α is determined by the total contribution from a large number of terms. To a first approximation, this contribution can be regarded as being averaged over a large number of impurities and be calculated using (11.3.14). However, averaging transforms a crystal into a homogeneous macroscopic medium for which

$$T_{\alpha\beta} = T(\boldsymbol{r}_\alpha - \boldsymbol{r}_\beta), \tag{11.6.2}$$

where $\boldsymbol{r}_\alpha$ is the α-th impurity coordinate that can now be considered to be varying continuously.

Substituting (11.6.2) into (11.3.7) and replacing the Kronecker delta symbol $\delta_{\alpha\beta}$ on the r.h.s. by a δ-function normalized in the appropriate way we obtain

$$D(\varepsilon) T(\boldsymbol{r} - \boldsymbol{r}') + \frac{cU_0}{V_0} \int G^0(\boldsymbol{r} - \boldsymbol{x}) T(\boldsymbol{x} - \boldsymbol{r}')\, d^3x = cU_0 V \delta(\boldsymbol{r} - \boldsymbol{r}'). \tag{11.6.3}$$

We introduce the Fourier transformation for the function $T(\boldsymbol{x})$

$$T_k = \frac{1}{V} \int T(\boldsymbol{x}) e^{-ikx} d^3x,$$

and rewrite (11.6.3) in $\boldsymbol{k}$-representation

$$[D(\varepsilon) - cU_0 G^0(\varepsilon, \boldsymbol{k})] T_k = cU_0, \tag{11.6.4}$$

where $G^0(\varepsilon, \boldsymbol{k})$ is the Green function of an ideal crystal determined in a standard way in $(\varepsilon, \boldsymbol{k})$-representation.

The matrix $\boldsymbol{T}$ has the following Fourier transform

$$T_k = cU_0[D(\varepsilon) - cU_0 G^0(\varepsilon, \boldsymbol{k})]^{-1}. \tag{11.6.5}$$

It is easy to see that (11.3.9), (11.3.12) are the first terms of the expansion (11.6.5) in powers of $cU_0/D(\varepsilon)$.

We make use of the homogeneity of a medium and rewrite (11.3.8) for the averaged Green function of a defect crystal in $\boldsymbol{k}$-representation

$$G(\varepsilon, \boldsymbol{k}) = G^0(\varepsilon, \boldsymbol{k}) + G^0(\varepsilon, \boldsymbol{k}) T_k G^0(\varepsilon, \boldsymbol{k}). \tag{11.6.6}$$

We substitute here (11.6.5) to obtain

$$G(\varepsilon, \boldsymbol{k}) = \frac{D(\varepsilon) G^0(\varepsilon, \boldsymbol{k})}{D(\varepsilon) - cU_0 G^0(\varepsilon, \boldsymbol{k})}. \tag{11.6.7}$$

It is clear that the additional poles of the Green function of a crystal with defects are found from the equation

$$D(\varepsilon) - cU_0 G^0(\varepsilon, \boldsymbol{k}) = 0. \tag{11.6.8}$$

If we take into account the three branches of the vibrations of a simple lattice, (11.6.8) will be replaced by

$$\mathrm{Det} \left\| \delta_{ik} - U_0^{ik}(0) - cU_0 G_0^{ik}(\varepsilon, \boldsymbol{k}) \right\| = 0. \tag{11.6.9}$$

However, returning to a scalar model, we take into account the explicit expression for the function $G^0(\varepsilon, \boldsymbol{k})$ and rewrite (11.6.7) in a canonical form for the Green function

$$G^0(\varepsilon, \boldsymbol{k}) = \frac{1}{\varepsilon - \omega_0^2(\boldsymbol{k}) - c\Pi(\varepsilon)}, \quad \Pi(\varepsilon) = \frac{U_0}{D(\varepsilon)}. \tag{11.6.10}$$

It is clear that (11.6.10) coincides with (11.6.1). Thus, for the range of frequencies distant from singular points of the spectrum, the expression (11.6.10) should be regarded as a good representation for the averaged Green function (Lifshits), 1964; Dzyub, 1964).

Note, finally, that the averaging has been performed over impurity configurations without taking into account the correlations in their mutual positions. Any effects generated by the fluctuational formation of impurity complexes, the distance between which is much smaller than the average distance $\sim a/c^{1/3}$, are not considered when we use (11.6.10). Sometimes (11.6.10) is regarded as the result of a selective summation (rather than a complete one) of a series of successive approximations such as the expansion (11.3.13).

Let us analyze (11.6.1) or (11.6.10) intending, first, to consider eigenvibrations such as plane waves, i. e., collective excitations of a crystal with point defects. These vibrations are characterized by the wave vector $\boldsymbol{k}$ and the frequency ω. The dispersion law $\omega = \omega(\boldsymbol{k})$ connecting the real ω and $\boldsymbol{k}$ is the primary characteristic of elementary excitations and determined by the Green function poles of a crystal in the $(\varepsilon, \boldsymbol{k})$-representation. Therefore, we first discuss the problem of the poles of the function (11.6.10), i. e., of the roots of the equations

$$\varepsilon - c\Pi(\varepsilon) = \omega_0^2(\boldsymbol{k}). \tag{11.6.11}$$

We come back to the above procedure of regularizing the Green function of stationary vibrations of an ideal lattice and note that the function $\Pi(\varepsilon)$ is complex.

Hence it follows that for a real wave vector $\boldsymbol{k}$ the solutions to the dispersion equation (11.6.11) for ω will be complex. The presence of an imaginary part of the vibration frequency is evidence that the wave with fixed $\boldsymbol{k}$ gets damped in time. Thus, a plane wave of displacements in a crystal with impurities has the form

$$u(\boldsymbol{r}, t) = u_0 \exp\left(-\frac{t}{\tau}\right) e^{i(\boldsymbol{kr} - \omega t)}, \tag{11.6.12}$$

where the damping time τ is determined by the imaginary part of the function $\Pi(\varepsilon)$. It is natural that separation of collective excitations of a crystal such as (11.6.12) is only physically meaningful for $\omega\tau \gg 1$.

Let the condition $\omega\tau \gg 1$ be satisfied, i. e., the damping is weak. Writing the solution to (11.6.11) in the form

$$\varepsilon = (\omega - i/\tau)^2 \simeq \omega^2 - 2i\omega/\tau \simeq \omega^2 - 2i\omega_0(\boldsymbol{k})/\tau,$$

it is then easy to calculate $\omega(\boldsymbol{k})$ and $\tau(\boldsymbol{k})$ in the main approximation in $\omega\tau$.

The dependence $\omega = \omega(\boldsymbol{k})$ (see the problem) obtained in this way plays the role of a dispersion law of crystal vibrations with point impurities, but the corresponding excitations (11.6.12) prove to be damped, i. e., "living" for a finite time. The lifetime of these excitations τ can be directly associated with a correction $\delta g(\varepsilon)$ for the vibration density (11.5.2) near a quasi-local frequency

$$\frac{1}{\tau} = \frac{\pi}{2}\varepsilon_* \omega \delta g(\omega^2), \quad \omega = \omega_0(\boldsymbol{k}). \tag{11.6.13}$$

It is clear that τ has a minimum at $\omega = \omega_q$. A small lifetime of the collective excitation with $\omega = \omega_q$ can be explained easily. The energy of a "homogeneous" plane wave (11.6.12) is spent to excite the continuous spectrum vibrations whose frequencies are close to a quasi-local one. However, the coordinate dependence of quasi-local vibrations differs from (11.6.12). Therefore, the collective vibration is damped.

For the maximum of (11.6.13) we have the estimate

$$\frac{1}{\tau} \sim c\frac{\omega_D^2}{\Gamma_q} \sim c\left(\frac{\Delta m}{m}\right)^{3/2}. \tag{11.6.14}$$

We introduce the concentration c_0 at which the average distance between impurities has the order of magnitude of a characteristic wavelength of a single quasi-local vibration $\lambda \sim 2\pi s/\omega_q$. To an order of magnitude,

$$c_0 \sim \left(\frac{m}{\Delta m}\right)^{3/2}. \qquad (11.6.15)$$

Under the condition $\Delta m \gg m$, we always have $c_0 \gg c_*$.

Consequently, the estimate (11.6.14) can be rewritten as $\omega\tau \sim c_0/c$; hence it follows that the condition $\omega\tau \gg 1$ is satisfied for $c \ll c_0$. Thus at small enough concentrations of impurity atoms ($c \ll c_0$), ordinary vibrational excitations (such as plane waves) are weakly damped at all frequencies.

11.7 Possible Rearrangement of the Spectrum of Long-Wave Crystal Vibrations

Now consider the dynamic properties of a crystal where the concentration of heavy impurities is not restricted by the inequality $c \ll c_0$ (but the condition $c \ll 1$ remains). If $c \gtrsim c_0$ then in the resonance region $\omega\tau \lesssim 1$, and the concept of collective excitations (11.6.12) with frequencies close to ω_q is physically meaningless. The wave (11.6.12) is damped practically in one period of vibrations.

In this frequency region, for $c_0 \lesssim c \ll 1$ the spectrum of crystal eigenvibrations can be characterized by the quantity $\operatorname{Im} G(\varepsilon, \boldsymbol{k})$ as a function of ε and $\boldsymbol{k}$, because it can be measured experimentally. We consider $\operatorname{Im} G(\varepsilon, \boldsymbol{k})$ as a function of $\boldsymbol{k}$ with a given ε (this is typical for ordinary optical experiments) and examine how the position of the maximum of this function changes depending on the value of $\omega = \sqrt{\varepsilon}$. The maxima we are interested in are in the space $\omega, \boldsymbol{k}$ on hypersurfaces whose points are given by a straightforward condition

$$\omega^2 - \omega_0^2(\boldsymbol{k}) - c\,\operatorname{Re}[\Pi(\omega^2)] = 0. \qquad (11.7.1)$$

We restrict ourselves to the long-wave isotropic approximation when $\omega_0(\boldsymbol{k}) = s_0 k$. As a result of simple calculations, we get the following frequency dependence of the modulus of a wave vector providing the maximum $\operatorname{Im} G$:

$$(s_0 k)^2 = \omega^2 \left(1 - c\frac{\varepsilon_*(\omega^2 - \omega_q^2)}{(\omega^2 - \omega_q^2)^2 + \Gamma^2}\right). \qquad (11.7.2)$$

The plot of $k = k(\omega)$ (Fig. 11.6) can be considered as the dependence of the wave vector on the frequency of the crystal eigenvibrations in this case.

The function $k = k(\omega)$ has extrema $k_{\max}$ and $k_{\min}$ between which the "anomalous dispersion" region is situated. The difference in heights of the maximum and minimum in Fig. 11.6 is equal, in order of magnitude, to

$$\frac{k_{\max} - k_{\min}}{k_0} = \frac{\Delta k}{k_0} \sim c\frac{\varepsilon_*}{\Gamma} \sim c\left(\frac{\Delta m}{m}\right)^{3/2} \sim \frac{c}{c_0}.$$

With increasing concentration the minimum in Fig. 11.6 decreases and, at concentrations $c \sim c_0$, $k_{\min}$ may reach zero. For long-wave vibrations, one should then expect phenomena such as total internal reflection in optics. In fact, for $c \gg c_0$, in addition to $\omega = 0$, there appear two more frequency values corresponding to $k = 0$. These frequencies lying somewhat higher than ω_q limit the range of frequencies at which the crystal has no collective excitations described by the wave vector.

Therefore, in the case of large concentrations of impurities with pronounced quasi-local frequencies the spectra of long-wave crystal vibrations change significantly.

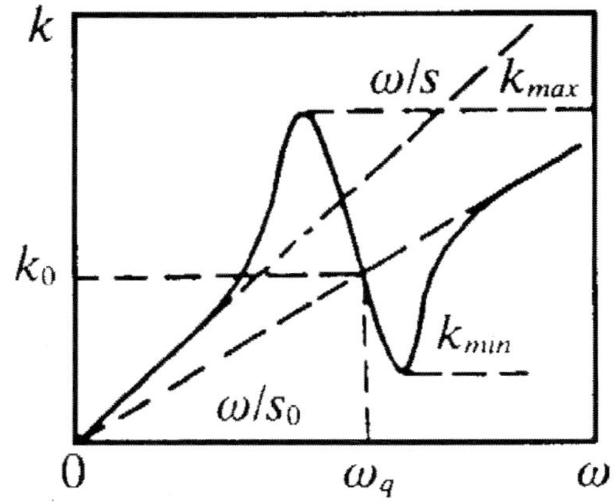

Fig. 11.6 The dependence $k = k(\omega)$ that provides the maximum of $\operatorname{Im} G(\varepsilon, \boldsymbol{k})$.

We consider $\operatorname{Im} G(\varepsilon, \boldsymbol{k})$ as a function of ε at fixed $\boldsymbol{k}$ (just this dependence is generally studied in neutron experiments). The maximum value of this function is determined by the condition (11.7.1), so that the frequency range where vibrations such as plane waves are absent is wide enough and satisfies the condition

$$\left|\omega^2 - \omega_q^2\right| \gg \Gamma \sim (\omega/\omega_D)^3 \varepsilon_* . \qquad (11.7.3)$$

The inequality (11.7.3) enables us to omit Γ in the real part of the function $\Pi(\omega^2)$ and the condition (11.7.1) will be replaced by a similar one:

$$[\omega^2 - \omega_0^2(\boldsymbol{k})][\omega^2 - \omega_q^2] = c\varepsilon_* \omega^2 . \qquad (11.7.4)$$

It follows from (11.7.4) that $\operatorname{Im} G$ as a function of ε at any fixed $\boldsymbol{k}$ has two maxima instead of one in an ideal crystal. The dependence of the frequencies on k providing the maximum $\operatorname{Im} G$ (Fig. 11.7a) resembles a typical diagram of frequency splitting near the resonance point, i. e., in the vicinity of the intersection of the dispersion curves $\omega = s_0 k$ and $\omega = \omega_q = \text{const}$ (Fig. 11.7b). The choice of the parameters $\Delta m/m$, ω_k, and c in the plot is matched to the experiment of Zinken et al. (1977). A hypothetical degeneration of frequencies arising in Fig. 11.7b is removed and the dependence $\omega = \omega(\mathbf{k})$ in the isotropic model is obtained by solving an algebraic equation

$$\omega n(\omega) = s_0 k, \tag{11.7.5}$$

where the function $n(\omega)$ is the coefficient of sound wave refraction in a crystal and is determined by (11.7.2) with $\Gamma = 0$:

$$n(\omega) = \left[1 - \frac{c\varepsilon_*}{\omega^2 - \omega_q^2}\right]^{1/2} = \sqrt{\frac{\omega^2 - \omega_0^2}{\omega^2 - \omega_q^2}}. \tag{11.7.6}$$

Here $\omega_0^2 = \omega_q^2 + c\varepsilon_*$. We note that $n(0) = \sqrt{1 + c(\Delta m/m)}$ and $n(\infty) = 1$. A plot of $\omega(k)$ is constructed in an obvious way (Fig. 11.7c).

The vertical dashed lines in Fig. 11.7c cut off the value of the wave vector k at which (11.7.6) becomes meaningless. Near $\omega = \omega_q$ these are the values of k at which the wavelengths are comparable with the distance between the impurities, and near $\omega = \omega_0$ these are the values of k at which the damping length is comparable with the wavelength ($\omega\tau \sim 1$).

Figure 11.7c shows the dispersion law asymptotes at small frequencies (straight line 1) $\omega = s_0 k/[1 + c(\Delta m/m)]^{1/2}, \omega \ll \omega_q$ and at high frequencies (straight line 2) $\omega = s_0 k, \omega_q \ll \omega \ll \omega_D$.

We note that the plot in Fig. 11.7c resembles the plot of the dispersion law of transverse optical vibrations of an ionic crystal in Fig. 3.7.

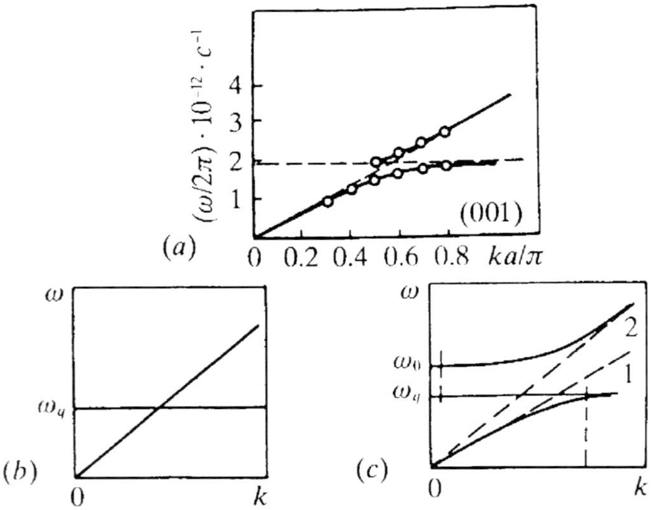

Fig. 11.7 A scheme of the phonon dispersion law for a crystal with large impurity concentration: (*a*) experimental observation of two branches; (*b*) the intersection of the sound dispersion law with the quasi-local frequency of homogeneously distributed impurities; (*c*) two branches of long-wave vibrations divided by a quasi-gap.

The most remarkable property of the plot is the presence of a forbidden range of frequencies (ω_q, ω) or a *quasi-gap*. For $c \gg c_0$ a new limiting frequency $\omega_0 = \omega_q\sqrt{1 + c(\Delta m/m)}$ is displaced from the frequency ω_q by a distance greatly exceeding the broadening due the concentration c of a quasi-local frequency $\delta\omega \sim c^{1/3}\omega_q$. The frequency ω_0 plays the role of a limiting frequency of optical vibrations (vibrations of a system of impurities relative to a crystal lattice). Thus, there are two branches of the spectrum of long-wave vibrations such as plane waves in a crystal with a large concentration of defects (Kosevich, 1965; Slutskin and Sergeeva, 1966; Ivanov,1970).

11.7.1
Problems

1. Find the frequency of a local vibration connected with an isotope-defect in a 1D crystal with interaction of nearest neighbors.

Hint. Make use of the dispersion law (2.1.6) and (11.1.12) or the vibration density (4.4.18) and (11.1.13).

Solution.

$$\omega = \frac{\omega_m}{\sqrt{1 - (\Delta m/m)^2}}, \qquad \Delta m < 0.$$

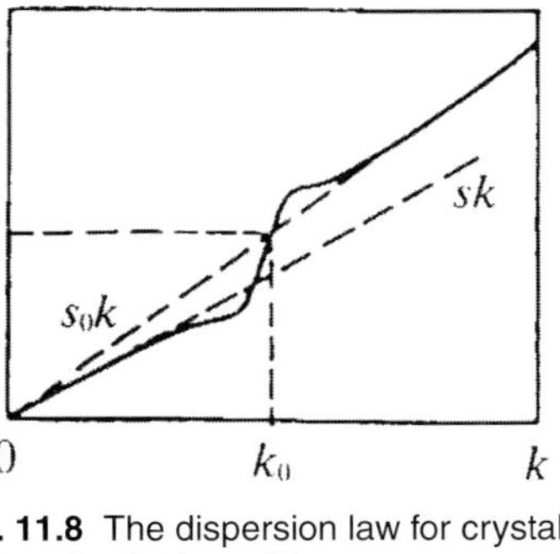

Fig. 11.8 The dispersion law for crystal vibrations with a small concentration of point impurities.

2. Find, in the isotropic approximation, the dispersion law for crystal vibrations with a small concentration of heavy impurities ($c \ll c_* = (m/\Delta m)^2 \ll 1$).

Hint. Take into account (11.6.13) and the fact that for $c \ll c_*$ for all frequencies $|c\Pi(\varepsilon)| \ll \varepsilon$ and $|c\Pi'(\varepsilon)| \ll 1$.

Solution.

$$\omega = s_0 k \left[1 + c\frac{\varepsilon_*(s_0^2k^2 - \omega_q^2)}{(s_0^2k^2 - \omega_q^2)^2 + \Gamma^2}\right]^{1/2}.$$

The plot of this dependence is shown in Fig. 11.8.

12
Localization of Vibrations Near Extended Defects

12.1
Crystal Vibrations with 1D Local Inhomogeneity

In any macroscopic specimen of a real crystal there are dislocations. A dislocation, being a quasi-one-dimensional structure, breaks the lattice regularity only in a small region near a certain line – its axis. Vibrations localized near the dislocation have the form of waves running along the dislocation line.

We are interested in small atomic vibrations near new equilibrium positions and thus in a zero approximation the dislocation axis may be considered to be fixed. We assume that the dislocation is a straight-line and its axis is perpendicular to the crystal symmetry plane. We direct the z-axis along the dislocation line and denote by $\boldsymbol{\xi}$ a 2D radius vector in the plane xOy: $\boldsymbol{r} = (\boldsymbol{\xi}, z)$, $\boldsymbol{\xi} = (x, y)$. Assume also that the dislocation does not change the substance density along its axis. The perturbation introduced by a dislocation is then connected with a local change in the matrix of atomic force constants. To analyze the long-wave crystal vibrations ($\lambda \gg a$), this perturbation is assumed to be concentrated on the dislocation axis. Using a scalar model we consider the displacement as a continuous function of the coordinates: $u = u(\boldsymbol{\xi}, z)$. The equation of long-wave lattice vibrations near the linear singularity can be represented as

$$\omega^2 u(\boldsymbol{\xi}, z) - \frac{1}{m} \sum_{\boldsymbol{n}'} \alpha(\boldsymbol{n} - \boldsymbol{n}') u(\boldsymbol{\xi}', z') = a^2 \delta(\boldsymbol{\xi}) \sum_{z'} U(z - z') u(0, z'), \qquad (12.1.1)$$

where $\delta(\boldsymbol{\xi})$ is a two-dimensional δ-function ($\delta(\boldsymbol{\xi}) = \delta(x)\delta(y)$) and the summation over the lattice sites can be replaced by the integration

$$\sum_{z} \cdots = \frac{1}{a} \int dz \cdots, \qquad \sum_{\boldsymbol{\xi}} \cdots = \frac{1}{a^2} \int dx\, dy \cdots,$$
$$\sum_{\boldsymbol{n}} \cdots = \frac{1}{V_0} \int dV \cdots . \qquad (12.1.2)$$

The Crystal Lattice: Phonons, Solitons, Dislocations, Superlattices, Second Edition. Arnold M. Kosevich

ISBN: 3-527-40508-9

The function $U(z)$ in (12.1.1) describing the force matrix perturbation is independent of frequency and can be regarded as an even function of z. It satisfies the obvious requirement

$$\sum_z U(z) = 0. \tag{12.1.3}$$

Using the property of homogeneity of the crystal along the Oz direction and applying a one-dimensional Fourier transformation relative to the z coordinate we obtain

$$u(\xi, z) = \frac{a}{2\pi} \int \chi_k(\xi) e^{ikz}\, dk, \quad \chi_k(\xi) = \sum_z u(\xi, z) e^{ikz}. \tag{12.1.4}$$

The equation for $\chi_k(\xi)$ is derived from (12.1.1)

$$\omega^2 \lambda_k(\xi) - \sum_{\xi'} \Lambda_k(\xi - \xi') \chi_k(\xi') = a^2 \delta(\xi) U_k \chi_k(0), \tag{12.1.5}$$

where

$$\Lambda_k(\xi) = \frac{1}{m} \sum_{n_z} \alpha(\boldsymbol{n}) e^{ikan_z}, \quad U_k = \sum_z U(z) e^{-ikz}. \tag{12.1.6}$$

Equation (12.1.5) is the equation of a 2D crystal vibrations with a point defect at the origin of the coordinates. The wave-vector component $k = k_z$ enters this equation as a parameter and determines the local perturbation intensity U_k. In the case $ak \ll 1$, the function U_k has an obvious expansion following from (12.1.6), (12.1.3)

$$U_k = -k^2 W_0, \quad W_0 = \frac{1}{2} \sum_z U(z) z^2. \tag{12.1.7}$$

To find the function $\chi_k(\xi)$, we use a 2D Fourier expansion

$$\chi_k(\xi) = \frac{a^2}{(2\pi)^2} \int \chi_k(\boldsymbol{\kappa}, k) e^{ik\xi}\, dk_x\, dk_y, \quad \chi_k(\boldsymbol{\kappa}, k) = \sum_\xi \chi_k(\xi) e^{ik\xi}, \tag{12.1.8}$$

where $\boldsymbol{\kappa}$ is a two-dimensional wave vector $\boldsymbol{\kappa} = (k_x, k_y)$.

The Fourier components $\chi(\boldsymbol{\kappa}, k)$ are determined from the relation

$$\left[\omega^2 - \omega_0^2(\boldsymbol{\kappa}, k)\right] \chi(\boldsymbol{\kappa}, k) = -k^2 W_0 \chi_k(0), \tag{12.1.9}$$

where the function $\omega_0^2(\boldsymbol{\kappa}, k)$ is the dispersion law of an ideal crystal.

To simplify calculations, we assume the axis Oz is a four-fold or six-fold symmetry axis. Then in the long-wave limit

$$\omega_0^2(\boldsymbol{\kappa}, k) = s_0^2 \kappa^2 + s^2 k^2. \tag{12.1.10}$$

The dependence of the vibration amplitude on ξ follows directly from (12.1.8)–(12.1.10)

$$\chi_k(\xi) = -(ak)^2 \frac{W_0}{(2\pi)^2} \chi_k(0) \int \frac{\cos(\boldsymbol{\kappa}\xi)\, dk_x\, dk_y}{\omega^2 - s^2 k^2 - s_0^2 \kappa^2}. \tag{12.1.11}$$

If we set $\xi = 0$ in (12.1.11), we obtain an equation for the vibration frequencies

$$1 + (ak)^2 \frac{W_0}{2\pi} \int_0^{k_0} \frac{\kappa d\kappa}{\omega^2 - s^2k^2 - s_0^2\kappa^2} = 0. \tag{12.1.12}$$

The upper limit of integration in (12.1.12) can be estimated as $\kappa_0 \sim 1/a$. The fact that the integration limit in (12.1.12) is determined by the order of magnitude only, and has the character of some "cut off" parameter, is connected with a model assumption of the point-like character of a perturbation in (12.1.5). The assumption (12.1.1) of a delta-like localization of the perturbation on the dislocation axis as well as the dispersion law (12.1.10) are valid for the long-wave vibrations only ($a\kappa \ll 1$).

At the same time the integration in (12.1.11), (12.1.12) should be extended to the whole interval of a continuous spectrum of frequencies. Since the integrand in (12.1.11) does not exhibit a decrease at infinity necessary for the integral to converge, we have to take into account the natural limit of integration over the quasi-wave vector ($a\kappa_0 \sim \pi$).

Simplifying (12.1.12) we obtain

$$1 - (ak)^2 \frac{W_0}{4\pi s_0^2} \log \frac{s_0^2\kappa_0^2}{s^2k^2 - \omega^2} = 0, \tag{12.1.13}$$

by omitting small terms of the order of magnitude

$$\frac{s^2k^2 - \omega^2}{s_0^2\kappa_0^2} \ll \frac{s^2k^2}{s_0^2\kappa_0^2} \sim (ak)^2 \ll 1.$$

As in Chapter 11, (12.1.13) has a solution for eigenfrequency squares ω^2 with a definite sign of W_0 only, namely, for $W_0 > 0$, so that necessarily $\omega^2 < s^2k^2$. If the condition $W_0 > 0$ is satisfied, (12.1.13) always has a solution (Lifshits and Kosevich, 1965)

$$\omega^2 = s^2k^2 - s_0^2\kappa_0^2 \exp\left\{-\frac{4\pi s_0^2}{(ak)^2 W_0}\right\}. \tag{12.1.14}$$

We recall that in (12.1.14), $W_0 > 0$ and $s_0^2\kappa_0^2 \sim \omega_D$.

The frequencies (12.1.14) have an exponential dependence on the perturbation intensity, i.e., on U_k that is characteristic for two-dimensional problems. Their definition has no critical value of the perturbation intensity at which the local vibration frequency starts splitting off and that is typical for point defects. The existence of the vibrations localized near a dislocation requires a definite sign of the perturbation ($W_0 > 0$), and the corresponding frequency is always separated by a certain finite gap $\delta\omega$ the origin of the spectrum of bulk crystal vibrations

$$\delta\omega = sk - \omega_d \sim \frac{\omega_D^2}{sk} \exp\left\{\frac{4\pi s_0^2}{(ak)^2 W_0}\right\}.$$

We have considered elastic waves traveling along the dislocation. Their existence is due to changes in the elastic moduli in the dislocation core, but the vibrations localized near the chain of impurity atoms also have frequencies such as (12.1.14). Such a chain is also a linear defect. The long-wave vibrations near this defect are described by the same equation (12.1.1), but with a different perturbation matrix

$$U(z) = U_0 a\delta(z) = -\left(\frac{\Delta m}{m}\right)\omega^2 a\delta(z). \tag{12.1.15}$$

It is clear that the problem of crystal vibrations with the perturbation (12.1.15) is reduced to finding a solution to the two-dimensional equation (12.1.5) in which the replacement $U_k \to U_0$ is necessary. Thus, the equation for local frequencies remains the same, replacing $k^2 W_0 \to (\Delta m/m)\,\omega^2$,

$$1 - \left(\frac{\Delta m}{m}\right)\frac{(a\omega)^2}{4\pi s_0^2}\log\frac{s_0^2\kappa_0^2}{s^2k^2 - \omega^2} = 0. \tag{12.1.16}$$

For $\Delta m > 0$, (12.1.16) has the following solution in a frequency region of the spectrum ($\omega^2 \ll (m\Delta m)\,\omega_D^2$),

$$\omega^2 = s^2k^2 - s_0^2\kappa_0^2\exp\left\{-\frac{4\pi}{(ak)^2}\left(\frac{s_0}{s}\right)^2\frac{m}{\Delta m}\right\}. \tag{12.1.17}$$

It is clear that (12.1.17) is not qualitatively different from (12.1.14).

Let us note that the frequencies (12.1.14), (12.1.17) do correspond to vibrations localized near the linear defect. According to (12.1.11), we perform the integration

$$\begin{aligned}\chi_k(\xi) = \; & -(ak)^2\frac{W_0}{(2\pi)^2}\chi_k(0)\int_0^\infty \frac{\kappa d\kappa}{\omega^2 - s^2k^2 - s_0^2\kappa^2} \\ & \times \oint \cos(\kappa\rho\cos\phi)d\phi,\end{aligned} \tag{12.1.18}$$

where $\rho^2 = x^2 + y^2$ and the upper limit of integration over κ is shifted to infinity since for $\rho \neq 0$ the integral in (12.1.18) converges.

The integrals on the r.h.s. of (12.1.18) can be taken from any reference book; therefore, we give only the final expressions for the vibration amplitude

$$\chi_k(\xi) \equiv \chi(\rho) = (ak)^2\frac{W_0}{2\pi s_0^2}K_0(\kappa_\perp\rho)\chi_k(0), \tag{12.1.19}$$

where $K_0(x)$ is a cylindrical zero-order Hankel function (of an imaginary argument); $\kappa_\perp$ is the inverse radius of localization of vibrations near the dislocation line

$$\kappa_\perp^2(k) = \frac{s^2k^2 - \omega^2}{s_0^2}. \tag{12.1.20}$$

The behavior of the function $K_0(x)$ at small and large values of its argument is well known:

$$K_0(x) = -\log\frac{x}{a}, \quad x \ll 1, \qquad K_0(x) = \sqrt{\frac{\pi}{2x}}\,e^{-x}, \quad x \gg 1.$$

Thus, the vibration amplitude distant from the dislocation axis has a characteristic exponential decrease confirming that the vibrations are localized near the dislocation. The limiting dependence $K_0(x)$ for $x \ll 1$ shows that the vibration amplitude has a logarithmic singularity as $\rho \to 0$. The equation (12.1.19) derived in the long-wave approximation is valid only for $\rho \gg a$; therefore, the extremely small distance ρ for which (12.1.19) is valid may not be smaller than a.

12.2
Quasi-Local Vibrations Near a Dislocation

The equation for the eigenfrequencies (12.1.2) is equivalent to the condition

$$1 - U_k G_2(\varepsilon, \mathbf{k}) = 0, \tag{12.2.1}$$

where $G_2(\varepsilon, \mathbf{k})$ is the Green function of the equation for 2D crystal vibrations ($\varepsilon = \omega^2$), with the spectrum of frequency squares (12.1.10) beginning with s^2k^2, where k is a fixed parameter. This function has an obvious definition

$$G_2(\varepsilon, \mathbf{k}) = \frac{a^2}{(2\pi)^2}\int G_0(\varepsilon, \mathbf{k})\, d^2k,$$

where $G_0(\varepsilon, \mathbf{k})$ is the Green function of an ideal crystal in a scalar model (4.5.10).

The dislocation localized waves have frequencies for which $\mathrm{Im}\, G_2(\varepsilon, \mathbf{k}) = 0$. But by examining quasi-local vibrations near a heavy impurity, we made it clear that in the frequency range where $\mathrm{Im}\, G_2 \neq 0$, resonance vibrational states may exist. In this case the frequencies of these vibrations are determined by

$$1 - U_k \,\mathrm{Re}\, G_2(\varepsilon, \mathbf{k}) = 0. \tag{12.2.2}$$

Using (12.1.12), it is easy to see that

$$\mathrm{Re}\, G_2(\varepsilon, \mathbf{k}) = \frac{a^2}{4\pi s_0^2}\log\frac{|s^2k^2 - \varepsilon|}{s_0^2k_0^2}. \tag{12.2.3}$$

It is clear that the r.h.s. of (12.2.3) tends to $-\infty$ at the point $\varepsilon = s^2k^2$ and determines the function symmetrical with respect to this point (Fig. 12.1). Finding from the plot a solution to (12.2.1), (12.2.2) we conclude that for $U_k < 0$ there are simultaneously solutions both to (12.2.1) for $\varepsilon < s^2k^2$ (dislocation waves) and to (12.2.2) for $\varepsilon > s^2k^2$ (quasi-local vibrations near the dislocation). But the latter have the physical meaning

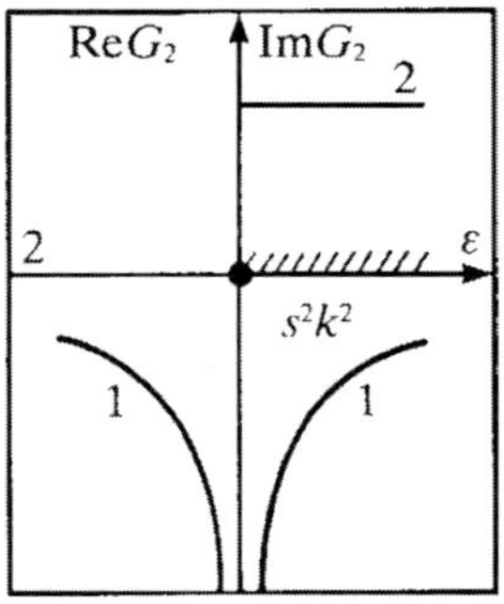

Fig. 12.1 The real (1) and imaginary (2) parts of the Green function as functions of ε at fixed k_z.

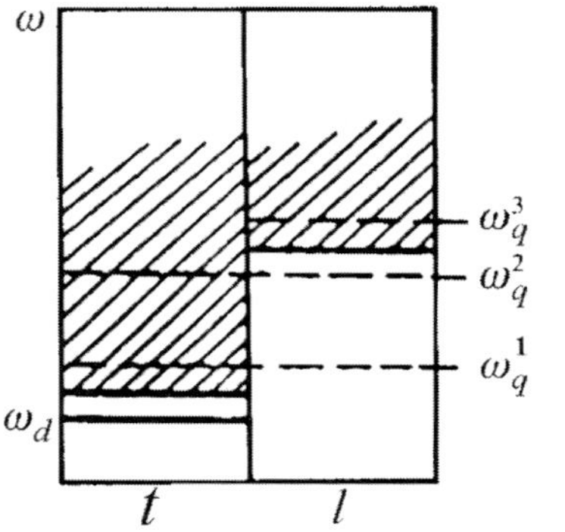

Fig. 12.2 The position of local ω_d and quasi-local ω_q frequencies of dislocational vibrations with fixed k_z.

of isolated frequencies only in the case when the damping of a corresponding resonance frequency (i. e., the imaginary part of this frequency) is small. We know that the damping is determined by $\text{Im}\, G_2(\varepsilon, \boldsymbol{k}) = \pi g^{(2)}(\varepsilon, \boldsymbol{k})$, where $g^{(2)}(\varepsilon, \boldsymbol{k})$ is a two-dimensional vibration density. In the considered range of frequency squares, $\pi g^{(2)}$ is practically constant (line 2 in Fig. 12.1) $\pi g^{(2)} = (a/s_0)^2 \sim 1/\omega_D^2$. Analyzing the width of the quasi-local vibration peak, we have

$$\Gamma = \pi g^{(2)} \left(\frac{d}{d\varepsilon} \,\text{Re}\, G_2 \right)^{-1} = g^{(2)} \left(\frac{2\pi s_0}{a} \right)^2 \left(\varepsilon - s^2 k^2 \right) \sim \varepsilon - s^2 k^2 \ , \qquad (12.2.4)$$

i. e., it has the order of magnitude of the square of a resonance frequency measured from the edge of the spectrum s^2k^2. This means that the quasi-local frequency near a dislocation is very weakly pronounced.

In conclusion, we consider in brief a short qualitative characteristic of dislocation localized vibrations of a real crystal lattice that has three polarizations of the displacement vector with three branches of the dispersion law. In the isotropic approximation, there exist two branches of transverse vibrations with the dispersion law

$\omega^2 = s_t^2(\kappa^2 + k^2)$ and one branch of longitudinal vibrations with the dispersion law $\omega^2 = s_l^2(\kappa^2 + k^2)$, so that $s_l > s_t$ always. If the value of k is fixed, there are two bands of continuous frequency values of the bulk vibrations (Fig. 12.2) $\omega > sk$.

With the corresponding sign of the perturbation U_k, the frequency lying near the boundary of the corresponding band may split off from the lower edge of each of the bands. One of these frequencies (the lowest one, ω_d) corresponds to the vibration localized near the dislocation. It arises near the edge of the transverse vibration band and the localized vibrations have the form of transverse waves running along the dislocation. The dislocation axis participates in these vibrations, bending and vibrating like a spanned string. As the quantity $s_t k - \omega_\mathrm{d}$ is exponentially small, the bending waves have a velocity that practically does not differ from that of s_t. The character of vibrations allows us to formulate the *elastic string model* often used in different applications of a dynamic theory of dislocations. In this model a dislocation line is considered as a heavy string vibrating in a slip plane. The ratio of linear tension to dislocation mass is such that the dispersion law of string-bending vibrations coincides practically with the dispersion law of transverse sound waves in a crystal $\omega = s_t k$.

The frequency "split off" from the boundary of the longitudinal vibrations spectrum (the frequency ω_d in Fig. 12.2) could be considered as discrete only if the interaction between different branches of vibrations is disregarded. But the linear defect violates the independence of different types of vibrations so that they are "mixed together". Since the frequency ω_d^2 is in the region of a continuous spectrum of transverse vibrations, it gets broadened and the corresponding vibration is transformed into a quasi-local one.

Finally, even in the case of an independent branch of vibrations, the quasi-local vibrations discussed in detail above are possible. These vibrations in Fig. 12.2 correspond to the frequencies ω_q^1 and ω_q^3. The quasi-local peak width at the frequency ω_q^1 has been evaluated, and the peak width at ω_q^2 cannot be smaller. Hence, only the frequencies of bending vibrations of a dislocation as a spanned string are actually singled out.

12.3 Localization of Small Vibrations in the Elastic Field of a Screw Dislocation

The elastic vibrations near the dislocation as a source of static stresses in a crystal can be included, using a simple anharmonic approximation, into the initial state of a vibrating crystal and small vibrations on the background of a distorted lattice can be considered.

We represent the vibrating crystal displacement in the form

$$u_z(\xi, z, t) = u_0(\xi) + u(\xi, z, t)\,,$$

where $u_0(\xi)$ is a static field of the screw dislocation coinciding with the axis Oz

((10.2.8) is convenient for this case)

$$u_0(\boldsymbol{\xi}) = \frac{b\theta}{2\pi} = \frac{b}{2\pi}\arctan\left(\frac{y}{x}\right), \tag{12.3.1}$$

$u(\boldsymbol{\xi}, z, t)$ are small dynamic displacements relative to a stationary crystal with a dislocation.

Apart from a static deformation, the equations of motion of a vibrating crystal with a dislocation will include anharmonicities (cubic and fourth-order ones). Then, restricting ourselves to the approximation linear in dynamic displacements, we get in a scalar model the following equation for the displacement u:

$$\begin{aligned} &\rho_0\frac{\partial^2 u}{\partial t^2} - \mu\frac{\partial^2 u}{\partial x_\alpha^2} - (\lambda + 2\mu)\frac{\partial^2 u}{\partial z^2} \\ &= A\frac{\partial u_0}{\partial x_\alpha}\frac{\partial^2 u}{\partial x_\alpha \partial z} + B\left(\frac{\partial u_0}{\partial x_\alpha}\right)^2\frac{\partial^2 u}{\partial x_\beta^2} + C\left(\frac{\partial u_0}{\partial x_\alpha}\right)^2\frac{\partial^2 u}{\partial z^2}, \end{aligned} \tag{12.3.2}$$

where λ and μ are the renormalized second-order elastic moduli taking into account anharmonicities ($\mu > 0$, $\lambda + 2\mu > 0$); A is the third-order elastic modulus; B and C are the fourth-order elastic moduli of a nonlinear elasticity theory (from general considerations, $A \sim B \sim C \sim \mu$).

Equation (12.3.2) should be supplemented with a certain boundary condition on the surface of a dislocation tube $r = r_0$ to describe the phonon reflection from the dislocation axis. We use further the fact that no phonons penetrate into the region of the dislocation core. But we try, using the same approximation in which (12.3.2) is written to take correctly into account in the boundary conditions the symmetry of lattice distortions along the screw dislocation axis. If $f(\theta, z)$ is some function characterizing these distortions on a dislocation tube, it should have screw symmetry

$$f(\theta + \theta_0, z) = f\left(\theta, z - \frac{b\theta_0}{2\pi}\right), \quad f(\theta, z + b) = f(\theta, z), \tag{12.3.3}$$

that takes into account static displacements (12.3.1) about the screw dislocation.

In view of (12.3.3), we write the solution (12.3.2) as

$$u = \chi(r)e^{ik(z+b\theta)+im\theta-i\omega t}, \tag{12.3.4}$$

where $m = 0, \pm 1, \pm 2, \ldots$.

It is typical that the solution (12.3.4) satisfying the screw symmetry (12.3.3) is nonperiodic with respect to the angle θ, and, therefore, is not a single-valued function of the coordinate $\boldsymbol{x}(\boldsymbol{\xi}, z)$. We remember that $u(\boldsymbol{x})$ is the atomic displacement relative to their equilibrium positions in a crystal with a dislocation, and $\boldsymbol{x}(\boldsymbol{\xi}, z)$ are the atomic positions (coordinates) in a nondeformed crystal. The atom coordinates in a crystal with a dislocation, i. e., in a statically deformed medium, are different from those in the initially nondeformed medium, namely, the coordinates $\boldsymbol{\xi}(x,y)$ are unchanged and the third coordinate (along the dislocation axis) is $Z = z + u_0(\theta) = z + (b\theta/2\pi)$. If we consider (12.3.4) as a function of atomic coordinates $\mathbf{X}(\boldsymbol{\xi}, Z)$ in a deformed medium, then it is a single-valued function. Ambiguity of the expression (12.3.4) for the displacement u as a function $\mathbf{x}(\boldsymbol{\xi}, z)$ is a direct result of the fact that (12.3.2) is written in the coordinate system connected with the initial undeformed medium.

Before writing the equation for $\chi(r)$ that follows from (12.3.2) we note that the spatial derivatives in (12.3.2) have different orders. It follows from (12.1.14) that the frequency square of a local vibration ω^2 differs insignificantly from $(s_0k)^2$. But then

$$\frac{\partial^2 u}{\partial x_\alpha^2} \sim \kappa_\perp^2 u \sim \frac{s_0^2k^2 - \omega^2}{(s_0k)^2}\frac{\partial^2 u}{\partial z^2} \ll \frac{\partial^2 u}{\partial z^2},$$

where $\kappa_\perp$ is the inverse radius of the vibration localization near the dislocation axis. Thus, it is meaningless to retain on the r.h.s. of (12.3.2) the term with a fourth-order elastic modulus B. Taking this into account, we get the following ordinary differential equation

$$\frac{d^2\chi}{dr^2} + \frac{1}{r}\frac{d\chi}{dr} - \left(\kappa_\perp^2 + \frac{\nu}{r^2}\right)\chi = 0, \tag{12.3.5}$$

where

$$\nu = \left(m + \frac{bk}{2\pi}\right)^2 + m\left(\frac{bk}{2\pi}\right)\frac{A}{\mu} + \left(\frac{bk}{2\pi}\right)^2\frac{A+C}{\mu}, \tag{12.3.6}$$

and the parameter $\kappa_\perp$ is determined by a relation such as (12.1.20)

$$\kappa_\perp^2(k) = \frac{s_l^2k^2 - \omega^2}{s_t^2}, \quad \rho s_l^2 = \lambda + 2\mu, \quad \rho s_t^2 = \mu. \tag{12.3.7}$$

Equation (12.3.5) is a Schrödinger equation for the radial part of the wave function in cylindrical coordinates where the Planck constant should be equal to unity, and the particle mass to $1/2$. It describes a particle located in the potential field $U = \nu/r^2$ and having the energy $E = -\kappa_\perp^2$.

If $m \neq 0$, it follows from (12.3.6) that in the long-wave approximation ($bk \ll 1$)

$$\nu \approx m^2 > 0. \tag{12.3.8}$$

But under the condition (12.3.8) and with a repulsive potential on the dislocation axis (12.3.5) has no discrete spectrum corresponding to finite (localized) states. Thus, in the long-wave approximation for $m \neq 0$ there are no localized vibrations near the dislocation.

The situation is quite different in the case $m = 0$ when

$$\nu = \left(\frac{bk}{2\pi}\right)^2\left(1 + \frac{M}{\mu}\right), \quad M = A + C.$$

For $M > -\mu$, as before, $\nu > 0$ and there are no localized vibrations near the dispersion, but if $M < -\mu$, then $\nu < 0$ and the situation changes. The (12.3.5) describes the particle motion in a 2D potential well, and there is an infinite number of discrete levels with the point of condensation at $E = 0$. These levels can be calculated quasi-classically.

Equation (12.3.5) corresponds to the Hamiltonian

$$H = p_x^2 + p_y^2 + U(r) \equiv p_r^2 - \frac{\tau^2}{r^2},$$

(here $\tau^2 = -\nu$). Thus, the Bohr quasi-classical quantization condition is

$$\int_{r_0}^{r_m} p_r \, dr \equiv \int_{r_0}^{r_m} \sqrt{\frac{\tau^2}{r^2} + E} \, dr = \pi \left(N + \frac{1}{2} \right), \tag{12.3.9}$$

where N is a natural number ($N \gg 1$); $r_m^2 = \tau^2 / |E| = (\tau / \kappa_\perp)^2$.

We perform a trivial integration in (12.3.9) and retain logarithmically large terms only ($\ln(r_m / r_0) \gg 1$)

$$\tau \ln \left(\frac{\lambda}{r_0} \tau \right) = \pi \left(N + \frac{1}{2} \right), \quad \lambda = \frac{1}{\kappa_\perp}. \tag{12.3.10}$$

The dispersion law for localized vibrations follows from (12.3.6), (12.3.7), (12.3.9)

$$\omega^2 = s_l^2 k^2 - |\nu| \left(\frac{s_t}{r_0} \right)^2 \exp \left\{ - \frac{2\pi^2 (2N+1)}{b |k| \sqrt{\frac{|M|}{\mu} - 1}} \right\}, \tag{12.3.11}$$

where the pre-exponential factor is determined to an order of magnitude only.

Let us note two important peculiarities of the dispersion law (12.3.11). First, the second term on its right-hand side depends less on k than in (12.1.14). Second, the eigenfrequencies of localized vibrations form an infinite geometric progression that describes the concentration of local frequencies toward the edge of the continuous spectrum. The latter should be taken into account in evaluating the statistical weight of local vibrations that determines their contribution to the low-temperature thermodynamic properties of a crystal.

If the signs of the perturbations on the dislocation axis and the ratio of M to μ are such that there are vibrations with frequencies (12.1.14), (12.1.11), then these frequencies are mixing. But these vibrations "do not interfere" with one another under the condition that the frequency (12.1.14) is lower than the first of the frequencies determined by (12.3.11) (with a given k).

12.4
Frequency of Local Vibrations in the Presence of a Two-Dimensional (Planar) Defect

A two-dimensional defect of the crystal breaks the lattice regularity along a certain surface inside a solid. If the character of the crystal structure distortions is the same along the indicated surface (more frequently, it is a plane), the break in the rigorous lattice periodicity is called a *stacking fault* (Fig. 12.3). The possible types of planar stacking fault are completely determined by the crystallography of a given lattice.

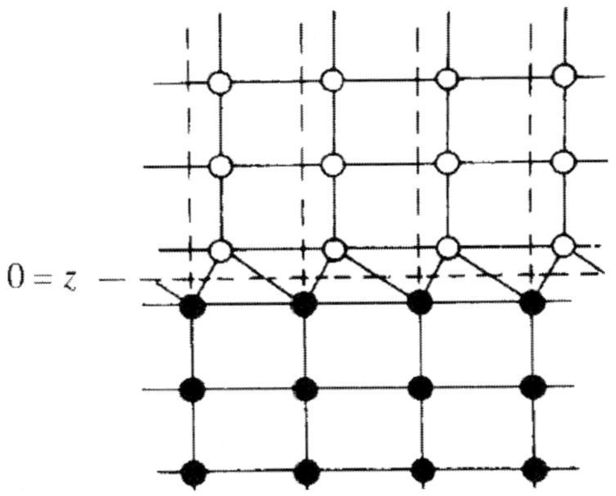

Fig. 12.3 Planar stacking fault that is perpendicular to the z-axis.

We consider an unlimited, extended planar defect, assuming it to be coincident with a plane $z = 0$ parallel to some crystallographic plane. We restrict ourselves to the case of a symmetric lattice and assume that the z-axis is a four-fold symmetry axis of an ideal crystal. In general, the symmetry group of the plane defect is smaller than the corresponding group of atomic planes for an unbounded crystal lattice. Thus, there is no four-fold symmetry axis perpendicular to the defect plane in Fig. 12.3.

We show first that the equations of vibrations localized near such a defect coincide with the equations of vibrations of a certain 1D crystal. The stacking fault does not, generally, change the atomic mass; thus, it may seem that its influence on lattice vibrations is described only by the corresponding force matrix perturbations (just such perturbations are usually taken into consideration in qualitatively discussing the problem). However, if the interatomic distances in the stacking fault differ from those in a nondefect crystal lattice ($h \neq a$ in Fig. 12.3) a local change in the mass density takes place. We intend to account for this fact.

The planar defect running through the whole crystal does not break the lattice homogeneity in a plane perpendicular to the z-axis. Then the perturbation potential U_* due to the change in the mass density along the defect layer can be written in the form analogous to (11.1.7)

$$U_*(\boldsymbol{n}, \boldsymbol{n}') = U_0 \delta_{n_z 0} \delta_{n'_z 0}, \quad U_0 = -\omega^2 \frac{\Delta \rho}{\rho},$$

where $\Delta\rho$ is the change in the mass density. Assuming such a form of the perturbation we unite two atomic layers (or crystallographic planes) on both sides of the plane $z = 0$ in Fig. 12.3 into one planar defect.

The perturbation of the force matrix $U(\boldsymbol{n},\boldsymbol{n}')$ in this crystal should be dependent on the differences $n_x - n'_x$ and $n_y - n'_y$ and obey the requirement

$$\sum_n U(n_x, n_y; n_z, n'_z) = \sum_n U(n_x, n_y; n'_z, n_z) = 0\,. \tag{12.4.1}$$

Therefore, an equation of crystal vibrations including both types of perturbations is

$$\begin{aligned} &\omega^2(\boldsymbol{n}) - \frac{1}{m}\sum_{n'} \alpha(\boldsymbol{n}-\boldsymbol{n}')u(\boldsymbol{n}') \\ &= \sum_{n'} U(n_x - n'_x, n_y - n'_y; n_z n'_z)u(\boldsymbol{n}') + U_0 u(n_x, n_y, 0)\delta_{n_z 0}\,. \end{aligned} \tag{12.4.2}$$

Since the crystal is structurally uniform in the $x0y$ plane, it is convenient to switch in the equations of motion from a site representation in this plane to a two-dimensional $\boldsymbol{k}$-representation, retaining the site representation along the the z-axis. We represent the amplitude of vibrations in the form

$$u(\boldsymbol{n}) = \chi_{\boldsymbol{\kappa}}(n_z)e^{ia(k_x n_x + k_y n_y)}\,, \quad \boldsymbol{\kappa} = (k_x, k_y),$$

and rewrite (12.3.2) in the $\boldsymbol{\kappa}$-representation

$$\omega^2 \chi_{\boldsymbol{\kappa}}(n_z) - \sum_{n'_z} \Lambda_{\boldsymbol{\kappa}}(n_z - n'_z)\chi_{\boldsymbol{\kappa}}(n'_z) = \sum_{n'_z} U_{\boldsymbol{\kappa}}(n_z, n'_z)\chi_{\boldsymbol{\kappa}}(n'_z) + U_0 \chi_{\boldsymbol{\kappa}}(0)\delta_{n_z 0}\,, \tag{12.4.3}$$

where

$$\Lambda_{\boldsymbol{\kappa}}(n_z) = \frac{1}{m}\sum_{n'_x n'_y} \alpha(\boldsymbol{n})e^{-ia(k_x n_x + k_y n_y)}\,,$$

$$U_{\boldsymbol{\kappa}}(n_z, n'_z) = \sum_{n'_x n'_y} U(n_x, n_y; n_z, n'_z)e^{-ia(k_x n_x + k_y n_y)}\,.$$

It is very important that the quantity $\boldsymbol{\kappa}$ appears in (12.4.3) as a parameter. If we assume this parameter to be fixed and are interested only in the dependence of the vibration amplitude on z, then omitting the index $\boldsymbol{\kappa}$ in all the quantities in (12.4.3), we obtain an equation for 1D crystal vibrations

$$\omega^2 \chi(n) - \sum_{n'} \Lambda(n - n')\chi(n') = \sum_{n'} V(n, n')\chi(n')\,, \tag{12.4.4}$$

where $n = n_z$ and the matrix $V(n, n')$ is equal

$$V(n, n') = U(n, n') + U_0 \delta_{n0}\delta_{n'0}.$$

The matrix $\Lambda(n)$ has the following property $\sum \Lambda(n) = \omega_0^2(\boldsymbol{\kappa}, 0)$, where the function $\omega_0(\boldsymbol{\kappa}, k_z)$ determines the dispersion relation of an ideal crystal.

In order to take into account a band structure of the dispersion relation and have at the same time a possibility to obtain simple analytical results of calculations we analyze the dynamics of the planar defect in a strongly anisotropic crystal (see Section 2.10). Let the crystal has possess a primitive body-centered tetragonal lattice. Assume that neighboring atomic layers perpendicular to the four-fold symmetry axis have weak interlayer bonds and the stacking fault is parallel to such layers. The dispersion relation for a crystal of this type is given in the nearest-neighbors approximation by (2.10.2) that determines the vibration band of the strongly anisotropic crystal.

Under the condition $\omega_1 \gg \omega_2$ the band width (for any value of $\boldsymbol{k}$) is determined by the weak interlayer interaction and is proportional to ω_2. The maximum frequency of the vibrations is determined by the strong interlayer interaction and is proportional to ω_1. As a result the band is very narrow and strongly elongated. This specific form of the frequency distribution allows us to propose a simple description of the crystal dynamics at low enough frequencies $\omega \ll \omega_1$ admitting $\omega \sim \omega_2$.

For the frequencies mentioned above we can use the long-wave approximation in respect to $\boldsymbol{\kappa}$-dependence, and the dispersion relation (2.10.2) acquires the form (2.10.4):

$$\omega^2 = \omega_0^2(\boldsymbol{\kappa}, k_z) \equiv s_1^2\kappa^2 + \omega_2^2 \sin^2 \frac{bk_z}{2}; \quad \kappa^2 = k_x^2 + k_y^2), \quad s_1^2 = \frac{1}{4}a^2\omega_1^2, \quad (12.4.5)$$

where b is the interatomic distance along the z-axis and s_1 is the sound velocity in the basal plane.

If the value of $\boldsymbol{\kappa}$ is fixed then (12.4.5) can be considered as a dispersion relation for 1D crystal vibrations of the optical type. Remembering the results of Section 11.1 we expect to find discrete frequencies of localized 1D modes outside the continuous spectrum of an ideal crystal, namely the discrete frequencies squared should lie either under $\omega_1^2(\boldsymbol{\kappa})$ or above $\omega_1^2(\boldsymbol{\kappa}) + \omega_2^2$. Assume the perturbation to be small enough and refer to Section 1.1 where it was shown that in such a case (1) possible discrete frequencies are located close to the edges of the continuous spectrum; (2) the planar defect can be assumed to be concentrated in the plane $z = 0$. The later means that the Kronecker delta δ_{n0} in the r.h.p. of (12.4.3) can be substituted with the Dirac delta-function and the total perturbation matrix V_0 should be assumed to have the following form

$$V(n, n') = b^2 V_0 \delta(z)\delta(z'), \quad V_0 = [U_1(\boldsymbol{\kappa}) + U_0(\omega)] \,. \qquad (12.4.6)$$

Of course, (12.4.6) can be used only for calculations of the localized frequencies close to boundaries of the continuous spectrum.

Due to the condition (12.4.1) and the broken symmetry of the problem, the function $U_1(\boldsymbol{\kappa})$ has the following general expansion in powers of $\boldsymbol{\kappa}$: $U_1(\boldsymbol{\kappa}) = -W_{\alpha\beta}k_\alpha k_\beta$, $(\alpha, \beta = 1, 2)$. We substitute (12.4.6) into (12.4.4)

$$\omega^2\chi(z) - \sum_{z'} \Lambda(n - n')\chi(z') = bV_0\chi(0)\delta(z)\,, \qquad (12.4.7)$$

and use the ordinary Fourier expansion

$$\chi(z) = \frac{b}{2\pi}\int \chi_k e^{ikz}\,dk\,, \quad \chi_k = \sum_z \chi(z)e^{-ikz}\,. \tag{12.4.8}$$

The equation for the Fourier component is then reduced to the relation

$$\left[\omega^2 - \omega_0^2(\boldsymbol{\kappa}, k)\right]\chi_k = V_0\chi(0)\,. \tag{12.4.9}$$

In the approximation taken for calculations, the dispersion law of the crystal concerned is given by (12.4.5).

From (12.4.8), (12.4.9) we just find the dependence of the vibration amplitude on the coordinate z,

$$\chi(z) = \frac{bV_0}{2\pi}\chi(0)\int_{-\frac{\pi}{2}}^{\frac{\pi}{2}} \frac{\cos kz\,dk}{\omega^2 - \omega_0^2(\boldsymbol{\kappa}, k)}\,. \tag{12.4.10}$$

Setting $z = 0$ in (12.4.10), we find an equation for the possible frequencies of such vibrations

$$\frac{bV_0}{2\pi}\int_{-\frac{\pi}{2}}^{\frac{\pi}{2}} \frac{dk}{\omega^2 - s_1^2\kappa^2 - \omega_2^2 \sin^2\dfrac{bk_z}{2}} = 1\,. \tag{12.4.11}$$

Before calculations of the integral in (12.4.11) look on the sketch of the vibration spectrum in Fig. 12.4. In Fig. 12.4. the hatched region corresponds to a band in the continuous spectrum of vibrations of an ideal crystal at small κ. The bottom $\omega_{\text{low}}(\boldsymbol{\kappa})$ and top $\omega_{\text{up}}(\boldsymbol{\kappa})$ boundaries of this band are given by the expressions

$$\omega_{\text{low}}(\boldsymbol{\kappa}) = s_1\kappa, \qquad \omega_{\text{up}}^2(\boldsymbol{\kappa}) = \omega_2^2 + s_1^2\kappa^2\,. \tag{12.4.12}$$

Notations proposed in (12.4.12) allow us to give the simple representation of the calculation in (12.4.11):

$$V_0 = \sqrt{\omega^2 - \omega_{\text{low}}^2(\boldsymbol{\kappa})}\sqrt{\omega^2 - \omega_{\text{up}}^2(\boldsymbol{\kappa})}\,. \tag{12.4.13}$$

Analyzing (12.4.12) and (12.4.13) we obtain localized states and their dispersion relations of the following types.

1. $V_0 > 0$. In this case a frequency of the localized wave lies above the band of the continuous spectrum $\left(\omega > \omega_{\text{up}}(\boldsymbol{\kappa})\right)$:

$$V_0^2 = [\omega^2 - \omega_{\text{low}}^2(\boldsymbol{\kappa})][\omega^2 - \omega_{\text{up}}^2(\boldsymbol{\kappa})]\,. \tag{12.4.14}$$

Considering small perturbations we suppose $|V_0| \ll \omega_2^2$. Under such a condition (12.4.14) has the following solution

$$\omega_{\text{d}}^2 = \omega_{\text{up}}^2(\boldsymbol{\kappa}) + \frac{1}{\omega_2^2}\left(\frac{\Delta\rho}{\rho}\omega_2^2 + \frac{\Delta\rho}{\rho}s_1^2\kappa^2 + W_{\alpha\beta}k_\alpha k_\beta\right)^2\,. \tag{12.4.15}$$

If $\Delta\rho < 0$ then there is a gap at $\boldsymbol{\kappa} = 0$ between the local frequency ω_d^2 and the band of the continuous spectrum:

$$\omega_\mathrm{d}^2 - \omega_\mathrm{up}^2 = \left(\frac{\Delta\rho}{\rho}\omega_2\right)^2 . \qquad (12.4.16)$$

It is interesting to note that this gap is proportional to the perturbation squared $(\frac{\Delta\rho}{\rho})^2$.

To continue our analysis at $\boldsymbol{\kappa} \neq 0$ we assume $W_{\alpha\beta} = W_0\delta_{\alpha\beta}$ for the sake of simplicity; then

$$V_0(\kappa) = -\frac{\Delta\rho}{\rho}\omega_2^2 - \left(\frac{\Delta\rho}{\rho}s_1^2 + W_0\right)\kappa^2 . \qquad (12.4.17)$$

One can see that the behavior of ω_d as a function of κ depends on the signs of the perturbation parameters $\Delta\rho$ and W_0. For $\Delta\rho < 0$

$$V_0 = \frac{|\Delta\rho|}{\rho}\omega_2^2 + \left(\frac{|\Delta\rho|}{\rho}s_1^2 - W_0\right)\kappa^2 . \qquad (12.4.18)$$

If $W_0 < 0$ a dispersion relation for the localized wave is characterized by curve (1) in Fig. 12.4 for all κ. At $W_0 > 0$, localized states can exist only for wave numbers $0 < \kappa < \kappa_0$ (curve (2) in Fig. 12.4). The value of κ_0 for which the dispersion curve corresponding to the localized wave touches the top boundary of the band can be found from the equation

$$V_0(\kappa_0) = 0.$$

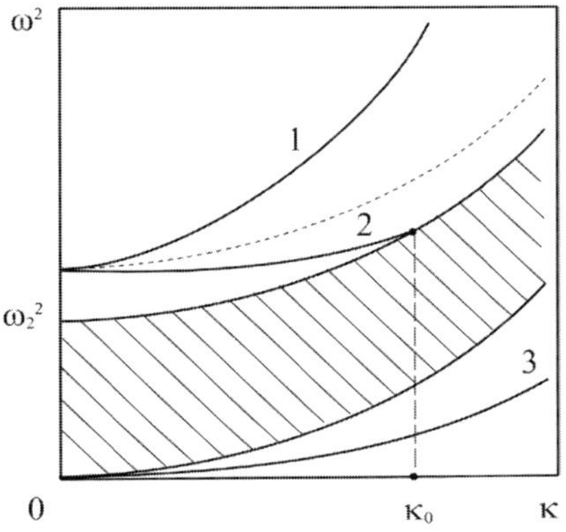

Fig. 12.4 Dispersion lines of localized waves: (1) vibrations exist at all wave vectors; (2) the dispersion line ends on the boundary of the frequency band; (3) the dispersion line for localized vibrations of the type of a surface wave.

2. $V_0 < 0$. Now (12.4.10) necessarily has a solution for $\omega < s_1\kappa$

$$\omega_\mathrm{d}^2 = s_1^2\kappa^2 - \left(\frac{aV_0}{b\omega_2}\right)^2 = s_1^2\kappa^2 - \left(\frac{1}{\omega_2}\right)^2\left(\frac{\Delta\rho}{\rho}s_1^2\kappa^2 + W_{\alpha\beta}k_\alpha k_\beta\right)^2 . \qquad (12.4.19)$$

The specific feature of this spectrum is that at fixed $\boldsymbol{\kappa}$ the frequency of the corresponding vibrations is separated from the spectrum of bulk crystal vibrations by a gap vanishing as $\boldsymbol{\kappa} \to 0$ as a power law. It follows from (12.4.11) that as $\boldsymbol{\kappa} \to 0$ not only the gap width but also its relative value tends to zero (curve 3 in Fig. 14.4). Using the simplification (12.4.17) we can obtain

$$\frac{s_1\kappa - \omega_\mathrm{d}}{\omega_\mathrm{d}} \sim \left(\frac{\kappa}{\omega_2}\right)^2 \left(\frac{\Delta\rho}{\rho} s_1^2 + W_0\right)^2 . \tag{12.4.20}$$

It is easily seen that these vibrations are localized near the planar defect. Since small wave vectors k play the main role in the integral in (12.4.10) we can take into account the approximation $bk \ll 1$ and simplify the integral:

$$\chi(z) = \frac{aV_0}{2\pi}\chi(0) \int_{-\infty}^{\infty} \frac{\cos kz\, dk}{\omega^2 - s_1^2\kappa^2 - s^2k^2} , \tag{12.4.21}$$

where s is the sound velocity along the z-axis ($2s = b\omega_2$).

Perform the integration in (12.4.21) and use (12.4.19):

$$\begin{aligned} \chi_{\boldsymbol{\kappa}}(z) &= \chi_{\boldsymbol{\kappa}}(0) \exp\left\{-\sqrt{s_1^2\kappa^2 - \omega_\mathrm{d}^2}\,\frac{|z|}{s}\right\} \\ &= \chi_{\boldsymbol{\kappa}}(0) \exp\left\{-\frac{1}{2}\left|V_0(\boldsymbol{\kappa})\frac{bz}{s^2}\right|\right\} \\ &= \chi_{\boldsymbol{\kappa}}(0) \exp\left\{\frac{1}{2}V_0(\kappa)\frac{a\,|z|}{s^2}\right\} . \end{aligned} \tag{12.4.22}$$

One should remember that the case $V_0(\boldsymbol{\kappa}) < 0$ is considered. As we see, the amplitude $\chi(z)$ exhibits an exponential decay typical for surface waves. The rate of decay of the amplitude depends on the direction of the vector $\boldsymbol{\kappa}$. If $V_0(\boldsymbol{\kappa})$ is not negative at all $\boldsymbol{\kappa}$, the localized wave with the dispersion relation (12.4.19) and the coordinate dependence (12.4.22) exist only for the directions $\boldsymbol{\kappa}$ for which $V_0(\boldsymbol{\kappa}) < 0$.

The scheme presented allows one in principle to describe the vibrations of the free surface of a crystal. Generally, the atomic interaction in a crystal decreases with increasing distance between the atoms. Thus, in analyzing the equations for the lattice vibrations, only the elements of the matrix $\alpha(\boldsymbol{n})$ with comparatively small $\boldsymbol{n}$ are taken into account. In a model where a small number of elements $\alpha(\boldsymbol{n})$ is nonzero, the free crystal surface is equivalent to a certain planar defect. Indeed, a crystal with a free surface can be obtained from an unbounded ideal crystal by cutting along a certain plane between two atomic layers and replacing the force matrix of an ideal crystal by another matrix of atomic force constants that results in the interaction between the atomic planes vanishing.

A scalar model is convenient for describing the bulk vibrations, but is unsatisfactory for analyzing the surface vibrations. To construct a consistent theory of long-wave surface vibrations it is natural to turn to elasticity theory making some assumptions on the properties of a planar defect that models the free crystal surface.

The vibrations localized near the planar defect are described as follows. The equations of elasticity theory for the bulk vibrations that occur with frequency ω have the form

$$\rho\omega^2 u^k + \nabla_l \lambda_{klmn} \nabla_m u^n = 0\,, \tag{12.4.23}$$

and the elasticity modulus tensor of a crystal λ_{iklm} in an isotropic approximation is reduced to the two Lamé coefficients λ and G

$$\lambda_{iklm} = \lambda\delta_{ik}\delta_{lm} + G(\delta_{il}\delta_{km} + \delta_{im}\delta_{kl})\,.$$

If a solid has a defect coincident with the plane $z = 0$, the "perturbed" Lamé coefficients and the mass density are given by

$$\begin{gathered} \lambda' = \lambda + Lh\delta(z)\,, \quad G' = G + Mh\delta(z)\,, \\ \rho' = \rho + \Delta\rho h\delta(z) = \rho\left[1 - (h-a)\delta(z)\right]\,, \end{gathered} \tag{12.4.24}$$

where h is the "thickness" of the planar defect; L and M are characteristics of its elastic properties; a is the interatomic distance along the z-axis in an ideal crystal.

Substituting (12.4.24) into (12.4.23), we get a system of equations generalizing the scalar equation (12.4.7). It is clear that such a description of the planar defect has a literal meaning only for $hk \ll 1$. Therefore, the singular functions describing the space perturbation localization in (12.4.24) must be treated with caution. Analyzing the vibrations near the linear defect, we have seen that using the δ-like perturbation in the long-wave approximation leads to introducing a finite upper limit of the integration over the wave vectors. Similarly, in the case of a perturbation of the type (12.4.24) caused by the planar defect, when we use the method of Fourier transformations along the z-axis, it is necessary to restrict the possible values of k_z by the limiting value $k_0 \sim 1/h$. In the case we are interested in it is determined by

$$hk_0 = \pi\,. \tag{12.4.25}$$

The solution of a similar elasticity theory problem has shown that the frequency of waves localized near the defect splits off from the edge of the lowest-frequency band of vibrations (in the isotropic approximation, from the boundary of the transverse vibration frequency). This frequency is at a distance of $\delta\omega \sim \kappa^3$ from the band boundary, in agreement with (12.4.20).

This approach may also be used to study surface waves in an elastic half-space (Rayleigh waves), assuming that they are localized vibrations near the planar defect.

If this planar defect is a free crystal surface, than as a result of the perturbation of elastic moduli the connection between elastic half-spaces $z > 0$ and $z < 0$ vanishes.

The latter can be provided in a straightforward way by setting

$$L = -\lambda\,, \quad M = -G\,, \quad h = a\,. \tag{12.4.26}$$

Substituting (12.4.26) into (12.4.24), taking (12.4.25) into account, and using the described method of finding the possible frequencies of surface vibrations, one can show that these waves have the dispersion relation $\omega = s_S \kappa$, where s_S coincides with the known velocity of Rayleigh waves (Kosevich and Khokhlov, 1970). We do not prove this statement here because the calculations are cumbersome.

13
Elastic Field of Dislocations in a Crystal

13.1
Equilibrium Equation for an Elastic Medium Containing Dislocations

Our initial definition of a dislocation

$$\oint_L dx_i \nabla_i u_k = -b_k \tag{13.1.1}$$

may be rewritten in a somewhat different form using the notation of a distortion tensor $u_{ik} = \nabla_i u_k$

$$\oint_L u_{ik}\, dx_i = -b_k. \tag{13.1.2}$$

In dislocation theory the elastic distortion tensor may be regarded as an independent quantity describing the crystal deformation. Similarly to the strain tensor ε_{ik} and the stress tensor σ_{ik} it is a single-valued function of the coordinates even in the presence of a dislocation.

Let us write down the condition (13.1.2) in a differential form. For this purpose we transform the integral around the contour L into an integral over any surface S spanning this contour:

$$\oint_L dx_i u_{ik} = \int dS_i e_{ilm} \nabla_l u_{mk}.$$

The constant vector $\boldsymbol{b}$ is represented in the form of an integral over the same surface by means of the 2D δ-function introduced in (10.3.4)

$$b_k = \int_S \tau_i b_k \delta(\xi)\, dS_i. \tag{13.1.3}$$

The Crystal Lattice: Phonons, Solitons, Dislocations, Superlattices, Second Edition. Arnold M. Kosevich

ISBN: 3-527-40508-9

As the contour L is arbitrary, the equality of these integrals means the equality of the integrand expressions:

$$e_{ilm}\nabla_l u_{mk} = -\tau_i b_k \delta(\xi). \tag{13.1.4}$$

This is just the desired differential form of (13.1.1). It is clear that on the dislocation line ($\xi = 0$), which is a line of singular points, a representation in the form of derivatives is meaningless.

If there are simultaneously many dislocations in a crystal with relatively small separations (but large compared to the lattice constant), their averaged consideration becomes important. This consideration is useful in problems where an exact field distribution between separate dislocations is of no interest and in which a theory operates with physical quantities averaged over small volume elements. It is clear that many dislocation lines should run through such "physically infinitesimal" volume elements.

The equation that expresses the main property of dislocation deformations is formulated by generalizing (13.1.4). We introduce the dislocation density tensor α_{ik} by requiring its integral on the surface spanned on any contour L to be equal to the sum of Burgers vectors $\boldsymbol{b}$ of all dislocation lines enveloped by this contour:

$$\int_S dS_i \alpha_{ik} = b_k. \tag{13.1.5}$$

The tensor α_{ik} replaces the expression on the r.h.s. of (13.1.4)

$$e_{ilm}\nabla_l u_{mk} = -\alpha_{ik}, \tag{13.1.6}$$

and describes a continuous dislocation distribution in a crystal.

It follows from (13.1.4), (13.1.6) that in the case of a discrete dislocation

$$\alpha_{ik} = \tau_i b_k \delta(\xi). \tag{13.1.7}$$

As is seen from (13.1.6), the tensor α_{ik} should satisfy the condition

$$\nabla_i \alpha_{ik} = 0, \tag{13.1.8}$$

which in the case of a single dislocation shows that the Burgers vector is constant along the dislocation line.

With such an interpretation of dislocations, the tensor u_{ik} becomes the primary quantity that describes the crystal deformation. The displacement vector $\boldsymbol{u}$ cannot then be introduced. Indeed, at $u_{ik} = \nabla_i u_k$ the l.h.s. of (13.1.4) would identically become zero over the whole crystal volume.

Equations (13.1.6) or (13.1.4), together with

$$\nabla_i \sigma_{ik} + f_k = 0, \tag{13.1.9}$$

and Hooke's law constitute a complete system of equilibrium equations of an elastic medium with dislocations.

In a scalar elastic field the role of a distortion tensor is played by the vector $\boldsymbol{h}$ that in a medium without vortices is determined by

$$\boldsymbol{h} = \operatorname{grad} u. \tag{13.1.10}$$

If vortex dislocations are present in the medium, the vector $\boldsymbol{h}$ obeys the equation

$$\operatorname{curl} \boldsymbol{h} = -\boldsymbol{\alpha}, \tag{13.1.11}$$

where $\boldsymbol{\alpha}$ is the density of vortices. For a separate vortex, $\boldsymbol{\alpha} = \boldsymbol{\tau} b \delta(\boldsymbol{\xi})$.

An analog of the stress tensor is the vector $\boldsymbol{\sigma} = G\boldsymbol{h}$ that is determined by the elastic scalar field equilibrium equation:

$$\operatorname{div} \sigma = -f, \tag{13.1.12}$$

where f is the analog of the force density.

13.2 Stress Field Action on Dislocation

Let us consider a dislocation loop D in a field of external (with respect to a dislocation) elastic stresses σ_{ik} and find a force acting on the dislocation. We calculate the work done by external forces δR for an infinitely small movement of the loop. If this work is given as

$$\delta R = \oint_{\mathcal{D}} \boldsymbol{F} \delta \boldsymbol{X} \, dl, \tag{13.2.1}$$

where $\delta \boldsymbol{X}$ is the displacement element of a dislocation line, $\boldsymbol{F}$ will determine the force with which an elastic stress acts on a unit length of the dislocation.

Let the dislocation displacement generate a certain change in the displacement vector $\delta \boldsymbol{u}$. The work of external stresses performed in a volume V with a dislocation loop is

$$\delta R = \oint \sigma_{ik} \delta u_k \, dS_i^{\infty}, \tag{13.2.2}$$

where S^{∞} is the surface enveloping the volume V.

In the given case it is convenient to use transformations of the integral (13.2.2) based on Gauss's theorem. Therefore the displacement $\boldsymbol{u}$ near a dislocation should be considered as a single-valued function of the coordinates. Then, the vector $\boldsymbol{u}$ will have a discontinuity along the surface $S_{\mathcal{D}}$ spanned on the dislocation loop. To exclude discontinuity points of the vector $\boldsymbol{u}$, we surround the loop $\mathcal{D}$ with a certain closed surface $S_{\mathcal{D}}$ outside which (in the volume V') the function $\boldsymbol{u} = \boldsymbol{u}(\boldsymbol{r})$ is continuous.

Then,

$$\begin{aligned}\delta R = & \int \nabla_i(\sigma_{ik}\delta u_k)\, dV' - \oint \sigma_{ik}\delta u_k\, dS_i^0 \\ = & \int \nabla_i\sigma_{ik}\delta u_k\, dV' + \int \sigma_{ik}\delta\varepsilon_{ik}\, dV' + \oint \sigma_{ik}\delta u_k n_i^0\, dS^0,\end{aligned} \qquad (13.2.3)$$

where the unit normal vector $\boldsymbol{n}^0$ is directed outwards with respect to the surface S^0.

Since the dislocation is not associated with additional volume force in a crystal, then $\nabla_i\sigma_{ik} = 0$ and the first integral on the r.h.s. of (13.2.3) vanishes. To calculate the last two integrals, we choose S^0 as the surface going along the upper and lower banks of the cut $S_{\mathcal{D}}$ (with a "clearance" h) and connected by an infinitely thin tube S_t of radius ρ that envelopes the line D (Fig. 13.1).

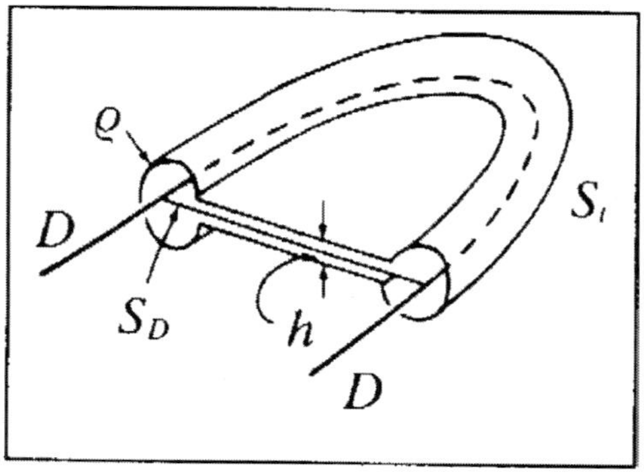

Fig. 13.1 Scheme of closed surface surrounding the cut $S_{\mathcal{D}}$.

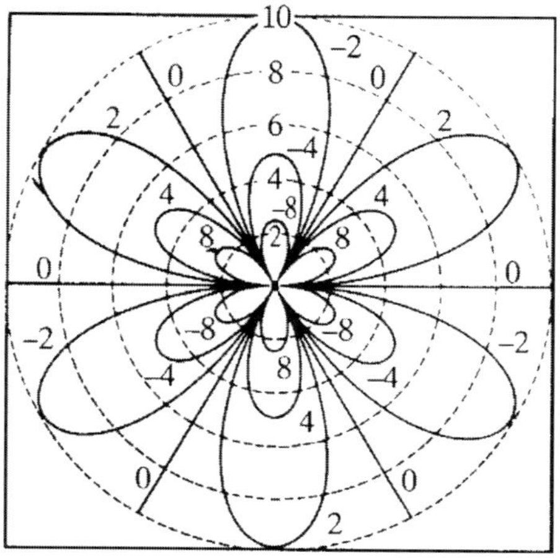

unit of distance: b
unit of dilatation: $1/400\,\pi$

Fig. 13.2 Dilatation field of a screw dislocation with an axis along [111] in a α-Fe crystal calculated by Chou (1965).

If we disregard the inhomogeneity of material generated by a dislocation and that is unimportant in evaluating the interaction between the dislocation and an elastic field, the volume integral remaining on the r.h.s. of (13.2.3) is simplified. Indeed, when the cut banks come closer ($h \to 0$) and the tube radius decreases ($\rho \to 0$), the integral over the volume V', because of continuity of σ_{ik} and ε_{ik}, transforms into an integral over the whole volume

$$\int_{V'} \sigma_{ik}\delta\varepsilon_{ik}\,dV \to \int_{V} \sigma_{ik}\delta\varepsilon_{ik}\,dV. \tag{13.2.4}$$

The surface integral in (13.2.3) can also be simplified by a corresponding limiting transition. We note that the integral over the surface of the tube S_t vanishes as $\rho \to 0$, since the dislocation field has the following obvious property:

$$\lim_{\rho\to 0} \rho u(\boldsymbol{r}) = 0.$$

On the cut banks still present in the integrals, the values of the continuous functions σ_{ik} in the limit are the same and the limiting values of $\boldsymbol{u}$ are different in a constant value $\boldsymbol{b}$. Thus, instead of (13.2.3), we get

$$\delta R = \int \sigma'_{ik}\delta\varepsilon'_{ik}\,dV + \frac{1}{3}\int \sigma_{ll}\delta\varepsilon_{kk}\,dV + b_i\delta\int_{S_D} \sigma_{ik}\,dS_k, \tag{13.2.5}$$

where σ'_{ik} is the stress deviator and ε'_{ik} is the strain deviator $[(\varepsilon'_{ik} = \varepsilon_{ik} - (1/3)\delta_{ik}\varepsilon_{ll})]$. The first two terms of the integral (13.2.3) due to the obvious identity in (13.2.5) follow from

$$\sigma_{ik}\varepsilon_{ik} = \left(\sigma_{ik} - \frac{1}{3}\delta_{ik}\sigma_{ll}\right) + \frac{1}{3}\sigma_{ll}\varepsilon_{kk} = \sigma'_{ik}\varepsilon'_{ik} + \frac{1}{3}\sigma_{ll}\varepsilon_{kk}.$$

In the last term on the r.h.s. in (13.2.5) the symbol of an infinitely small variation δ is taken outside the integral sign since the stress distribution σ_{ik} is assumed to be given.

If the element of the dislocation line dl is displaced by $\delta\boldsymbol{X}$ the area of the surface element S_D changes by

$$\delta S_i = e_{imn}\delta X_m\tau_n\,dl. \tag{13.2.6}$$

The relation (13.2.6) should be used in transforming the last integral in (13.2.5).

If the displacement $\delta\boldsymbol{X}$ is in the dislocation slip plane, (13.2.6) characterizes the changes in the crystal completely. If the displacement is perpendicular to the slip plane, additional conditions have to be taken into account that follow from the medium continuity. When the dislocation moves without breaking the medium continuity the local relative variation of the volume given by (10.3.4) exists. Using this formula, a formal replacement can be made to transform the second integral in (13.2.5)

$$\delta\varepsilon_{kk}\,dV \to e_{imn}b_m\tau_n\delta X_i\,dl. \tag{13.2.7}$$

Taking into account (13.2.6), (13.2.7), we single out in the work δR the part connected with inelastic medium deformation that accompanies the dislocation migration:

$$\delta R = \int \sigma_{ik} \delta \varepsilon_{ik} \, dV + \oint e_{imn} \tau_m \left(\sigma_{nk} - \frac{1}{3} \delta_{nk} \sigma_{jj} \right) b_k \delta X_i \, dl. \qquad (13.2.8)$$

The first integral in (13.2.8) equals the increase in elastic energy of a medium in a solid volume. The linear integral over the dislocation loop

$$\delta R_D = \oint e_{ilm} \tau_l \sigma'_{mk} \delta X_i \, dl \qquad (13.2.9)$$

will determine the work done for a dislocation displacement. Comparing (13.2.1), (13.2.9) it is clear that the force acting per unit length is

$$F_i = e_{ilm} \tau_i \sigma'_{mk} b_k. \qquad (13.2.10)$$

A formula such as (13.2.10) where the stress deviator σ'_{ik} is replaced by the stress tensor σ_{ik} was first obtained by Peach and Koehler (1950). The necessity of taking into account an inelastic change in the medium volume in climbing of a dislocation, i. e., the necessity to replace σ_{ik} by $\sigma_{ik} - (1/3)\delta_{ik}\sigma_{ll}$, was indicated by Weertman (1965).

As simple examples of using (13.2.10), we consider the forces acting on screw and edge dislocations. Let the Oz-axis be parallel to the dislocation line ($\tau_z = -1$). In the case of a screw dislocation, $b_z = b$ and

$$F_x^{\text{scr}} = b\sigma_{yz}, \quad F_y^{\text{scr}} = -b\sigma_{xz}. \qquad (13.2.11)$$

In the case of an edge dislocation, we direct the Oz-axis along its Burgers vector ($b_x = b$). Then,

$$F_x^{\text{edg}} = b\sigma_{xy}, \quad F_y^{\text{edg}} = -b\sigma'_{xx} = \frac{1}{3} b(2\sigma_{xx} - \sigma_{yy} - \sigma_{zz}). \qquad (13.2.12)$$

It is interesting to determine the projection of the force (13.2.10) on the slip plane of the corresponding dislocation element. Let $\boldsymbol{\kappa}$ be a vector that is perpendicular to the dislocation line in the slip plane. Then, this projection (we denote it as f) is equal $f = \boldsymbol{\kappa F} = e_{ikl} k_i \tau_k \sigma_{lm} b_m$ or

$$f = n_l \sigma_{lm} b_m, \qquad (13.2.13)$$

where $\boldsymbol{n} = [\boldsymbol{k\tau}]$ is the vector normal to the slip plane.

Since the vectors $\boldsymbol{n}$ and $\boldsymbol{b}$ are mutually perpendicular, by choosing two of the coordinate axes along these vectors, we see that the force f is determined by one of the components σ_{ik} only. If the dislocation is a plane curve lying in its slip plane and is in a homogeneous elastic stress field then the force f is the same for all dislocation line elements, independent of their position on the slip plane.

13.3 Fields and the Interaction of Straight Dislocations

If the dislocation is linear, the dependence on the distance of elastic stresses around it can easily be elucidated in a general case. In cylindrical coordinates r, φ, z (with the z-axis along the dislocation line) the deformation will be dependent only on r and φ. The integral (13.1.2) should not change, in particular, with an arbitrary similarity transformation of the integration contour in the plane xOy. It is obvious that this is possible only if all the elements of the tensor u_{ik} are inversely proportional to the distance $u_{ik} \sim 1/r$. The strain tensor ε_{ik} and also the stress tensor σ_{ik} will be proportional to $1/r$.

As an example of calculations of the elastic deformation generated by dislocations, we consider the dislocation around straight screw and edge dislocations in an isotropic medium. The physical meaning of these and other problems referring to an isotropic medium is conventional, since real dislocations are basically inherent only to crystals, i. e., to an anisotropic medium. These problems, however, are of interest as illustrations.

We start with a screw dislocation along which $\boldsymbol{\tau} \parallel \boldsymbol{b}$ (it is clear that only a straight dislocation may have pure screw character). We choose the axis z along the dislocation line, the Burgers vector then has the components $b_x = b_y = 0, b_z = b$. It follows from symmetry considerations that the displacement $\boldsymbol{u}$ is parallel to the z-axis and is independent of the coordinate z. Since in an isotropic medium $\sigma_{ik} = 2G\varepsilon_{ik}$ for $i \neq k$, the equilibrium equation (13.1.9) without bulk forces ($f = 0$) is reduced to a 2D harmonic equation for u_z:

$$\Delta u_z = 0, \quad \Delta \equiv \frac{\partial^2}{\partial x^2} + \frac{\partial^2}{\partial y^2}. \tag{13.3.1}$$

The solution (13.2.2) satisfying (13.1.1) has the form [1]

$$u_z = \frac{b}{2\pi}\varphi \equiv \frac{b}{2\pi}\arctan\frac{y}{x}. \tag{13.3.2}$$

The solution (13.3.2) is equivalent to (10.2.8).

The tensors ε_{ik} and σ_{ik} have the following nonzero components in cylindrical coordinates:

$$\varepsilon_{z\varphi} = \frac{b}{4\pi r}, \qquad \sigma_{x\varphi} = \frac{Gb}{2\pi r}, \tag{13.3.3}$$

where G is the shear modulus. Thus, the deformation around a screw dislocation in an isotropic medium is pure shear.

We recall that the field of a screw dislocation in an isotropic elastic medium coincides with the vortex field in a scalar model. In cylindrical coordinates it looks like:

$$h_\varphi = \frac{b}{2\pi r}, \qquad \sigma_\varphi = \frac{Gb}{2\pi r}. \tag{13.3.4}$$

1) In all problems with straight dislocations we take the vector $\boldsymbol{\tau}$ in the negative direction of the z-axis ($\tau_z = -1$).

A screw dislocation generates around itself a pure shear strain field due to the isotropic medium. To show the role of anisotropy in the elastic field variation we show the distribution of dilatation deformation (the distribution of $\varepsilon_{kk} = \operatorname{div} \boldsymbol{u}$) around the screw dislocation parallel to [111] in a cubic crystal. In Fig. 13.2, the levels of constant div $\boldsymbol{u}$ value forming a rosette that exhibits the anisotropy of a α-Fe monocrystal are given.

Calculation of elastic fields around a straight edge dislocation is much more complicated (in contrast to a screw dislocation, a purely edge dislocation can be curvilinear if the dislocation line is in a plane perpendicular to the vector $\boldsymbol{b}$). Let the z-axis be directed along the dislocation axis and the x-axis along the Burgers vector: $b_x = b, b_y = b_z = 0$. It follows from the symmetry of the problem when $\varepsilon_{zz} = 0$ that the deformation vector lies in the plane xO_y and is independent of z.

Since the solution to this seemingly simple elasticity theory problem is cumbersome, we do not repeat it here in detail.

We write down the final results for the displacement field[2]

$$\begin{aligned} u_x &= \frac{b}{2\pi}\left\{\arctan\frac{y}{x} + \frac{1}{2(1-\nu)}\frac{xy}{x^2+y^2}\right\}, \\ u_y &= -\frac{b}{4\pi(1-\nu)}\left\{(1-2\nu)\log\sqrt{x^2+y^2} + \frac{x^2}{x^2+y^2}\right\}, \end{aligned} \tag{13.3.5}$$

where ν is Poisson's ratio.

The first term in the expression for u_x is written in a form analogous to (13.3.2). Thus, the condition (13.1.1) is satisfied.

The stress tensor calculated on the basis of (13.3.5) has the Cartesian components

$$\begin{gathered} \sigma_{xx} = -bM\frac{y(3x^2+y^2)}{(x^2+y^2)^2}, \qquad \sigma_{yy} = -bM\frac{y(x^2-y^2)}{(x^2+y^2)^2}, \\ \sigma_{xy} = bM\frac{x(x^2-y^2)}{(x^2+y^2)^2}, \end{gathered} \tag{13.3.6}$$

or in polar coordinates (the corresponding coordinates r, φ in the plane xOy):

$$\sigma_{rr} = \sigma_{\varphi\varphi} = -bM\frac{\sin\varphi}{r}, \qquad \sigma_{r\varphi} = bM\frac{\cos\varphi}{r}, \tag{13.3.7}$$

with $M = G/2\pi(1-\nu)$.

It follows from the condition $\varepsilon_{zz} = 0$ that $\sigma_{zz} = \nu(\sigma_{xx} + \sigma_{yy}) = \nu(\sigma_{rr} + \sigma_{\varphi\varphi})$. Therefore, the mean hydrostatic pressure generated by the edge dislocation in an isotropic medium is

$$p_0 = -\frac{1}{3}\sigma_{kk} = -\frac{1}{3}(1+\nu)(\sigma_{rr} + \sigma_{\varphi\varphi}) = \frac{2(1+\nu)}{3}bM\frac{\sin\varphi}{r}. \tag{13.3.8}$$

2) The solution to this problem can be found in the literature on dislocation theory.

In an isotropic medium the levels of constant pressure p_0 are at the same time the levels of constant dilatation $\varepsilon_0 \equiv \varepsilon_{kk} = -p_0/K$ (K is the compression modulus). Therefore, the distribution of an elastic dilatation of the deformation field of an edge dislocation is characterized by the levels

$$\frac{\sin\varphi}{r} \equiv \frac{y}{x^2+y^2} = \text{constant}. \tag{13.3.9}$$

The curves (13.3.9) represent a family of circles where centers are on the Oy-axis and pass through the coordinate origin (the dislocation axis).

In a crystal the dilatation distribution around the edge dislocation has levels that are different from circles (13.3.9), but preserving the same character on the whole (Fig. 13.3).

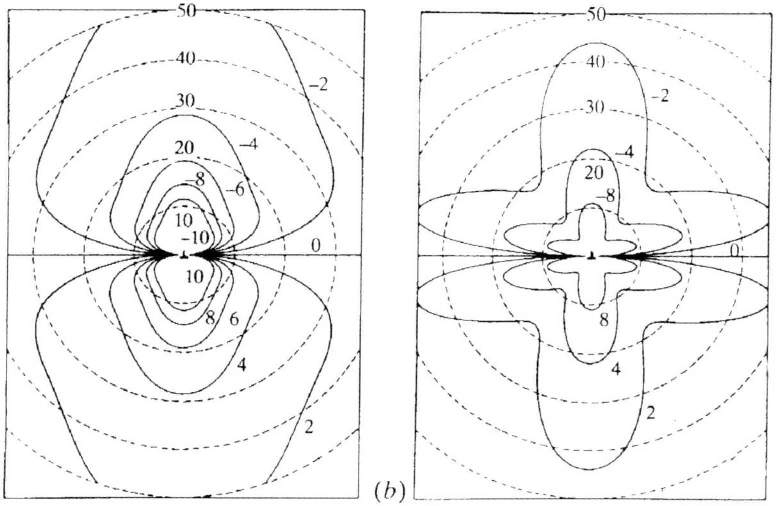

Fig. 13.3 The dilatation field of an edge dislocation in plane (110): (*a*) the crystal α-Fe; (*b*) the crystal Li calculated by Chou (1965).

The expressions for the elastic fields of straight dislocations allow us to describe the interaction of parallel dislocations easily.

To determine the force acting from one dislocation on another and parallel to it, we use (13.2.11), (13.2.12) and the expression for the stress tensor of an elastic field around the second dislocation. As seen from Fig. 13.2, the elastic field around a screw dislocation in a crystal differs considerably from that in an isotropic medium. Therefore, a detailed description of the interaction of screw dislocations in the isotropic model is meaningless, although it is justified for edge dislocations, and therefore we shall now describe the interaction of edge dislocations in an isotropic medium.

If a dislocation coincides with the z-axis, it acts on a second dislocation running through the point (x, y) in the (xOy) plane with a force whose projection on the slip

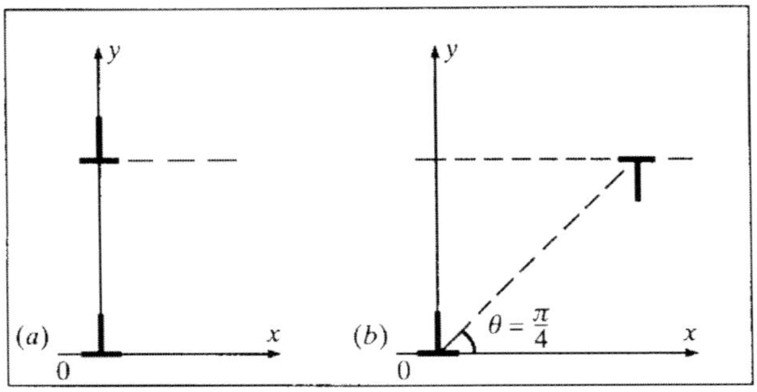

Fig. 13.4 Stable equilibrium configurations of a system of two parallel edge dislocations: (*a*) of a single sign, (*b*) of opposite signs.

plane is

$$F_x = Mb_1b_2\frac{\cos\varphi\cos 2\varphi}{r}, \qquad M = \frac{G}{2\pi(1-\nu)}. \tag{13.3.10}$$

Just this projection is of great interest, since in the slip plane only migration of the dislocation may have the character of a mechanical motion. The vanishing of this force projection corresponds to the equilibrium configuration of two dislocations relative to gliding. It follows from (13.3.10) that F_x vanishes at $\varphi = \pi/2$ and $\varphi = \pi/4$. It is easy to verify that the first of these positions (Fig. 13.4a) corresponds to the stable equilibrium condition of two dislocations of the same sign ($b_1b_2 > 0$), and the second one (Fig. 13.4b) to that of two dislocations of different sign ($b_1b_2 < 0$). Obviously, the first configuration (Fig. 13.4a) remains in stable equilibrium if the number of dislocations is greater than two. This is the reason for the dislocation wall (Section 10.5, Figs 10.8) to be an equilibrium and stable structure.

If a dislocation pair is formed by two dislocations of opposite sign (Fig. 13.4b) separated by a small distance (e. g., dislocations are positioned in the neighboring slip planes) such a pair is called a *dislocation dipole*. This system is stable enough to occur as a single dislocation entity.

Edge dislocations that are in the same glide plane (whose trace is the O_x-axis) interact in a very simple way

$$F_x = \frac{Mb_1b_2}{x}. \tag{13.3.11}$$

The formula (13.3.11) is also valid in an anisotropic medium when the constant M is connected in a different way with the elastic moduli, but remains positive.

We consider a set of a large number of identical edge dislocations placed parallel to one another in the same slip plane. The interaction of each pair of them will then be determined by the force (13.3.11).

We choose the x-axis along the Burgers vector and assume that the dislocations are distributed on the part (a_1, a_2) of the x-axis. For definiteness, we can assume that at the point $x = a_2$ there is an obstacle holding the dislocations and the whole set of them undergoes the action of a certain external stress σ^e_{xy}, that "presses" the dislocations to the obstacle ($\sigma^e_{xy} < 0$). Under these conditions we have a planar array of parallel dislocations on the same glide plane that is called a *dislocation pile-up*. The conditions of mechanical equilibrium of the pile-up are determined as follows.

We assume that there is such a large number of dislocations per unit length of the x-axis that their distribution can be described by the linear density $\rho(x)$, which is a continuous function of the coordinate x. Then, $\rho(x)\,dx$ is the Burgers vector of dislocations going through the points of the interval dx.

We single out a dislocation at the point x and write the condition of its mechanical equilibrium. The remaining dislocations of the pile-up act on a separated dislocation with a force that is found by taking account of (14.3.11)

$$F(x) = bM\,\text{P.V.}\int_{a_1}^{a_2} \frac{\rho(\xi)\,d\xi}{x-\xi}, \qquad (13.3.12)$$

P.V. means the principal value of the integral. The action of a dislocation on itself should be excluded.

Externally, the dislocation is acted upon by the force $b\sigma^e_{xy}$ so that in equilibrium

$$\text{P.V.}\int_{a_1}^{a_2} \frac{\rho(\xi)\,d\xi}{\xi - x} = \omega(x), \qquad \omega(x) = \frac{\sigma^e_{xy}(x)}{M}. \qquad (13.3.13)$$

If the function $\omega(x)$ is given, the relation is an integral equation for the equilibrium distribution $\rho(x)$. It belongs to the type of singular integral equation with a Cauchy kernel.

In conclusion, we calculate the energy of the elastic field generated by the straight dislocation in a crystal. The free energy of a dislocation of unit length is given by the integral

$$U + \frac{1}{2}\int \varepsilon_{ik}\sigma_{ik}\,dx\,dy. \qquad (13.3.14)$$

For a screw dislocation the integral (13.3.14) equals

$$U^{\text{scr}} = \frac{1}{2}\int 2\varepsilon_{z\varphi}\sigma_{z\varphi}2\pi r\,dr = \frac{Gb^2}{4\pi}\int \frac{dr}{r} \qquad (13.3.15)$$

where the integration over r should be performed within $(0,\infty)$. However, the integral (13.3.15) diverges logarithmically at both limits. The divergence at $r = 0$ is connected with the inapplicability of elasticity theory formulae at atomic distances. Therefore, as the low integration limit we must take the value of r_0 of the order of atomic distances ($r_0 \sim a \sim b$). Also, we have to omit the energy of the dislocation core U_0 whose maximum value is easy to estimate. In fact, in a dislocation core with cross sectional area $\sim b^2$ the relative atomic displacement is of the order of unity. Thus,

$$U_0 \sim Gb^2. \qquad (13.3.16)$$

The upper integration limit (13.3.15) is determined by the quantity R that is of the order of the dislocation length or crystal dimension. Then,

$$U^{\text{scr}} = \frac{Gb^2}{4\pi} \log \frac{R}{r_0}. \tag{13.3.17}$$

Comparing (13.3.16), (13.3.7), we see that (13.3.17) determines the main part of the dislocation energy under the condition $\log R/r_0 \gg 1$. The formula (13.3.17) determines the dislocation energy with logarithmic accuracy.

We note that the scalar field vortex energy is exactly equal to (13.3.17).

For an edge dislocation the integral (13.3.14) equals

$$\begin{aligned} U^{\text{edg}} &= \frac{1}{2} \int (\varepsilon_{rr}\sigma_{rr} + \varepsilon_{\varphi\varphi}\sigma_{\varphi\varphi} + 2\varepsilon_{r\varphi}\sigma_{r\varphi}) r\, dr\, d\varphi \\ &= b^2 M^2 \frac{1-\nu}{G} \int_{r_0}^{R} \frac{dr}{r} \int_0^{2\pi} \sin^2\varphi\, d\varphi = \frac{Gb^2}{4\pi(1-\nu)} \log \frac{R}{r_0}. \end{aligned}$$

It is natural that the energy of unit length of an edge dislocation has the same order of magnitude as the screw dislocation. Moreover, it can be shown (although it is quite obvious) that for any weakly distorted dislocation line (with radius of curvature $R \gg b$) its energy per unit length will have the order of magnitude

$$U \sim \frac{Gb^2}{4\pi} \log \frac{R}{r_0}. \tag{13.3.18}$$

However, it should be noted that the theoretical "large parameter" $\log(R/r_0)$ is, in fact, small. Indeed, taking the relation $R/r_0 \sim 10^5 - 10^6$ (and it is often less), we obtain that $\ln(R/r_0)$ is not different from 4π. Therefore, for rough estimates the energy per unit length of an isolated dislocation can be taken approximately equal to $U \equiv Gb^2$. Such an estimate also remains valid for an anisotropic medium.

Analyzing straight dislocation fields, we considered a screw and an edge dislocation separately. This is so because in an isotropic medium any straight dislocation with tangent vector $\boldsymbol{\tau}$ can be represented as consisting of two independent dislocations: a purely screw dislocation, the displacements around which are parallel to $\boldsymbol{\tau}$, and a purely edge dislocation, the displacements around which are perpendicular to $\boldsymbol{\tau}$. It is interesting to know whether the displacement field of a straight dislocation in an anisotropic medium can be divided into a screw part where all displacements are parallel to the dislocation line and an edge part where all displacements are perpendicular to this line. It turns out that this division is possible if the dislocation line is a two-fold symmetry axis or if the dislocation line is perpendicular to the crystal symmetry plane.

13.4 The Peierls Model

For a quantitative study of the role of crystal lattice discreteness in a dislocation deformation, Peierls (1940) suggested a semi-microscopic model that on the one hand, takes into account the crystal translation symmetry, and on the other hand allows one to obtain a direct limiting transition to the elasticity theory results.

We return to (13.3.12) and recall the geometrical meaning of the quantity $\rho(x)$. We assume that the shift generated by dislocations along the slip plane is absent at $x = \infty$. By the definition of an edge dislocation the integral

$$u(x) = \int_x^\infty \rho(\xi)\, d\xi \qquad (13.4.1)$$

equals the atomic layer displacement at the point x above the slip plane relative to that under this plane. It follows from (13.4.1) that

$$\rho(x) = -\frac{du}{dx}. \qquad (13.4.2)$$

Thus, knowing the relative crystal displacements on both sides of the slip plane we can, using (13.4.2), find a corresponding dislocation density.

We now formulate a semi-microscopic model of an edge dislocation. Let nondeformed crystal atoms form a simple tetragonal lattice with the constant b along the x-axis and the constant a along the y-axis (Fig. 13.5a). We consider two parts of this crystal, one of which (A) contains an extra atomic plane compared to the other part (B). We form a single crystal of these two parts, making the upper and lower parts come closer at an interatomic distance a and matching the left and right edges of these parts. To make the latter, with fixed right edges of parts A and B, it is necessary to perform a relative shift along the x-axis by the value b of their left edges. After joining the two parts of a crystal, the atoms form a single lattice with an edge dislocation (Fig. 13.5b).

We denote by $u(x)$ the relative displacement of the atoms on two sides of the slip plane in the direction of the x-axis. At an infinitely large distance from the dislocation, the lattice should be ideal. This means that

$$u(\infty) = 0, \qquad u(-\infty) = b. \qquad (13.4.3)$$

The function $u(x)$ should have a form qualitatively similar to the plot in Fig. 5.1. We set $u(x_0) = b/2$, i. e., assume the dislocation center to be at the point $x = x_0 = X$. For an exact definition of the function $u(x)$ the upper and lower parts of a crystal can be regarded as two continuous elastic solids. However, the tangential stress in the slip plane (in a plane where elastic half-spaces link) that prevents the deformation is treated as a periodic function of the local relative displacement $u(x)$ with period b.

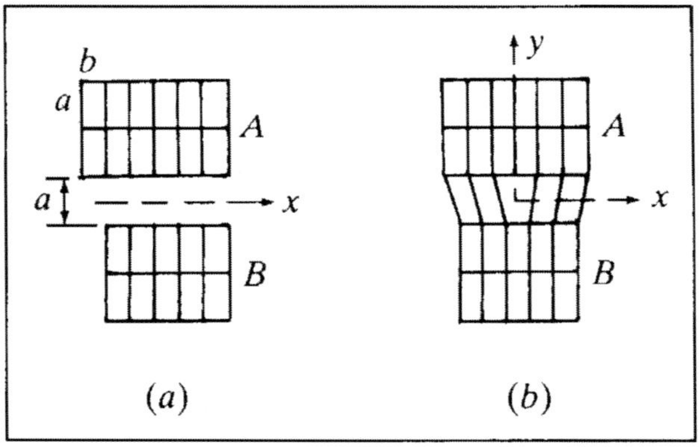

Fig. 13.5 A scheme of the formation of a monocrystal with an edge dislocation: (*a*) crystal part *A* contains an extra atomic plane. (*b*) Parts *A* and *B* are combined in a monocrystal with a dislocation.

Peierls chooses a simple periodic dependence, namely

$$\sigma^P_{xy}(x) = \mu_1 \sin\left[2\pi \frac{u(x)}{b}\right], \tag{13.4.4}$$

where the coefficient μ_1 has the order of magnitude of crystal elastic moduli. The procedure determining it will be indicated below.

The local stresses (13.4.4) should be coincident with the macroscopic stresses generated by an inhomogeneous distribution of the displacement $u(x)$ along the slip plane. We have already seen that an inhomogeneous relative displacement $u(x)$ is equivalent to a set of edge dislocations with the density (13.4.2) and, hence, it causes the shear stresses (13.3.12). Substituting (13.4.4) into (13.3.12) and equating the result to σ^P_{xy}, we get a singular integral equation for $u(x)$:

$$M\,\text{P.V.}\int_{-\infty}^{\infty}\left(\frac{du}{d\xi}\right)\frac{d\xi}{x-\xi} = -\mu_1 \sin\left[2\pi\frac{u(x)}{b}\right]. \tag{13.4.5}$$

The solutions to (13.4.5) should have the property (13.4.3).

First, we shall find the relation between the coefficient μ_1 and the crystal elastic moduli. It is clear that far from the dislocation core ($x \gg b$), where the displacements $u(x)$ are small, the solution to (13.4.5) should coincide with the known solution of elasticity theory if we take $u_{yx} = u/a$. It follows from (13.4.5) that for $x \to \infty$ when $u \ll b$,

$$2\pi\mu_1 u(x) = \frac{Mb}{x}\int_{-\infty}^{\infty}\left(\frac{du}{d\xi}\right)d\xi \equiv \frac{Mb^2}{x}. \tag{13.4.6}$$

We now calculate the derivative $\partial u_x/\partial y$ using (13.3.5) for $x \to \infty$ and compare it with (13.4.6). We get $\mu_1 = bG/\pi a(3-2\nu)$. Thus, the Peierls equation takes the form

$$\text{P.V.} \int_{-\infty}^{\infty} \left(\frac{du}{d\xi}\right) \frac{d\xi}{x-\xi} + \frac{2(1-\nu)}{3-2\nu} \frac{b}{a} \sin \frac{2\pi u}{b} = 0. \tag{13.4.7}$$

A very important property of (13.4.7) that distinguishes it from the Frenkel–Kontorova equation for dislocations in a 1D crystal is the very slow variation of the quantity $u(x)$ with distance x. While in a 1D crystal the decay of $u(x)$ as $x \to \infty$ is exponential, the solution to (13.4.7) decreases as a hyperbolic law (13.4.6). The solution corresponding to the conditions (13.4.3) is

$$u(x) = \frac{b}{2}\left(1 - \frac{2}{\pi} \arctan \frac{x-X}{l}\right), \tag{13.4.8}$$

where

$$l = \frac{3-2\nu}{4(1-\nu)} a. \tag{13.4.9}$$

The quantity l may conventionally be called a *dislocation half-width*. This notation is tentative, because the displacement $u(x)$ varies more slowly in space, which leads to the dislocation width being dependent on a specific form of the periodic function in (13.4.7). In the isotropic approximation the dislocation half-width depending on Poisson's ratio can range from $l = (3/4a)$ (for $\nu = 0$) to $l = a$ (for $\nu = 1/2$).

The Peierls model allows one to estimate the critical stress necessary to displace the dislocation in the slip plane. This was made by Nabarro (1947) using the following considerations. The stress σ_{xy}^P corresponds to the following potential energy per interval b of the x-axis

$$U(x) = -b \int_0^{u(x)} \sigma_{xy}^P \, du = -b\mu_1 \int_0^{u(x)} \sin \frac{2\pi u}{b} \, du = \frac{\mu_1 b^2}{\pi} \sin^2 \frac{\pi u}{b}. \tag{13.4.10}$$

We now take into account the discrete structure of a slip plane and associate each atomic layer with the coordinate

$$x = x_n \equiv nb, \qquad n = 0, \pm 1, \pm 2, \ldots, \tag{13.4.11}$$

choosing the layer numeration origin in the vicinity of the dislocation center. If we now substitute a discrete coordinate (13.4.11) into (13.4.10), then we obtain the n-th atomic layer energy in a lattice distorted by the presence of a dislocation. Hence, the dislocation energy related to local stresses along the slip plane is equal to

$$E = \sum_{n=-\infty}^{\infty} U(x_n) = \frac{\mu_1 b^2}{\pi} \sum_{n=-\infty}^{\infty} \sin^2 \frac{\pi u(x_n)}{b}. \tag{13.4.12}$$

Using the explicit form of (13.4.8) for the function $u(x)$, it is easy to obtain

$$\sin^2 \frac{\pi u(x)}{b} = \cos^2 \left(\arctan \frac{x-X}{l}\right) = \frac{l^2}{(x-X)^2 + l^2}. \tag{13.4.13}$$

We recall that X is the dislocation center coordinate that is an independent characteristic of the dislocation and noncoincident in general with the coordinate of any atomic layer.

Finally, we substitute (13.4.13) into (13.4.12) using the Poisson summation formula and obtain as a main approximation with regard to the small parameter $b/(2\pi l) \ll 1$:

$$E = \mu_1 b^2 + 2\mu_1 b^2 \exp\left(-\frac{2\pi\lambda}{b}\right) \cos\left(2\pi \frac{X}{b}\right). \tag{13.4.14}$$

It is clear that the dislocation energy is a periodic function of the coordinate of its center. Differentiating (13.4.14) with respect to the coordinate X, we find the force acting on a dislocation in the crystal:

$$F = -\frac{\partial E}{\partial X} = 4\pi\mu_1 b \exp\left(-\frac{2\pi l}{b}\right) \sin\left(2\pi \frac{X}{b}\right). \tag{13.4.15}$$

The formula (13.4.15) determines a *Peierls force*. The maximum value of this force F_m determines the shear stresses $\sigma_s = F_m/b$ to be applied to a crystal for the dislocation to begin moving in the slip plane. To evaluate σ_s, we set $l = a$ and $a = b$. It appears then that $\sigma_s \sim 2\pi\mu_1 10^{-2} \sim 10^{-2}G$. If $a = (3/2)b$, then $\sigma_s \sim 10^{-4}G$. Just these limiting values are generally given in analyzing dislocation glide.

13.5
Dislocation Field in a Sample of Finite Dimensions

Point-defect fields decrease slowly with distance. Thus, analyzing the fields of elastic point defects (Section 9) it is necessary to consider the effects associated with the presence of free external crystal surfaces. But dislocation fields decrease with distance even slower. Therefore, the finite dimensions of a crystal can result in effects that, when disregarded, could distort the result of solving dislocation problems.

As an example we consider a screw dislocation positioned along the axis of a specimen of finite dimensions having the form of a cylinder of radius R. Let the cylinder length L be much larger then the radius ($L \gg R$), then it is possible to exclude from our discussion a stress distribution near the cylinder ends. The stresses (13.3.3) resulting from (13.3.1) satisfy the boundary conditions $\sigma_{rr} = \sigma_{rz} = \sigma_{r\varphi} = 0$ on a free side surface of a specimen $r = R$. But these stresses generate a nonzero twisting moment on cylinder ends

$$M_z^0 = \int r\sigma_{z\varphi}\, dS = 2\pi \int_0^R \sigma_{z\varphi} r^2\, dr = \frac{1}{2} GbR^2. \tag{13.5.1}$$

Thus, if the cylinder ends are free the solution (13.3.3) does not satisfy the boundary conditions on the faces (even if the latter are infinitely distant). Thus, the true solution to the cylinder equilibrium equations should involve, apart from (13.3.3), additional stresses compensating the twisting moment (13.5.1), i. e., creating an average moment $M_z = -M_z^0$. These stresses (and also the corresponding displacements) are easily obtained from a rod torsion theory. It is known that for a rod twisted under the action of the moment M_z, there arises a displacement vector component u_φ providing a torsion angle that is homogeneous along the rod length:

$$\frac{d\theta}{dz} = \frac{\partial}{\partial z}\left(\frac{u_\varphi}{r}\right) = \frac{1}{r}\frac{\partial u_\varphi}{\partial z} = \frac{M_z}{C} = \text{constant},$$

where $C = (1/2)\pi G R^4$ is the torsional rigidity of a rod.

Taking this into account, it is easy to write an equilibrium distribution of additional displacements and stresses in a cylinder with a screw dislocation along its axis

$$u_\varphi = -\frac{brz}{\pi R^2}, \qquad \delta\sigma_{z\varphi} = -\frac{Gb}{2\pi}R^2 r. \tag{13.5.2}$$

The second term on the r.h.s. of (13.5.2) shows the role of finite dimensions of the specimen in generating the elastic stresses around the dislocation. With increasing cylinder radius R, the contribution of the last term decreases for $r \ll R$. The torsion angle also decreases:

$$\frac{d\theta}{dz} = -\frac{b}{\pi R^2}.$$

However, the total torsion of the cylindrical specimen

$$\delta\theta = -\frac{bL}{\pi R^2}, \tag{13.5.3}$$

for $L \gg R$ may be essential. Torsional deformation is observed in long and thin thread-like crystals with a screw dislocation (whiskers).

Now consider a superstructure formed by a system of a large number of screw dislocations oriented along the axis of the cylinder of radius R and we will be interested in the macroscopic properties of such a superstructure. Then, in the main approximation the distribution of the dislocations can be assumed continuous, characterized by the density $n_0 = 1/S_0$, where S_0 is the average area of the xy plane per dislocation.

The dislocation creates a stress (13.3.3) in an unbounded media (denote it as σ_0). But because the stress field of the screw dislocations is similar to the electric field of linear charges, the stresses at a distance r from the axis of the cylinder are created by all the dislocations intersecting an area $S = \pi r^2$ around the axis of the cylinder and are equal to the stresses around one dislocation lying along the axis of the cylinder ($x = y = 0$) and carrying the total "charge" (the Burgers vector bS/S_0) of all those dislocations:

$$\sigma_\varphi \equiv \sigma_\varphi = \frac{S}{S_0}\sigma_0 = \frac{1}{2}Gbn_0 r. \tag{13.5.4}$$

These stresses, first, create a force acting on a dislocation lying a distance r from the axis of the cylinder:

$$F_r = b\sigma_\varphi = \frac{1}{2}Gb^2 n_0 r\,. \tag{13.5.5}$$

Second, the dislocation field (13.5.4) creates the following moment of torque on the end of the cylinder:

$$M_z = \int r\sigma_\varphi \, dS = 2\pi \int_0^R \sigma_\varphi r^2 \, dr = \frac{\pi}{4} GbR^4 . \tag{13.5.6}$$

One sees that (13.5.6) differs from (13.5.1) with the multiplier $(1/2)\pi R^2 n_0$. Consequently, in order to calculate strains and stresses produced by the dislocation system in the cylinder it is sufficient to multiply (13.5.2) by this multiplier:

$$u_\varphi = -\frac{1}{\pi} b n_0 r z , \qquad \delta\sigma_\varphi = -\frac{1}{2} G b n_0 r . \tag{13.5.7}$$

The stresses described by the second relation in (13.5.7) can be looked upon as a certain external field in relation to the dislocation system. Comparing (13.5.7) with (13.5.4) and (13.5.5) one can see that the interaction force of a given dislocation with remaining continuously distributed dislocations is exactly compensated by these "external stresses". This means that the expected repulsion of discrete dislocations calculated according to formula (13.5.5) with the use of (13.3.3) is eliminated on average when the boundary conditions and symmetry of the problem with a continuous distribution of dislocations are correctly taken into account. In other words, the equilibrium state of such a dislocation system is stabilized by the twisting of the sample.

Another more trivial example of the influence of boundary conditions for a free crystal surface on the distribution of the elastic fields around the dislocation is the problem of a screw dislocation parallel to the plane of a free surface of an isotropic medium. Let the plane zOy coincide with a surface of the solid, and the dislocation parallel to the z-axis have the coordinates $x = x_0, y = 0$. The stress field leaving the free medium surface is described by the sum of dislocation fields and its mirror reflection in the plane yOz as if they were in an unbounded medium

$$\begin{aligned} \sigma_{zx} &= -\frac{Gb}{2\pi}\left[\frac{y}{(x-x_0)^2+y^2} - \frac{y}{(x+x_0)^2+y^2}\right] , \\ \sigma_{xy} &= \frac{Gb}{2\pi}\left[\frac{x-x_0}{(x-x_0)^2+y^2} - \frac{x+x_0}{(x+x_0)^2+y^2}\right] . \end{aligned} \tag{13.5.8}$$

The dependence of stresses (13.5.8) on the distance x_0 is typical for any linear dislocation near the crystal surface and is very important in discussing various boundary problems.

13.6 Long-Range Order in a Dislocated Crystal

According to (13.3.5), far from the dislocation an unexpected dependence of u_y on r ($u_y \sim b \ln r$) is observed. This dependence is quite natural, if we consider the displacement as a strain (or stress) field "potential" generated by a linear source. The vector $\boldsymbol{u}$ has a simple physical meaning in a crystal with an isolated dislocation. It determines the displacement of an atom in a dislocated crystal lattice with respect to its equilibrium position in the same lattice without a dislocation. Thus, the displacements of atoms extremely distant from the linear dislocation axis increase logarithmically with increasing crystal dimension.

Although the relative displacements of the neighboring atoms, given by the deformation tensor proportional to $1/r$, are vanishingly small as $r \to \infty$, it is necessary to specify more accurately the notion of crystal order in a dislocated lattice since the displacement vector increases at large distances from the dislocation. Since dislocations can generate arbitrarily large atomic displacements without breaking the crystal structure, it is unnecessary to require in describing the crystalline ordering in a real crystal that the same periodic spatial atomic network is preserved in the whole space.

To determine the crystal order, we consider the scheme of a deformed crystal (Fig. 7.7) used to explain the effect of lattice deformation on the phonon spectrum. A system of atoms has no space periodicity, and the "unit cells" in its different parts differ in form and dimension, but it is nevertheless considered as a deformed crystal. To make this system ordered, we introduce a curvilinear net describing at each point of the space a quite definite crystal structure. Going along this net, it is possible to establish a relation between a local short-range order and that in any part of the crystal. The point defects do not break the general structure of the net.

In a crystal that has several dislocations (Fig. 13.6), in addition to local crystal net distortions, there are "topological" faults of the spatial lattice that significantly deform the long-range order.

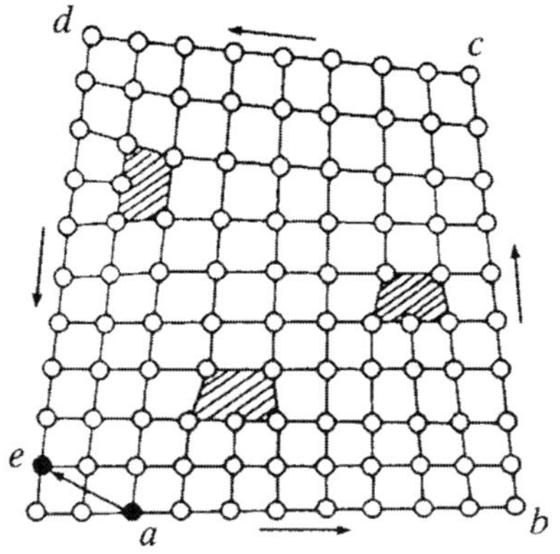

Fig. 13.6 A dislocated crystal with topological long-range order.

The increase in the density of the opposite-sign dislocations makes the whole structure amorphous without breaking the short-range order. Thus, an *amorphous* solid can be considered as the limiting state of a crystal with a large number of dislocations such that the average distance between them is of a comparable order of magnitude as the lattice period.

However, we are interested in the crystal state, whose existence implies that the long-range order distortions induced by dislocations are small. These distortions can be displayed during the following process of observation. Basing our treatment on the local short-range order in a crystal, we pass around a macroscopic contour that would be closed in an ideal crystal. For instance, if we pass an external contour in a counterclockwise direction (Fig. 13.6) starting from the point a, it will not be closed and we will arrive at the point e noncoincident with the contour origin. The impossibility of closing a similar contour characterizes the fault in the long-range order in the crystal.

The situation when continuous "short-range ordering" in passing around a closed contour cannot result in a common short-range order at all contour points has been called (in a theory of crystal singularities) *frustration*.

In the given case the degree of "frustration" in the structure produced is characterized by the vector $(\boldsymbol{ae})$, which is equal to the total Burgers vector $\boldsymbol{B}$ of all dislocations enveloped by tracing around a contour. Thus, a quantitative measure of breaking the long-range order in a certain part of a dislocated crystal is the total Burgers vector of dislocations coupled with a "closed" macroscopic contour that envelopes the crystal part concerned.

We denote by L the length of the contour $abcde$ (Fig. 13.6). For the system under consideration, the inequality $B \ll L$ is satisfied. If a similar inequality is satisfied for any macroscopic contour in a dislocated crystal then the long-range order fault can be assumed to be unessential.

For the atomic system concerned, in determining the crystal order we replace the literal translational symmetry of an ordered system by a set of the following structural properties.

1. The system has a short-range order given entirely by the crystallographic structure of substance. To formulate this property, it is necessary to exclude the vicinities of point defects or of certain specific lines–dislocation cores.

2. During a continuous motion from a spatial point to the neighboring one, it is possible to establish a correspondence between equivalent atoms and crystallographic directions in the unit cells positioned close to one another.

3. The total Burgers vector coupled with any closed macroscopic contour inside the system remains small compared to the length of this contour. A closed contour is constructed by using the short-range order at each of its intermediate points, and the Burgers vector coupled with it equals the displacement of an atom at the end of the contour relative to an atom at its beginning.

It is natural to consider the atomic system characterized by these properties as a crystal with long-range order. A real crystal has long-range order only in this sense.

We note that the suggested definition of crystal long-range order does not give the information about the smallness of atomic displacements with respect to their equilibrium positions in the initial crystal lattice. Sometimes, the long-range order determined in this way is called *topological order*.

The vanishing of topological long-range order means the destruction of a crystal structure. When the absence of long-range order in a solid substance is not accompanied by the loss of shear elasticity, we assume that a solid is in an amorphous state. If the long-range order vanishes with increasing temperature simultaneously with shear elasticity (the substance starts leaking), we can speak of crystal *melting*. Unfortunately, there is no systematic theory of 3D crystal melting.

The situation is opposite in a 2D case. In a 2D crystal, a dislocation is a point defect, whose 2D displacement vector field $\boldsymbol{u}(u_x, u_y)$ is not different from that of a linear edge dislocation in a 3D crystal (13.3.5). The energy of this defect can be obtained by multiplying the linear dislocation energy per unit length by the lattice constant a:

$$E = \varepsilon_0 \ln \frac{R}{a}, \qquad \varepsilon_0 = \frac{aGb^2}{4\pi(1-\nu)}. \tag{13.6.1}$$

Since the energy ε_0 is comparable with the atomic energy of a 2D crystal and the dependence on crystal dimension (13.6.1) is very weak, a thermo-fluctuational initiation of the dislocation can be assumed in a 2D crystal.

With given temperature T, the thermodynamic equilibrium condition is determined by the minimum of the free energy of the solid F. We consider the change in F associated with the appearance of one dislocation and note that the dislocation contribution to the configuration entropy in a 2D crystal is

$$\delta S = \ln \frac{R^2}{a^2}. \tag{13.6.2}$$

Then

$$\delta F = E - T\delta S = (\varepsilon_0 - 2T) \ln \frac{R}{a}. \tag{13.6.3}$$

It is clear that for $T < (1/2)\varepsilon_0$ the appearance of a dislocation increases the free energy of the system and, thus, the dislocation initiated fluctuationally will inevitably escape from the crystal. In total thermodynamic equilibrium there are no isolated free dislocations.

For $T > (1/2)\varepsilon_0$, the free energy of the system decreases with the appearance of a dislocation and thus the process of their initiation in a 2D crystal becomes advantageous thermodynamically. Hence, with increasing temperature at $T = T_0 = (1/2)\varepsilon_0$ there occurs a phase transition of a 2D crystal into a state with an arbitrary number of free dislocations (Berezinsky, 1970, Kosterlits, Taules, 1973). As a result, the initial long-range order in a 2D crystal is broken.

It is easy to follow what mechanism is responsible for breaking the 2D crystal long-range order. At low $(T < T_0)$ but finite temperatures a certain equilibrium density of dislocation dipoles exists in a crystal. This is the case because the energy of an opposite-sign dislocation pair, in contrast to (13.6.2) is independent of crystal dimension. Indeed, the dipole energy can be written as $E_D = 2E + E_{int}$, where the interaction energy of a dislocation pair E_{int} is determined by (13.6.7) shown in solving Problem 2. In this formula it is necessary to put $b_1 = b = -b_2$. Thus,

$$E_D = 2\varepsilon_0 \left[\ln \frac{r}{a} + \sin^2 \varphi\right] + 2U_0, \tag{13.6.4}$$

where the angle φ determines the dipole orientation and U_0 is the dislocation core energy.

Since the energy (13.6.4) is bounded, there exists a finite density of dislocation dipoles of the given dimension r, proportional to the Boltzmann factor $\exp(-\beta E_D)$, where $\beta = (1/T)$. We calculate the mean square of the dislocation dipole dimension:

$$\langle\langle r^2 \rangle\rangle = \int_a^R r^2 e^{-\beta E_D} r\, dr\, d\varphi \left\{ \int_a^R e^{-\beta E_D} r\, dr\, d\varphi \right\}^{-1} \tag{13.6.5}$$

$$= \int_a^R r^{3-2\beta\varepsilon_0}\, dr \left\{ \int_a^R r^{1-2\beta\varepsilon_0}\, dr \right\}^{-1} = \frac{1-\beta\varepsilon_0}{2-\beta\varepsilon_0} \frac{R^{4-2\beta\varepsilon_0} - a^{4-2\beta\varepsilon_0}}{R^{2-2\beta\varepsilon_0} - a^{2-2\beta\varepsilon_0}}.$$

Since we are interested in the limit $R \to \infty$ $(R \gg a)$ the mean square (13.6.5) will have different forms for $4 < 2\beta\varepsilon_0$ when $T < T_0$, and with $4 > 2\beta\varepsilon_0$ when $T > T_0$. In the first case $T < T_0$, in the limit $R \to \infty$ we have

$$\langle\langle r^2 \rangle\rangle = \frac{1-\beta\varepsilon_0}{2-\beta\varepsilon_0} a^2 = \frac{2T_0 - T}{T_0 - T} \left(\frac{a^2}{2}\right)^2. \tag{13.6.6}$$

Thus, the average dimension of equilibrium dislocation dipoles in an ordered 2D crystal is limited, but increases with increasing temperature and goes to infinity at the phase transition point $T = T_0$.

In the second case $(T > T_0)$, in the limit $R \to \infty$ we get $\langle\langle r^2 \rangle\rangle = \infty$. Hence, it follows that when $T > T_0$ the dislocation dipoles are broken down, on average, and a phase with free 2D dislocations is occurs. A more detailed description of the properties of the resulting phase requires taking into account the dislocation interaction and a knowledge of how to treat disordered systems.

13.6.1 Problems

1. Calculate the shear stresses generated by an infinitely extended dislocation wall (Fig. 10.8c) at large distances.

Hint. Write down the total field of all dislocations

$$\sigma_{xy}(x,y) = bMx \sum_{n=-\infty}^{\infty} \frac{x^2 - (y-nh)^2}{[x^2 + (y-nh)^2]}$$

and make use of the Poisson summation formula (9.4.3).

Solution.

$$\sigma_{xy} = 4\pi^2 bM \frac{x}{h^2} e^{-2\pi x/h} \cos\left(2\pi \frac{y}{h}\right), \qquad x \gg h.$$

2. Find the interaction energy of two linear edge dislocations lying in the parallel glide planes.

Hint. Make use of the fact that the desired energy is equal to the work done to remove one dislocation in the glide plane from infinity (from crystal surface) to the line of its location in the field of a fixed second dislocation. Thus, the interaction energy per unit length of the dislocation equals

$$e_{12} = \int_{r\cos\varphi}^{R} F_x dx \quad \text{at} \quad R \to \infty,$$

where F_x is determined by an equation such as (13.3.10).

Solution.

$$E_{21} = -b_1 b_2 M \left(\ln \frac{r}{R} + \sin^2 \varphi\right). \qquad (13.6.7)$$

3. Describe the distribution of straight dislocations near the obstacle under the action of a homogeneous compressive stress.

Hint. Make use of (13.3.13), setting $\sigma^e_{xy} = \sigma_0 =$constant, $a_2 = 0$, and find $a_1 = -L$ from the condition $\int_{-L}^{0} \rho(x)\,dx = N$.

Solution.

$$\rho(x) = \frac{\sigma_0}{\pi bM}\sqrt{\frac{L+x}{|x|}}, \quad L = \frac{2NbM}{\sigma_0} \qquad (-L < x < 0).$$

14
Dislocation Dynamics

14.1
Elastic Field of Moving Dislocations

Let us now elucidate what form the total system of equations of elasticity theory takes that determines strains and stresses in a crystal when dislocations perform a given motion.

Equation (13.1.6) is independent of whether the dislocations are at rest or in motion. However, in the dynamic case the distortion tensor should change with time and this will be determined by the character of dislocation motion.

If during the deformation of a medium the dislocations remain stationary the following equality is valid

$$\frac{\partial v_k}{\partial x_k} = \frac{\partial u_{ik}}{\partial t}, \tag{14.1.1}$$

where $v = v(r, t)$ is the velocity of displacement of a medium element with a coordinate r at time t.

If the dislocations move and their density changes with time, (14.1.1) is incompatible with (13.1.6). Thus, we replace it with

$$\frac{\partial v_k}{\partial x_k} = \frac{\partial u_{ik}}{\partial t} - j_{ik}, \tag{14.1.2}$$

where the tensor j_{ik} should be chosen so that these equations are compatible. The compatibility condition of (13.1.6), (13.1.2) has the form

$$\frac{\partial \alpha_{ik}}{\partial t} + e_{ilm}\nabla_l \frac{\partial j_{mk}}{\partial t} = 0\,. \tag{14.1.3}$$

To associate the tensor j_{ik} with dislocation motion, we note that the condition (14.1.3) can be regarded as the differential form of the conservation law of the Burgers vector in a medium. Indeed, we integrate both sides of (14.1.3) over the surface

The Crystal Lattice: Phonons, Solitons, Dislocations, Superlattices, Second Edition. Arnold M. Kosevich

ISBN: 3-527-40508-9

spanned on a certain closed line L and introduce the total Burgers vector $\boldsymbol{b}$ of the dislocations enveloped by the contour L. Using Stokes's theorem, we obtain

$$\frac{db_k}{dt} = -\oint_L dx_i j_{ik}.$$

It is clear from this equality that the integral on its r.h.s. determines the Burgers vector "flowing" per unit time through the contour L, i. e., carried over by dislocations that cross the line L. Thus, it is natural to consider j_{ik} as the *dislocation flux density* tensor, and (14.1.3) as an equation for the continuity of dislocation flow.

In particular, it is obvious that in the case of an isolated dislocation loop, the tensor j_{ik} has the form

$$j_{ik} = e_{ilm}\tau_l V_m b_k \delta(\xi)\,, \qquad (14.1.4)$$

where $\boldsymbol{V}$ is the dislocation line velocity at a given point. The vector of the flux $j_{ik}\,dl_i$ through an element $d\boldsymbol{l}$ of the contour is then proportional to $d\boldsymbol{l}\,[\boldsymbol{\tau V}] = \boldsymbol{V}\,[d\boldsymbol{l}\boldsymbol{\tau}]$, i. e., to the projection velocity $\boldsymbol{V}$ onto the direction perpendicular both to $d\boldsymbol{l}$ and $\boldsymbol{\tau}$. It follows from geometrical considerations that only this projection results in the dislocation crossing the element $d\boldsymbol{l}$.

If the dislocation distribution is described by the continuous functions α^s_{ik} (14.1.4) is generalized by

$$j_{ik} = e_{ilm}\sum_s \alpha^s_{ik} V^s_m,$$

where the index s denotes the densities of dislocations of various types (for instance, dislocations of different sign in the case of a system of parallel edge dislocations) and the vector $\boldsymbol{V}^s$ is equal to the average velocity of dislocations of the type s at a given point in the crystal.

The tensor j_{ik} has an independent meaning and is the principal characteristic of dislocation motion.

Finally, we obtain a total system of differential equations describing the elastic fields in a crystal with moving dislocations. This system consists of (13.1.6), (14.1.2) and the equations of motion of a continuous medium:

$$\rho\frac{\partial v_i}{\partial t} = \nabla_k\sigma_{ki},\quad \sigma_{ik} = \lambda_{iklm}u_{lm},\quad e_{ilm}\nabla_l u_{mk} = -\alpha_{ik},$$
$$\nabla_i v_k - \frac{\partial u_{ik}}{\partial t} = -j_{ik}\,. \qquad (14.1.5)$$

In these equations the tensors α_{ik} and j_{ik} are the given functions of the coordinates (and time) characterizing the dislocation distribution and motion.

The compatibility conditions for (14.1.5) are the conservation laws

$$\nabla_i\alpha_{ik} = 0,\quad \frac{\partial\alpha_{ik}}{\partial t} + e_{ilm}\nabla_l j_{mk} = 0\,. \qquad (14.1.6)$$

Using the definition for the tensor of the dislocation flow density and the system of (14.1.5), the dynamics of dislocations in an elastic medium can be developed (Kosevich, 1962; Mura, 1963).

The relation between the trace of the tensor j_{ik} ($j_0 \equiv j_{kk}$) and the equation for continuity of a continuous medium is of special interest. The convolution j_0 is involved in the equation obtained from (14.1.5):

$$\operatorname{div} v - \frac{\partial \varepsilon_{kk}}{\partial t} = -j_0. \tag{14.1.7}$$

It is easy to explain the physical meaning of (14.1.7). Indeed, the convolution ε_{kk} is a relative elastic change in the medium element volume obviously related with a corresponding relative change in its density (ρ):

$$\varepsilon_{kk} = -\frac{\delta\rho}{\rho}. \tag{14.1.8}$$

Substituting (14.1.8) into (14.1.7) and using the linearity of the theory, we get the relation

$$\frac{\partial\rho}{\partial t} + \operatorname{div} \rho v = -\rho j_0. \tag{14.1.9}$$

If, with a moving dislocation, the medium elements move without breaking continuity, the l.h.s. of (14.1.9) vanishes due to the continuity equation and (14.1.10)

$$j_0 \equiv j_{kk} = 0. \tag{14.1.10}$$

For linear dislocations (14.1.10) has a simple interpretation. Indeed, in the case of an isolated linear dislocation, the convolution j_0 is proportional to $[\boldsymbol{b\tau}]\,\boldsymbol{V}$, i. e., proportional to the projection of the dislocation velocity onto the directional perpendicular to the vectors $\boldsymbol{\tau}$ and $\boldsymbol{b}$, or in other words, onto the direction perpendicular to the dislocation glide plane. Thus, (14.1.10) implies that with medium continuity preserved the vector of the dislocation velocity $\boldsymbol{V}$ lies in the glide plane, so that the mechanical motion of a dislocation can only take place in this plane.

If dislocation motion is accompanied by the formation of some discontinuities, for instance, by a macroscopic cluster of vacancies along some part of the dislocation line, the l.h.s. of (14.1.9) is nonzero and equals the velocity of a relative inelastic increase or decrease of the mass of some elementary volume of the medium.

The action of this mechanism amounts to a macroscopic clustering of vacancies or interstitial atoms along the dislocation line. We denote by $q(\boldsymbol{r})$ a relative increase of the specific volume of a medium at the point $\boldsymbol{r}$ per unit time. According to the formulae (14.1.4), (14.1.9), the motion of an isolated dislocation should then generate the following value of q:

$$q(\boldsymbol{r}) = \boldsymbol{V}\,[\boldsymbol{b\tau}]\,\delta(\boldsymbol{\xi}), \tag{14.1.11}$$

when $\boldsymbol{\xi}$ is a 2D radius vector measured from the dislocation axis.

One of the methods of solving the system of equations (14.1.5) consists in replacing an infinitely small displacement of the dislocation line element by an infinitely small dislocation loop. This method is only fruitful when the time dependence of the deformation field is studied.

We differentiate the equation of elastic medium motion with respect to time using (14.1.2). We then obtain a dynamic equation of elasticity theory for the velocity vector $\boldsymbol{v}$:

$$\rho \frac{\partial^2 v_i}{\partial t^2} - \lambda_{iklm} \nabla_k \nabla_l v_m = \lambda_{iklm} \nabla_k j_{lm} \,. \tag{14.1.12}$$

The role of the force density in this equation is played by the vector $f_i^D = \lambda_{iklm} \nabla_k j_{im}$. The solution to (15,1.12) can be written as

$$v_i(\boldsymbol{r}, t) = \int dV' \int\limits_{-\infty}^{t} dt' \, G_{ik}(\boldsymbol{r} - \boldsymbol{r}', t - t') f_k^D(\boldsymbol{r}', t') \,,$$

where $G_{ik}(\boldsymbol{r}, t)$ is the Green tensor of the elasticity theory dynamic equation. This formula solves completely the problem of finding the displacement velocity and determines the time dependence of the displacements.

We recall that the Green dynamic tensor for an isotropic medium is written explicitly as

$$\begin{aligned} G_{ik}(\boldsymbol{r}, t) = {} & \frac{1}{4\pi\rho r} \left[\frac{n_i n_k}{s_l^2} \delta\left(t - \frac{r}{s_l}\right) + \frac{\delta_{ik} - n_i n_k}{s_t^2} \delta\left(t - \frac{r}{s_t}\right) \right] \\ + {} & \frac{t(\delta_{ik} - 3 n_i n_k)}{4\pi\rho r^3} \left[\Theta\left(t - \frac{r}{s_t}\right) - \Theta\left(t - \frac{r}{s_l}\right) \right] , \end{aligned}$$

where s_l and s_t are the longitudinal and transverse sound velocities; $\boldsymbol{r} = \boldsymbol{n} r$, $\Theta(x)$ is the Heaviside discontinuity function.

The present Green tensor resembles the Green function for an electromagnetic field in the medium. The equations for the dynamics of an elastic field with dislocations are not very different from the dynamic equations for an electromagnetic field with charges and currents. These are the differential equations in partial derivatives of the type of wave equations with sources. Since nonstationary (accelerated) motion of the sources generates a radiation field, the accelerated motion of dislocations induces radiation of elastic (sound) waves. The description of acoustic radiation of dislocations can be made by a standard scheme developed in field theory, but is somewhat complicated by the tensor character of the deformation fields.

14.2 Dislocations as Plasticity Carriers

Dislocation motion is accompanied (apart from changes of elastic deformation) by changes in the crystal not associated directly with stresses, i. e., *plastic deformation* (Fig. 14.1). By a plastic deformation we understand a residual crystal deformation that does not vanish after the process that generates it is over.

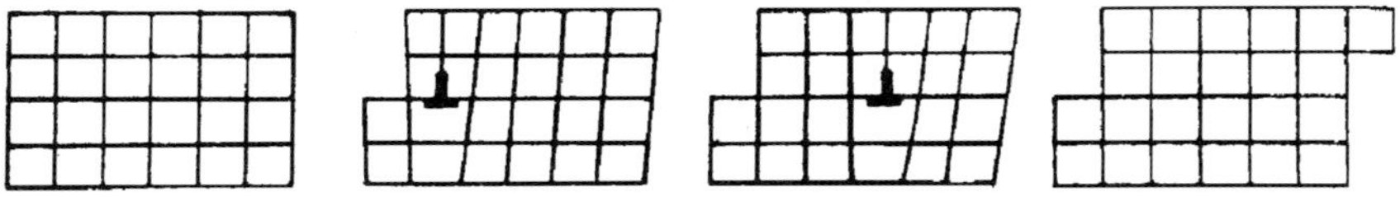

Fig. 14.1 Scheme of a residual atomic rearrangement resulting from the motion of an edge dislocation.

Let an edge dislocation pass through a crystal from left to right. As a result, part of the crystal above the glide plane is displaced by one lattice period. Since the lattice at any point inside the specimen is regular after the dislocation has moved through it, the crystal remains unstressed. Unlike the elastic deformation associated directly with the thermal state of a solid, plastic deformation is a function of the process. (In considering stationary dislocations the question of distinguishing between elastic and plastic deformation does not arise. We are interested in stresses that are independent of the previous crystal history.)

It is clear that the dislocation displacement is inevitably related to the occurrence of plastic deformation: dislocations are *elementary plasticity carriers*.

For a quantitative description of dislocation elasticity, it is necessary to return to (14.1.2). We introduce the vector of the total geometric displacement of the points of the medium $\boldsymbol{u}$ measured from their position before the deformation starts. The time derivative of the vector $\boldsymbol{u}$ determines the displacement velocity of the medium element $v = \partial \boldsymbol{u}/\partial t$. Therefore,

$$\frac{\partial w_{ik}}{\partial t} = \nabla_i v_k, \tag{14.2.1}$$

where w_{ik} is the total distortion tensor, $w_{ik} = \nabla_i u_k$.

Using (14.2.1) we rewrite (14.1.2) as

$$\frac{\partial}{\partial t}(w_{ik} - u_{ik}) = -j_{ik}.$$

The difference $w_{ik} - u_{ik}$ determines the part of the total distortion tensor that is not associated with elastic stresses and is generally called the plastic distortion of a solid. We denote this quantity by u_{ik}^{pl} and obtain

$$\frac{\partial u_{ik}^{\mathrm{pl}}}{\partial t} = -j_{ik}. \tag{14.2.2}$$

Thus, the variation in the plastic distortion tensor at a certain point of the medium for small time δt is equal to

$$\delta u_{ik}^{\mathrm{pl}} = -j_{ik}\delta t .$$

A similar relation for the plastic strain tensor $\varepsilon_{ik}^{\mathrm{pl}}$ has the form

$$\delta\varepsilon_{ik}^{\mathrm{pl}} = -\frac{1}{2}(j_{ik} + j_{ki})\delta t . \tag{14.2.3}$$

The relation between dislocation flow density and plastic deformation velocity, i. e., the relations equivalent to (14.2.2) or (14.2.3) were indicated by Kroener and Rider (1956).

If, during the dislocation motion, the elements of the medium move without breaking the continuity, then $j_{kk} = 0$. It then follows from (14.2.2) that $\varepsilon_{kk}^{\mathrm{pl}} = 0$. Thus, we come to the assertion, known in plasticity theory, that a purely plastic deformation taking place without breaking the medium continuity does not result in a hydrostatic compression (which should be connected with internal stresses).

It is clear that (14.2.2), (14.2.3) are valid for any mechanism responsible for dislocation migration, in particular, their nonconservative motion (climb). In the last case, as shown above, the crystal volume changes locally, but the general scheme of plasticity is not broken.

To evaluate the role of dislocation climb it is expedient to imagine a semi-microscopic scheme for this process. In Fig. 14.2, the edge of an extra atomic half-plane coincident with the edge dislocation axis is shown. When the interstitial atom gets attached to the edge of a half-plane inserted in a crystal, it becomes a "regular" atom, resulting in a protuberance of atomic dimensions on the dislocation line. The dislocation line itself is determined up to atomic dimensions, so that its position can only be affected by the "condensation" of a macroscopic number of interstitial atoms. The change in position of an extra half-plane edge under the condensation of a great number of interstitial atoms (I), when an extended row of extra atoms is formed on the dislocation, is shown in the middle of Fig. 14.2. The appearance of an extra row of atoms results in the displacement of a corresponding part of the dislocation line by the value a in a direction perpendicular to the slip plane. The dislocation appears to go down by one atomic layer; it goes to the next slip plane.

The condensation of vacancies (V) on the edge dislocation leads to similar results, the difference being that the dislocation displacement is oppositely directed – the dislocation line goes up to a higher slip plane.

Both the glide and climb occur under the action of elastic stresses. The glide, when viewed as a fast mechanical motion, differs from climb in its threshold character: it only begins when the stress exceeds a definite value (the initiation stress). But in a number of crystals (e. g., in many metals and quantum crystals) comparatively small stresses are needed to start dislocation motion. If the crystal has a regular structure the dislocations at the first stage of their motion glide easily and often travel large

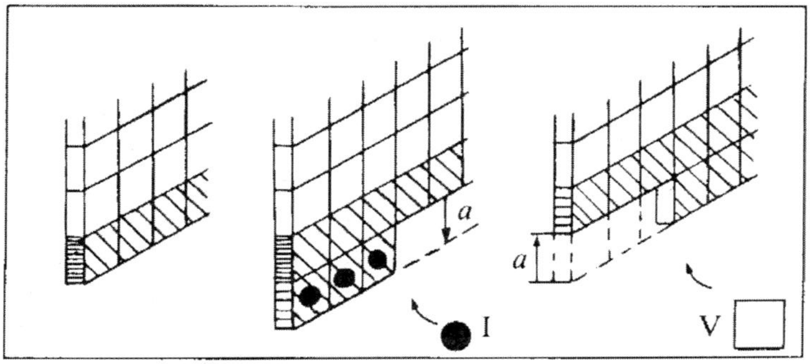

Fig. 14.2 Climb of an edge dislocation via elongation of an extra atomic half-plane under the condensation of interstitial atoms (I) or shortening of an extra half-plane under the condensation of vacancies (V).

distances in their slip planes. However, the process of easy gliding is short since obstacles decelerating the dislocation motion arise in a crystal.

The obstacles decelerating the dislocation are different. These may be defects such as impurities, fixing the dislocation line at some points only. But sometimes the dislocation stops gliding over an extended part of its line. This occurs when the dislocation is retarded by clusters of impurities or macroscopic heterogeneous inclusions. Obstacles of any kind retard the dislocation motion. As a result, the velocity of crystal plastic deformation is determined by factors such as climb and glide, and whether the dislocation can overcome the obstacles.

14.3 Energy and Effective Mass of a Moving Dislocation

The system of equations (14.1.5) determines the elastic distortion tensor u_{ik} and the vector v_α of the medium displacement velocity using a given distribution of dislocations and flows. The tensors α_{ik} and j_{ik} and, hence, the dislocation motion are assumed to be given. For a system of equations to be completely closed and to determine a self-consistent evolution of dislocations and elastic field, it is necessary to take into account the changes in the dislocation density and their flow under the action of elastic fields. In other words, it is necessary to know the equation of dislocation motion.

Since separate dislocation loops represent lines of elastic field singularities, the equation for dislocation motion is an equation for the motion of an elastic field singularity. The physical idea of obtaining the equation of motion of a field singularity (in the given case, a dislocation) is well known.

A dislocation is a source of internal stress, which creates stress and strain fields in a crystal free of external loads. A definite elastic energy that can be regarded as the dislocation energy is associated with this field. When the dislocation moves, the elastic field induced by it also moves, but it always has inertia, because the dynamic elastic field energy is different from the static field energy. The inertia of the elastic field of the dislocation can be interpreted as the inertia of the dislocation, which can be described by an effective mass. With such an approach, the energy and the mass of the dislocation and, hence, the equation of the dislocation motion are of a pure field origin.

Accordingly, to evaluate the dislocation dynamics, it is necessary to know the dislocation energy. To calculate the latter it is necessary to express the elastic energy of the field caused by a moving dislocation loop (or a system of loops) through the instantaneous coordinates and velocities of dislocation line elements. In accordance with the electromagnetic field theory of moving charges, this procedure is generally possible using an approximation that is quadratic in velocities (Section 3.6). Since the structure of the dynamical equations of an elastic field with dislocations is qualitatively the same as that of Maxwell's equations, the difficulties in carrying out the above procedure may be due to additional calculations only. The latter are rather cumbersome (even in an isotropic approximation) for the general case of a dislocation loop moving arbitrarily in a crystal. Taking this into account, we try, by analyzing a simple example and its almost obvious generalizations, to explain the general features of deriving the equation for dislocation motion.

We consider the uniform motion of a linear screw dislocation in an isotropic medium. We choose the z-axis along the dislocation line and the x-axis parallel to the direction of the dislocation velocity $\boldsymbol{V}$. As in the static case (Section 3.14), the elastic field of a screw dislocation is completely described by a single nonzero component of the displacement vector $u_z = u(x,y,t)$ and the equations of the elastic medium motion reduce to the 2D wave equation

$$\frac{\partial^2 u}{\partial t^2} = s^2 \left(\frac{\partial^2 u}{\partial y^2} + \frac{\partial^2 u}{\partial y^2} \right), \quad s^2 = \frac{G}{\rho}. \tag{14.3.1}$$

The solution to (14.3.1) should have a standard dislocation singularity on the line $x = Vt$, $y = 0$, and, thus it is convenient by changing the variables

$$\xi = \frac{x - Vt}{\sqrt{1-\beta^2}}, \beta^2 = \frac{V^2}{s^2} = \frac{\rho V^2}{G},$$

to write it as

$$\Delta u^0 = 0, \quad \Delta \equiv \frac{\partial^2}{\partial \xi^2} + \frac{\partial^2}{\partial y^2}, \tag{14.3.2}$$

where $u^0(\xi, y) = u(x,y,t)$.

Equation (14.3.2) coincides with (13.3.1) and its solution has a dislocation singularity at the point $\xi = y = 0$. Thus the function $u^0(\xi, y)$ is identical to the corresponding static solution (13.3.2) if the following replacement $x \to \xi$ is made.

The total field energy per the dislocation unit length is

$$
\begin{aligned}
E = & \frac{1}{2}\int\left[\rho\left(\frac{\partial u_z}{\partial t}\right)^2 + G\left(u_{xz}^2 + U_{yz}^2\right)^2\right] dx\,dy \\
= & \frac{1}{2}\int\left[\frac{1+\beta^2}{1-\beta^2}\left(\frac{\partial u^0}{\partial \xi}\right)^2 + \left(\frac{\partial u^0}{\partial y}\right)^2\right] dx\,dy \qquad (14.3.3) \\
= & \frac{1}{2}G\sqrt{1-\beta^2}\int\left[\frac{1+\beta^2}{1-\beta^2}\left(\frac{\partial u^0}{\partial \xi}\right)^2 + \left(\frac{\partial u^0}{\partial y}\right)^2\right] d\xi\,dy.
\end{aligned}
$$

Using polar coordinates in the plane $\xi O y$ we obtain instead of (14.3.3) the following expression:

$$
\begin{aligned}
E = & \frac{1}{2}G\sqrt{1-\beta^2}\int\left(\frac{1+\beta^2}{1-\beta^2}\sin^2\varphi + \cos^2\varphi\right) u_{\varphi z}^2 r\,dr\,d\varphi \\
= & \frac{\pi G}{\sqrt{1-\beta^2}}\int \varepsilon_{\varphi z}^2 r\,dr\,.
\end{aligned}
$$

Finally, we substitute the expression for $\varepsilon_{\varphi z}$ that follows from (13.3.3):

$$E = \left(\frac{b}{2\pi}\right)^2 \frac{\pi G}{\sqrt{1-\beta^2}}\int\frac{dr}{r} = \frac{Gb^2}{4\pi}\frac{\log(R/r_0)}{\sqrt{1-\beta^2}}\,, \qquad (14.3.4)$$

where the parameters R and r_0 are chosen as in the static case.

Expressing in (14.3.4) the shear modulus G through the transverse sound wave velocity ($G = \rho s^2$), we arrive at the following formula for the energy per unit length of a screw dislocation

$$E = \frac{m^* s^2}{\sqrt{1 - v^2/s^2}}, \qquad m^* = \frac{\rho b^2}{4\pi}\log\frac{R}{r_0}. \qquad (14.3.5)$$

In spite of the "relativistic" form of (14.3.5) this formula is valid only for small dislocation velocities, assuming $V \ll s$. Such a "pseudodepletion" of (14.3.5) results from the fact that the dislocation energy has the field as origin and involves, in particular, the interaction energy of different elements of the same dislocation line. To take into account the interaction of distant dislocation loop elements in the case of nonstationary motion the elastic dislocation field at a certain time should be expressed through its coordinates and velocities at the same time. In the general case, because the elastic waves are retarded, this expression does not exist. As mentioned above for a nonstationary motion this is possible within the approximation quadratic in V/s.

Thus an expression for E, valid for any time dependence of the screw dislocation velocity V will be obtained using the condition $V \ll s$ and replacing (14.3.5) by

$$E = m^* s^2 + \frac{1}{2}m^* V^2\,. \qquad (14.3.6)$$

The first term in (14.3.6) coincides with the eigenenergy per unit length of the dislocation at rest (13.3.17), the second is assumed to be its kinetic energy. Therefore, m^* can be called the *unit length effective mass* of a screw dislocation. As in the case of a rest energy, our assumption is justified, i. e., when a real dislocation moves in a crystal, some atoms in the vicinity of the dislocation axis at a distance of the order of r_0 from it also start moving. This generates an additional dislocation inertia due to the usual atomic mass. The order of magnitude of the atomic mass inside tube of radius $r_0 \sim b$ per unit length of the dislocation can be estimated as $\rho r_0^2 \sim \rho b^2$. Comparing this estimate with (14.3.5) for m^*, the dislocation mass can be regarded with logarithmic accuracy as the field mass.

The inertial properties of an arbitrary weakly deformed dislocation loop are characterized by some tensor of line density of the effective mass m^*_{ik} dependent on a point on the dislocation line. It can be concluded that at a point where the radius of curvature of the dislocation loop $R_{\text{curv}} \gg b$, an estimate of the order of magnitude of the effective mass is the same as that of the dislocation rest energy (13.3.18):

$$m^* \sim \frac{\rho b^2}{4\pi} \log \frac{R}{r_0}. \qquad (14.3.7)$$

In the case of translational motion of the linear dislocation, R_{curv} is the dislocation length. If the dislocation vibrates, R_{curv} equals the wavelength of the dislocation bending vibrations.

A very important physical conclusion follows from the previous comments concerning the estimation of the parameter R_{curv}. By writing the dislocation energy in the form of (14.3.6), we introduced the effective mass of unit length of the dislocation, but characterized in fact the motion of the dislocation. This mass is not a local characteristic of the dislocation. We recall that with the retardation of electromagnetic waves in a 2D electron crystal taken into account, the mass of a vibrating atom has transformed into a nonlocal characteristic of the inertial properties of the crystal. Analogously, the inertial properties of a dislocation loop should be characterized by a nonlocal mass density. This means that the energy of a moving dislocation loop can be written as

$$E = E_0 + \frac{1}{2} \oint \oint \mu_{ik}(l, l') V_i(l) V_k(l')\, dl\, dl', \qquad (14.3.8)$$

where E_0 is the quasi-static dislocation energy, which is dependent only on the instantaneous form and the instantaneous position of the dislocation and plays the role of a rest energy. The second term in (14.3.8) should be considered as the kinetic energy of the dislocation. $\boldsymbol{V}(l)$ is the velocity of a dislocation line element dl, and the double integration is over the length of the entire dislocation loop. Then $\mu_{ik}(l, l')$ plays the role of a *nonlocal density* of the *effective mass of the dislocation*.

Even in an isotropic medium, the effective mass density of the dislocation is anisotropic. In view of the above expression for the effective mass of a screw dislocation (14.3.5), it is easy to understand the general form of the tensor function of two variables $\mu_{ik}(l,l')$ for a dislocation moving in its slip plane in an isotropic medium. This is, as follows (Kosevich, 1962):

$$\mu_{ik}(l,l') = \frac{1}{2}\rho b^2 \left[\tau\tau'\delta_{ik} - \tau_i'\tau_k\right] \left(1 + \left(\frac{s_t}{s_l}\right)^2 \sin^2\theta\right) K(l,l'), \tag{14.3.9}$$

$$b\cos\theta = \boldsymbol{nb}, \quad \boldsymbol{\tau} = \boldsymbol{\tau}(l), \quad \boldsymbol{\tau}' = \boldsymbol{\tau}(l'),$$

where $\boldsymbol{n}$ is the unit vector in the direction of a straight line linking two points l and l' on the dislocation: $\boldsymbol{n} = \left[\boldsymbol{r}(l) - \boldsymbol{r}(l')\right]/(r - r')$. The scalar function of two variables $K(l,l')$ resembles the formula for the self-induction coefficient of a linear conductor coincident with the dislocation loop. It is a purely geometrical characteristic of the dislocation loop D, independent of the crystal properties and even of those physical properties of the medium that are related with this line. This function characterizes the self-action of any linear singularity of the classical field in a 3D space if the dynamic field equations are differential equations in partial derivatives, such as wave equations with sources. Its general form is

$$K(l,l') = \frac{1}{4\pi}\iint \frac{\gamma(\boldsymbol{\xi})\gamma(\boldsymbol{\xi}')\, d^2\xi\, d^2\xi'}{|\boldsymbol{r}(l,\boldsymbol{\xi}) - \boldsymbol{r}(l',\boldsymbol{\xi}')|}, \tag{14.3.10}$$

where $\boldsymbol{\xi}$ is the two-dimensional radius vector measured from the dislocation axis in a plane perpendicular to it; $\gamma(\boldsymbol{\xi})$ is some smooth function localized in the vicinity of the dislocation axis and describing the Burgers vector "distribution" in the cross section of the dislocation core:

$$\alpha_{ik} = \tau_i b_k \gamma(\boldsymbol{\xi}), \quad \int \gamma(\boldsymbol{\xi})\, d^2\xi = 1. \tag{14.3.11}$$

The necessity of introducing such an arbitrary function is "payment" for changing to a continuum description of the dislocation as a specific crystal defect. If we replace the function $\gamma(\boldsymbol{\xi})$ by the delta-function $\gamma(\boldsymbol{\xi}) = \delta(\boldsymbol{\xi})$, a two-fold integral (14.3.10) will become meaningless. Thus, by introducing this function we avoid a nonphysical divergence in elasticity theory. Since the dependence on the dislocation-core radius is involved in the eigenenergy and the effective mass logarithmically, then arbitrariness in choosing the function $\gamma(\boldsymbol{\xi})$ does not affect the results. Generally, this function is assumed to be constant and nonzero inside a cylinder of radius r_0.

To obtain the expression (14.3.10) by direct calculation, it is necessary to calculate the self-induction coefficient of a linear conductor and also to solve problems on dislocation vortices in a scalar crystal model.

14.4 Equation for Dislocation Motion

Knowing the effective mass of the dislocation makes it possible to write its equation of motion. The equation of motion of an element of a dislocation loop located at the point l on the dislocation line can be written similarly to Newton's second law:

$$\oint \mu_{ik}(l,l')W_k(l')dl' = F_i^0(l) + e_{ikm}\tau_k(l)\sigma'_{mn}(l)b_n + S_i(l,\boldsymbol{V}), \qquad (14.4.1)$$

where $\boldsymbol{W}(l) = \partial \boldsymbol{V}/\partial t$ is the dislocation acceleration; $\boldsymbol{F}^0$ is the force of a quasi-static self-action of a distorted dislocation, generated by the energy density (13.3.18). In a *line tension approximation* used with certain reservations the order of magnitude of this force is evaluated when $R \gg b$ as

$$F_0 = \frac{T_D}{R}, \qquad T_D \sim GB^2. \qquad (14.4.2)$$

The last term on the r.h.s. of (14.4.1) is the retardation force experienced by the dislocation in a real crystal (this has not been discussed yet). We distinguish two physically different parts in the value of this force.

The first part of the force $\boldsymbol{S}$ is composed of forces that arise because the crystal structure is discrete and the dislocation core is constructed of atoms. Among these forces there is a static component (the Peierls force) and a component dependent on the dislocation motion. The last force is related to lattice distortions moving together with the dislocation on its axis and the induced reconstruction of the dislocation core in motion. The dislocation-core reconstruction involves, in particular, the formation and migration (along the dislocation) of so-called *steps* (kinks) that link parts of the same dislocation positioned in the neighboring "valleys" of the Peierls potential relief. A similar defect (kink) was observed by us in studying solutions to the sine-Gordon equation that, as a good model, can be applied to the kink dynamics on the dislocation. In a number of cases the kink kinetics is the main mechanism of dislocation motion.

The second part of the force $\boldsymbol{S}$ is due to various dynamic mechanisms responsible for the energy dissipation of a moving dislocation. First, these are the microscopic processes of interaction between the dislocation and phonons and other elementary crystal excitations. Then, there are the macroscopic processes of energy losses in the dynamic elastic dislocation field connected with the dispersion of the elastic moduli of a real crystal. Both processes are greatly dependent on the defect structure of a real crystal.

As well as with the equation of motion of a dislocation element (14.4.1) we can consider the equation of motion of the entire dislocation loop:

$$\oint m_{ik}(l)W_k(l)\,dl = e_{ikm}b_n \oint \tau_k\sigma'_{mn}\,dl + \oint S_i(l)\,dl, \qquad (14.4.3)$$

where

$$m_{ik} = \oint \mu_{ik}(l,l')\,dl'. \qquad (14.4.4)$$

When the integration is over the whole loop, the first term on the r.h.s. of (14.4.1) is omitted, as the total force of the dislocation static self-action is zero.

It follows from (14.4.3) that just $m_{ik}(l)$ can be considered as the value of the effective mass of a unit dislocation for the motion of the entire dislocation loop. But the mass per unit length introduced in this way is not a local property of a given point on the dislocation loop and is dependent on the dimensions and form of the entire loop. Using (14.4.4) and the definition of the tensor μ_{ik} it is easy to obtain the estimate (14.3.7) for the effective mass per unit length of the dislocation.

The given estimate for m_{ik} makes it possible to justify our assumption of a purely field origin of the dislocation mass. When a real dislocation moves in a crystal, some of the atoms in the vicinity of the dislocation axis (at distances of the order of r_0) also start moving. This leads to the appearance of an additional dislocation energy associated with the ordinary mass of these atoms. The order of magnitude of atomic masses inside a tube of radius $r_0 \sim b$ per unit length of the dislocation can be estimated as $\rho r_0^2 \sim \rho b^2$. Comparing this estimate with (14.3.7), we see that for $\log R_{\text{curv}}/r_0 \gg 1$, taking into account the mass of atoms moving near the dislocation axis practically does not affect the dislocation energy and, up to a logarithmic accuracy, the dislocation mass can be assumed to be the field mass.

For a linear dislocation, when the vectors $\boldsymbol{\tau}$ and $\boldsymbol{b}$ remain unchanged along the dislocation line, the tensor m_{ik} is

$$m_{ik} = \frac{\rho b^2}{4\pi}(\delta_{ik} - \tau_i \tau_k)\left(1 + \left(\frac{s_t}{s_l}\right)^4 \sin^2\theta\right) \ln\frac{R_n}{r_0}, \qquad (14.4.5)$$
$$b^2 \cos^2\theta = (\boldsymbol{\tau b})^2,$$

where R_m is the dislocation length ($R_m \gg r_0$). In conclusion, we note that the inertia term in the equation of motion is essential only for nonstationary motion of the dislocation when its acceleration is too great. If the dislocation acceleration is small, the action of retarding forces involving dissipative forces will be dominant. Their values and the dependence on the dislocation velocity mainly determine the character of an almost stationary motion of the dislocation.

The equation of dislocation motion is often applied in formulating a *string model* where the dislocation line is considered as a heavy string possessing a certain tension and lying in a "corrugated" surface. The corrugated surface relief is described by the Peierls potential and the valleys on it correspond to the potential minima on the slip plane that are occupied by a straight dislocation in equilibrium (Fig. 14.3).

Let the Ox-axis be directed along the equilibrium position of a straight dislocation and the transverse dislocation displacement η be along the Oy-axis. In the string model, this displacement as a function of x and t is described by the following equation of motion:

$$m\frac{\partial^2\eta}{\partial t^2} - T_D\frac{\partial^2\eta}{\partial x^2} + b\sigma_P \sin 2\pi\frac{\eta}{a} = b\sigma, \qquad (14.4.6)$$

where m is the mass per unit length of the dislocation; T_D is the line tension; σ_p is the Peierls stress; σ is the corresponding component of the stress tensor caused by external loads.

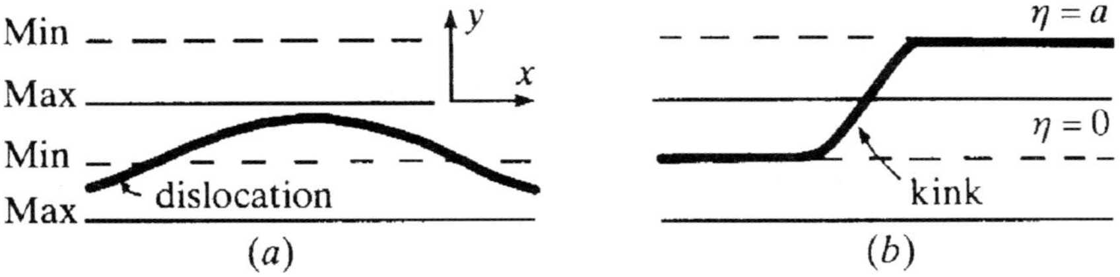

Fig. 14.3 Two types of dislocation motion in the field of the Peierls potential: (*a*) the dislocation vibrates in one potential valley; (*b*) the dislocation forms a kink moving along the x-axis.

It is clear that (14.4.6) is an inhomogeneous sine-Gordon equation, so that its solitons are associated with the dislocation kinks.

Unfortunately, the local equation of motion for the dislocation element (14.4.6) cannot be derived consistently from the equation of motion (14.4.1). In a crystal there are no specific interactions decreasing so fast in space that it be possible to pass from the integral equation (14.4.1) to the differential equation in partial derivatives (14.4.6). The elastic stresses causing the interaction of various parts of the same dislocation decrease very slowly with distance, so that such dislocation parameters as the mass m per unit length or the linear tension coefficient T_D are not local characteristics of the dislocation. Although the string model is limited, it is often used to demonstrate the physical phenomena generated by the dislocation bending vibrations. The model is attractive primarily because it is simple and enables good results to be obtained. However, the conclusions made from an analysis of (14.4.6) need to be re-examined.

We shall take the r.h.s. of (14.4.6) as an external force. In the absence of an external field ($\sigma = 0$), then the dislocation motion has the character of free oscillations corresponding to the normal modes of string vibrations and having the dispersion law

$$\omega^2 = \omega_0^2 + c_0^2 k^2, \tag{14.4.7}$$

where

$$\omega_0^2 = 2\pi b\sigma_P/(ma), \qquad c_0^2 = T_D/m. \tag{14.4.8}$$

Let us discuss the dispersion law (14.4.7). In analyzing the waves localized near the dislocation (Section 12.2) it was noticed that its bending vibrations should have a propagation velocity that does not actually differ from that of transverse sound vibrations s_t. To confirm this conclusion with (14.4.7) we assume $c_0 = s_t$, which does not contradict the estimates (14.3.7), (14.3.8). But the dispersion law frequencies (14.4.7) satisfy the inequality

$$\omega > c_0 k = s_t k. \tag{14.4.9}$$

The wave running along the dislocation with wave vector $\boldsymbol{k}$ and frequency (14.4.8) cannot be localized near the dislocation because it inevitably excites bulk vibrations whose frequencies satisfy the same condition (14.4.9). Thus, the string vibration with the dispersion law (14.4.7), if it exists, has the character of a quasi-local vibration. Unfortunately, it cannot be pronounced. Indeed, it is easily seen that the connection of the dislocation line vibrations with the vibrations of crystal lattice atoms is characterized by the same parameter that determines the eigenfrequency ω_0.

We note that the periodic potential in whose valleys the dislocation vibrates is created by the crystal atoms. Thus, it is more natural in (14.4.6) to write the periodic term in a different way, i. e.,

$$\frac{\partial^2 \eta}{\partial t^2} - c_0^2 \frac{\partial^2 \eta}{\partial x^2} + \frac{a\omega_0^2}{2\pi} \sin\left(2\pi \frac{\eta - u_y}{a}\right) = 0, \tag{14.4.10}$$

where u_y is the long-wave atomic displacement in the vicinity of the dislocation core.

In studying small vibrations, we have instead of (14.4.10)

$$\frac{\partial^2 \eta}{\partial t^2} - c_0^2 \frac{\partial^2 \eta}{\partial x^2} + \omega_0^2 \eta = \omega_0^2 u_y, \tag{14.4.11}$$

where there is no small coupling parameter between the dislocation and sound displacements that permits consideration of free dislocation vibrations in the valley of a periodic potential relief.

Thus, examining the eigenvibrations of a dislocation-string leads to the conclusion that (14.4.6) is not a satisfactory model for describing the motion of a free dislocation.

However, this model is quite applicable for investigating the dislocation vibrations under the action of an oscillating external force. We consider the dislocation vibrations under the conditions when the Peierls forces can be neglected (e. g., at high temperatures) and the retarding force of the dislocation that is proportional to its velocity becomes important. In this case the dislocation motion obeys an equation similar to (14.4.6), but the term appearing from the Peierls potential can be omitted and the dissipative term $B(\partial\eta/\partial t)$, where B is the multiplier depending on the nature of dissipative forces, is added. As a result we get

$$m\frac{\partial^2 \eta}{\partial t^2} + B\frac{\partial \eta}{\partial t} - T_D \frac{\partial^2 \eta}{\partial x^2} = b\sigma_0 e^{-i\omega t}. \tag{14.4.12}$$

We assume that the dislocation is fixed at the two points $\pm l/2$. This fixing can be produced by a strong interaction of the dislocation core with a poorly mobile impurity in a crystal. Then, the solution for the forced vibrations of a dislocation segment is the function

$$\eta = \frac{b\sigma_0 e^{-i\omega t}}{k^2 T_D}\left(\frac{\cos kx}{\cos \frac{1}{2}kl} - 1\right), \tag{14.4.13}$$

where $k^2 = (m\omega^2 + iB\omega)/T_D$.

Equation (14.4.12) and its solution (14.4.13) are the basis for a dislocation theory of internal friction worked out in detail by Granato and Lucke (1956).

14.5 Vibrations of a Lattice of Screw Dislocations

Consider a superstructure formed by a lattice of parallel screw dislocations. By dislocation lattice we mean a system of parallel screw dislocations oriented along the z-axis and intersecting the xy plane at discrete periodically arranged points, forming a 2D lattice, the unit cell of which has area S_0: $S = NS_0$, where S is the cross-sectional area of the sample in the xy plane and N is the number of dislocations. The coordinates of these points in the equilibrium lattice are

$$\boldsymbol{x}(\boldsymbol{n}) = \boldsymbol{R}_n + \sum_n \boldsymbol{d}_\alpha n_\alpha, \qquad \boldsymbol{n} = (n_1, n_2, 0), \tag{14.5.1}$$

where $\boldsymbol{d}_\alpha$ ($\alpha = 1, 2$) are the basic translation vectors of the lattice ($d_\alpha \sim d$ is the distance between neighboring dislocations: $S_0 = d^2$).

We intend to use a simple one-component scalar model of vibrations in which it is assumed that all the atoms are displaced only in one direction. The basis of using such a model is the fact that a static screw dislocation in the isotropic media produces the scalar field of displacements w along the z-axis. The model gives a correct description of the elastic field created in an isotropic medium by parallel screw dislocations. The solution of such a problem in a real vector displacement scheme can in principle be found analytically, but it permits obtaining the dispersion relation of the dislocation lattice only in implicit form. For the sake of simplicity we restrict ourselves to the scalar model.

If there are rectilinear screw dislocations directed along the z axisx, then the elastic field is more conveniently described not by the displacement w along the z-axis but by the distortion and velocity of the displacements as functions of the coordinate and time. Following (13.1.10) and (13.1.12), for describing the shear field of screw dislocations we introduce a distortion vector $\boldsymbol{h}$ and a stress vector $\boldsymbol{\sigma} = G\boldsymbol{h}$ (G is the shear modulus):

$$\boldsymbol{h} = \operatorname{grad} w, \quad h_i = \nabla_i w = \frac{\partial w}{\partial x_i}, \quad i = 1, 2, 3, \tag{14.5.2}$$

and a velocity v: $v = \partial w / \partial t$. Equation (13.1.11) conserves its form

$$\operatorname{curl} \boldsymbol{h} = -\boldsymbol{\alpha}, \tag{14.5.3}$$

and the density of dislocations $\boldsymbol{\alpha}$ is equal to

$$\boldsymbol{\alpha} = \boldsymbol{\tau} b \sum_n \delta(\boldsymbol{x} - \boldsymbol{R}_n),$$

where b is the modulus of the Burgers vector and $\boldsymbol{\tau}$ is the tangent vector to the dislocation; for a static dislocation it is conveniently chosen as $\boldsymbol{\tau}(0, 0, -1)$.

The wave equation for the elastic field in the medium between dislocations takes the usual form (14.3.1):

$$\operatorname{div} \boldsymbol{h} - \frac{1}{s^2} \frac{\partial v}{\partial t} = 0\,. \tag{14.5.4}$$

If the dislocations move (vibrate), then (14.5.3), (14.5.4) do not change, but a new variable of the dislocation structure appears: $\boldsymbol{u} = (u_x, u_y, 0)$ (of course, $\boldsymbol{u} = \boldsymbol{u}(\boldsymbol{n}, t)$ for the $\boldsymbol{n}$-th dislocation), which determines the instantaneous coordinate of an element of the dislocation:

$$x_n = \boldsymbol{R}(\boldsymbol{n}) + \boldsymbol{u}(\boldsymbol{n}, t) .$$

The time dependence of the displacement vector $\boldsymbol{u}$ gives the velocity $\boldsymbol{V}$ of an element of the dislocation ($V_\alpha = \partial u / \partial t$) that generates a dislocation flux. The dislocation flux density vector $\boldsymbol{j}$ arises in the dynamical equation (14.1.3):

$$\frac{\partial \boldsymbol{h}}{\partial t} = \operatorname{grad} v + \boldsymbol{j} . \tag{14.5.5}$$

In the case under consideration the flux density is given by the formula following from (14.1.4):

$$j_\alpha = b\varepsilon_{\alpha\beta} \sum_n V_\beta(\boldsymbol{n}) \delta\left(\boldsymbol{x} - \boldsymbol{R}(\boldsymbol{n})\right), \quad \alpha = 1, 2 , \tag{14.5.6}$$

where the matrix $\varepsilon_{\alpha\beta}$

$$\varepsilon_{\alpha\beta} = \begin{pmatrix} 0 & 1 \\ -1 & 0 \end{pmatrix} . \tag{14.5.7}$$

Collecting together (14.5.3)–(14.5.5) we obtain the total set of equations describing the elastic field in the sample if the distribution of dislocations and their fluxes are known. To close this system it is necessary to write equations of motion for the dislocations under the influence of the elastic fields. The simplest form of such an equation can be obtained using (14.4.1) and (14.4.4) for rectilinear dislocations:

$$m \frac{\partial V_\alpha}{\partial t} = f_\alpha + S_\alpha, \quad \alpha = 1, 2 , \tag{14.5.8}$$

here m is the effective mass of a unit length of the dislocation (with the order of magnitude (14.3.7), where R is the distance between dislocations in our case), and f is equal to

$$f_\alpha = b\varepsilon_{\alpha\beta}\sigma_\beta = bG\varepsilon_{\alpha\beta}h_\beta . \tag{14.5.9}$$

In the case of a curved dislocation line expression (14.5.9) includes the self-force from different elements of the same dislocation, which is proportional to the curvature of the dislocation line at the given point. In the analysis of small vibrations the curvature of the dislocations can not be taken into account, and the force (14.5.9) includes only the stresses created by the other dislocations.

Usually $\boldsymbol{S}$ is the force due to the discreteness of the lattice, including dissipative forces. As we are interested in the dispersion relation for small vibrations, we neglect the latter and take the force in the form equivalent to that on (14.4.11), namely,

$$\boldsymbol{S} = -m\omega_0^2 \boldsymbol{u}, \tag{14.5.10}$$

where ω_0 is the frequency of vibrations of the dislocation string in a valley of the Peierls relief.

Let us investigate the long-wavelength vibrations of the dislocation lattice, assuming the wavelength of the vibrations is much larger than the lattice period $d(dk \ll 1)$. In this approximation the distribution of the dislocations can be assumed continuous, characterized by a density $n(\boldsymbol{x}, t)$. In equilibrium $n = n_0 = 1/S_0$. The dynamics of the lattice is governed mainly by the average dislocation flux density, which in the linear approximation has the nonzero components

$$j_\alpha(\boldsymbol{x}, t) = bn_0\varepsilon_{\alpha\beta}V_\beta(\boldsymbol{x}t), \quad \alpha = 1, 2, \tag{14.5.11}$$

where $\boldsymbol{V}$ is the average velocity of the dislocations. The velocity $\boldsymbol{V}$ must be determined by the equation of motion of the dislocations (14.5.8). We write the equation of motion with the use of (14.5.9) and (14.5.10):

$$\frac{\partial V_\alpha}{\partial t} + \omega_0^2 u_\alpha = \frac{Gb}{m}\varepsilon_{\alpha\beta}h_\beta. \tag{14.5.12}$$

We differentiate (14.5.12) with respect to time and use (14.5.5) and (14.5.11). After elementary calculations we get

$$\frac{\partial V_\alpha}{\partial t} + (\omega_0^2 + \omega_{pl}^2)V_\alpha = \frac{Gb}{m}\varepsilon_{\alpha\beta}\frac{\partial v}{\partial x_\beta}, \tag{14.5.13}$$

where

$$\omega_{pl}^2 = \frac{Gb^2 n_0}{m}. \tag{14.5.14}$$

The frequency ω_{pl} is the analog of the plasma frequency.

We now differentiate (14.5.4) with respect to time and again use (14.5.5):

$$\left(\Delta - \frac{1}{s^2}\frac{\partial^2}{\partial t^2}\right)v = -bn_0\varepsilon_{\alpha\beta}\nabla_\alpha V_\beta. \tag{14.5.15}$$

Here Δ is the Laplacian operator.

The pair of equations (14.5.13), (14.5.15) describe the collective dynamics of the dislocation lattice and the elastic field. It is easy to show that the equation for the "longitudinal" vibrations of the lattice separates. Indeed, for the variable $P = \operatorname{div}\boldsymbol{V} = \nabla_\alpha V_\alpha$ it follows from (14.5.13) that

$$\frac{\partial^2 P}{\partial t^2} + \omega_l^2 P = 0, \qquad \omega_l^2 = \omega_0^2 + \omega_{pl}^2, \tag{14.5.16}$$

where ω_l plays the role of the frequency of the longitudinal vibrations of the lattice.

The longitudinal component of the average velocity of the dislocations is derivable from a potential: $V_\alpha^{(1)} = \nabla_\alpha Q, \alpha = 1, 2$. Taking that into account, we see that the "longitudinal" vibrations of the elastic field $v = v(z)$ do not depend on the lattice vibrations. Thus one branch of collective vibrations (we call it the branch of "longitudinal" vibrations) corresponds to independent oscillations of the elastic filed $v = v(z,t)$ with the dispersion relation $\omega = sk_z$ and to compression–rarefaction oscillations of the dislocation lattice $P = P(z,t)$ with the dispersion relation $\omega = \omega_l$.

To describe to "transverse" vibrations we introduce the variable

$$M = bn_0(\text{curl}\,\boldsymbol{V})_z = bn_0\varepsilon_{\alpha\beta}\nabla_\alpha V_\beta. \tag{14.5.17}$$

The equation for this variable follows from (14.5.13):

$$\frac{\partial^2 M}{\partial t^2} + \omega_l^2 M = -\omega_{pl}^2 \frac{\partial^2 v}{\partial x_\alpha^2}. \tag{14.5.18}$$

The "transverse" collective vibrations are described by (14.5.18) and the following equation obtained from (14.5.15) for the function $v(x,y,t)$:

$$\left(\frac{1}{s^2}\frac{\partial^2}{\partial t^2} - \frac{\partial^2}{\partial x_\alpha^2}\right) v = M. \tag{14.5.19}$$

The compatibility conditions for (14.5.18) and (14.5.19) give the dispersion relation for a wave with wave vector $\boldsymbol{k}(k_x, k_y, 0)$:

$$\omega^4 - (\omega_l^2 + s^2k^2)\omega^2 - \omega_0^2 s^2 k^2 = 0. \tag{14.5.20}$$

Equation (14.5.20) has two roots for ω^2, which correspond to low-frequency and high-frequency oscillations. Without writing the trivial expressions for these solutions in quadratures, we note the following:

Low-frequency branch. For $sk \ll \omega_0$ the dispersion relation has the form

$$\omega = \left(\frac{\omega_0}{\omega_l}\right) sk. \tag{14.5.21}$$

The vibrations are characterized by a transverse sound velocity, the value of which is less than the sound velocity s in the medium without the dislocations.

High-frequency branch. For $sk \gg \omega_l$ the inertial dislocation lattice is not entrained in the motion, and one observes only vibrations of the elastic field with the usual sound dispersion relation $\omega = sk$. Finally, in the long-wavelength limit ($sk \ll \omega_0$) we obtain

$$\omega^2 = \omega_l^2 + \left(\frac{\omega_{pl}}{\omega_l}\right) s^2(k_x^2 + k_y^2). \tag{14.5.22}$$

In comparing the graphs of the two branches of the dispersion relation, one must be particularly careful in rendering the low-frequency branch. The point is that dispersion relation (14.5.20) is valid for $\lambda \gg a$ (or $ak \ll 1$). At large k the dispersion relation of the lattice manifests a periodic dependence on the quasi-wave vector with the reciprocal lattice period $\boldsymbol{G}$: $\omega(\boldsymbol{k}) = \omega(\boldsymbol{k} + \boldsymbol{G})$. Therefore the dispersion relation obtained is actually valid in all small neighborhoods of any 2D reciprocal lattice

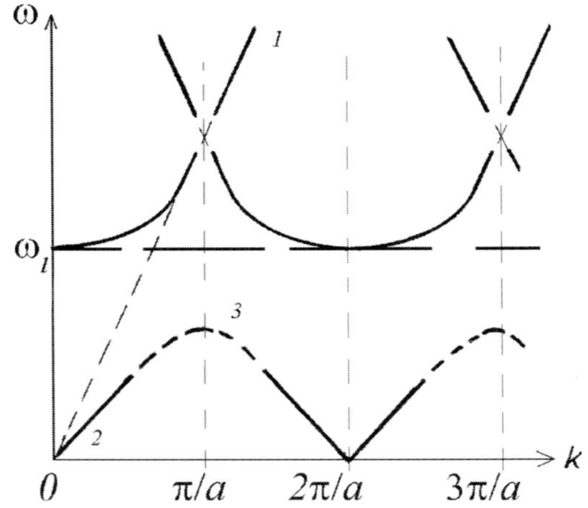

Fig. 14.4 Diagram of the dispersion relation : (1) $\omega = sk$, (2) plot of (14.5.21), (3) expected form of the graph in the short-wavelength region.

vector g, i. e., for $a\,|k - g| \ll 1$. Consequently, we are justified in drawing only the part of the graphs shown by the heavy solid lines 1 and 2 in Fig. 14.4 for a certain "good" direction in the reciprocal lattice. The continuation of the graph of the lower branch at $k \sim \pi/a$ and also the indicated crossing of the graphs of the upper branch at $k = (p + 1/2)\pi/a, p = 1, 2, 3, \ldots$ can be described only on the basis of a study of the dynamics of the discrete dislocation lattice. That is a subject for a separate study. We can only state that the graph of the lower branch is closed by the curves illustrated schematically by the dotted lines 3 in Fig. 14.4. Whether or not there is a band of forbidden frequencies between the upper and lower branches (gap in the spectrum) one cannot say on the basis of a long-wavelength treatment. However, one can say that the frequency spectrum has a limiting frequency ω_l that marks the edge of the upper branch of vibrations, which can certainly be manifested in the acoustic resonance properties of a crystal with a dislocation lattice.

Bibliography

Andreev A.F., Lifshits I.M., Zh. Eksper. Teor. Fiz. **56**, 2057 (1969) (in Russian).

Berezinsky V.L., Zh. Eksper. Teor. Fiz. **61**, 1144 (1971) (in Russian).

de Boer J., Physica **14**, 139 (1948).

Brout R., Visher W., Phys. Rev. Lett. **9**, 54 (1962).

Chebotarev L.V., private communication (1980).

Chou Y.T., Acta Metall. **13**, 251 (1965).

Debye P., Ann. Phys. **43**, 49 (1914).

Dzyub I.P., Fiz. Tverd. Tela **6**, 3691 (1964) (in Russian).

Goldstone J., Nuovo Cimento **19**, 154 (1961).

Granato A., Lucke K.J., Appl. Phys. **27**, 789 (1956).

Ivanov M.A., Fiz. Tverd. Tela **12**, 1895 (1970) (in Russian).

Kagan Yu., Iosilevskii Ya., Zh. Eksper. Teor. Fiz. **42**, 259 (1962); **44**, 284 (1963) (in Russian).

Kontorova T.A., Frenkel Ya.I., Zh. Eksper. Teor. Fiz. **8**, 80 (1938); **8**, 1340 (1938) (in Russian).

Kosevich A.M., Zh. Eksper. Teor. Fiz. **43**, 637 (1962) (in Russian); Soviet Phys. – JETP (English Transl.) **16**, 455 (1963).

Kosevich A.M., Pis'ma Zh. Eksper. Teor. Fiz. **1**, 42 (1965) (in Russian).

Kosevich A.M., Khokhlov V.I., Fiz. Tverd. Tela **12**, 2507 (1970) (in Russian).

Kosterlitz J.M., Thouless D.J., J. Phys. **C6**, 1181 (1973).

The Crystal Lattice: Phonons, Solitons, Dislocations, Superlattices, Second Edition. Arnold M. Kosevich

ISBN: 3-527-40508-9

Kroener E., Rieder G., Z. Phys. **145**, 424 (1956).

Lifshits I.M., Zh. Eksper. Teor. Fiz. **17**, 1076 (1947) (in Russian).

Lifshits I.M., Zh. Eksper. Teor. Fiz. **22**, 475 (1952) (in Russian).

Lifshits I.M., Usp. Fiz. Nauk **83**, 617 (1964) (in Russian).

Lifshits I.M., Kosevich A.M., Rep. Progr. Phys. **29**, pt. 1, 217 (1966).

Lifshits I.M., Peresada V.I., Uchen. Zapiski Khark. Univers. Fiz.-Mat. Fakult. **6**, 37 (1955).

Mura T., Philos. Mag. **8**, 843 (1963).

Nabarro F.R.N., Proc. Phys. Soc. A **59**, 256 (1947).

Nicklow H.G., Wakabayashi N., Smith H.G., Phys. Rev. B **5**, 4951 (1972).

Ott H., Ann. Phys. **23**, 169 (1935).

Peach M.O., Koehler J.S., Phys. Rev. **80**, 436 (1950).

Peierls R.E., Proc. Roy. Soc. **52**, 34 (1940).

Pushkarov D.I., Zh. Eksper. Teor. Fiz. **64**, 634 (1973) (in Russian); Soviet Phys. – JETP (English Transl.) **37**(2), 322 (1973).

Rytov S.M., Akust. Zhurn. (in Russian) **2**, 71 (1956).

Slutskin A.A., Sergeeva G.G., Zh. Eksper. Teor. Fiz. **50**, 1649 (1966) (in Russian).

Walker C.B., Phys. Rev. **103**, 547 (1956).

Waller I., Dissertation, Uppsala (1925).

Warren J.L., Wenzel R.G., Yarnell J.L., Inelastic Scattering of Neutrons (IAEA, Viena, 1965).

Weertman J., Philos. Mag. **11**, 1217 (1965).

Woods A.D.B., Cochran W., Brockhouse B.N., Cowley R.A., Phys. Rev. **131**, 1025 (1963).

Yarnell J.L., Warren J.L., Koenig S.H., in: Lattice Dynamics (Pergamon Press, Oxford, 1965), p. 57.

Zinken A., Buchenau U., Fenzl H.J., Schrober H.R., Solid State Commun. **22**, 693 (1977).

Index

The Crystal Lattice: Phonons, Solitons, Dislocations, Superlattices, Second Edition. Arnold M. Kosevich

ISBN: 3-527-40508-9